FLOW CYTOMETRY IN DRUG DISCOVERY AND DEVELOPMENT

FLOW CYTOMETRY IN DRUG DISCOVERY AND DEVELOPMENT

Edited by

VIRGINIA LITWIN
Covance Central Laboratory Services, Inc.

PHILIP MARDER
Redram Consulting, LLC

WILEY

A JOHN WILEY & SONS, INC., PUBLICATION

Published by John Wiley & Sons, Inc., Hoboken, New Jersey
Published simultaneously in Canada

Library of Congress Cataloging-in-Publication Data is available.

ISBN: 978-0-470-43356-0

Printed in Singapore

1 0 9 8 7 6 5 4 3 2 1

PHILIP MARDER ***(1948–2010)***

DEDICATION

This book is dedicated to my dear friend and coeditor, Philip Marder.

Phil was one of the early leaders in the use of flow cytometry pertaining to drug discovery and development. He participated in ISAC, GLIFCA, Indy Flow, and AAPS, and was a key contributor to both the flow cytometry and the Pharma communities. In 1973, he joined Eli Lilly where he went on to build a strong team of research scientists: biomarker assays developed by this team supported the full spectrum of the drug development process, including in vitro studies, toxicology, and clinical trials.

Midway through this editing project, Phil was diagnosed with glioblastoma multiforme. To say that he approached this illness with grace, intelligence, and courage would be a considerable understatement. Those of us who followed his blog were impressed by his ability to continue to enjoy so many different aspects of his life throughout the treatment process. Among these would be the arrival of a much loved puppy by the name of Chaim, and the professional quality reviews he provided for many of the fine (and not so fine) restaurants from Indianapolis to Chicago with many stops in between. Most impressive of all, however, was the connection he maintained with his strong and loving family, and, of course, they with him.

Completely in character, Phil continued to be involved with the details of this project until November 2009. It was truly a pleasure to collaborate with him on this book that should be seen as a fitting tribute to his substantial contributions to flow cytometry and drug development.

Our very best wishes to his family and friends.

CONTENTS

PREFACE

The recent contraction within the pharmaceutical sector and the push to decrease the timelines and costs associated with drug development have, fortuitously, coincided with a marked advancement in the field of flow cytometry. As a result, flow cytometry is now a critically important analytical tool for drug development. *Flow Cytometry in Drug Discovery and Development* chronicles the unique application of flow cytometry in drug development and provides examples of how it can be applied along the drug development pathway from drug discovery to clinical testing.

The Part I of this book provides the reader with essential background information regarding both flow cytometry and drug development. In the chapters in Part II, the focus is on the way in which flow cytometry serves as an essential tool for each stage of the drug development life cycle. The organization of this part follows along the lines of the drug development process: drug discovery/preclinical toxicology/clinical testing. In addition, the reader will be introduced to distinct therapeutic areas such as oncology, anti-infectives, and autoimmunity/inflammation that are highlighted in several chapters.

During the drug discovery phase, the life cycle of a new drug begins with the identification of a therapeutic need and potential molecular targets that may interfere with the disease process. The next steps in the long and costly process of bringing a new drug to market are target validation, high-throughput compound screening, and multiple rounds of compound selection followed by characterization. With each successive round of selection and characterization, more rigorous standards are applied until, ultimately, a small pool of lead compounds, or potential drug candidates, is identified from the initial compound library. Chapters 4–7 address the application of flow cytometry in these early stages of drug development. The development of high-throughput flow cytometry and its application in compound screening are addressed in Chapter 4. During the earlier stages of compound selection and characterization in oncology drug discovery, cell cycle analysis using flow cytometry plays a critical role, as illustrated in Chapter 5. A wider variety of flow cytometric methods are critical during the later stages of preclinical compound selection and characterization as described in Chapter 6. The process of conducting formal GLP toxicology studies with a small pool of highly characterized potential drug candidates is also introduced in Chapter 6. The transition from preclinical testing to clinical testing is highlighted in Chapter 7 that emphasizes the use of panels of phosphospecific antibodies in cancer therapeutics.

Clinical trial testing of novel therapeutic compounds includes biomarker measurements for efficacy and pharmacodynamic (PD) effect, safety assessment, and pharmacokinetic (PK) evaluation. Chapters 8–11 describe how flow cytometry is used to achieve each of these objectives. Chapter 8 reviews the historical and continued importance of CD4 T-cell monitoring as an efficacy marker in the evaluation of antiretroviral therapy for HIV infections. Chapter 9 outlines the translational medicine approach to PD biomarker evaluation with an emphasis on inflammatory and autoimmune disorders. Chapter 10 focuses on safety concerns associated with the clinical administration of biotherapeutic compounds and the advantages of flow cytometric analysis in determining antidrug antibody responses. Flow cytometric methods can also be used for PK analysis of protein- and peptide-based therapeutics as described in Chapter 11.

The importance of validation and regulatory compliance in drug development is highlighted by the fact that an entire section of the book is dedicated to this topic (Part III, Validation and Regulatory Compliance). Different stages of the drug development life cycle are subjected to different regulatory requirements. For data to be acceptable for submission to regulatory agencies, laboratory instrumentation and analytical methods must be validated according to applicable standards.

The last, Part IV, focuses on how multiparametric flow cytometry data will be analyzed and applied in near future. At present, one of the most pressing needs in flow cytometry is for advanced data analysis tools, capable of rapidly evaluating the large, complex data sets generated by multiparametric methods. Chapter 14 presents a new paradigm for data analysis, called probability state modeling, which is capable of visualizing and analyzing multiparametric files. The last chapter brings together one of the newest, potentially most valuable applications of flow cytometry, phospho-flow, with one of the most exciting and potentially most valuable aspects of patient treatment, personalized medicine.

The contributors to the book include both flow cytometry thought leaders from academia and industry and flow cytometry users from the pharmaceutical sector. This diverse mix of expertise is typical of the collaboration among disciplines that increasingly has been a hallmark of the flow cytometry community, and part of what makes working in this challenging and exciting discipline a rewarding experience.

FOREWORD

Almost 10 years ago, Phil Marder asked me how ISAC could expand its interest in high-content screening and in particular the role that flow cytometry could play in this emerging area. Over the next decade, Phil Marder was responsible for developing and advancing high-content screening tools within the pharma industry. This book is the result of that drive of Phil Marder to show how much impact flow cytometry can have on this arena. It is with great sadness that Phil did not live to see this volume published. Without doubt, it reflects the knowledge, skills, and foresight of Phil Marder who saw the impact of flow cytometry on screening well before anyone else.

What this book does effectively is to establish a well-defined set of assays, processes, and regulations that must be considered in the performance of robust screening by flow cytometry. Specific assays are identified and discussed with sufficient details to be reproduced in screening mode. The variety of screens presented covers a wide spectrum from cell culture assays, to human and animal studies and to the role that management and manipulation of data play in today's scientific world.

Flow cytometry has been used for over 40 years to define intricate pathways and responses from single cells. Its power as the leader in single cell analysis is unquestioned by most, but many scientists fail to see its capability at the systems level. Indeed, it is this capacity that is now opening up new opportunities that were previously not possible. The ability to multiplex both cells and beads has made it to possible to track cellular phenotype and function in multiple dimensions simultaneously. In drug screening, very complex assays can be automated and analyzed in minutes rather than days. The need now is to advance the analytical tools required for the very complex opportunities.

This book brings out an entirely new set of opportunities for this mature technology. It opens up a Pandora's box of assay systems and approaches in a way that may well expand the field well beyond what was imaginable a few years ago.

J. Paul Robinson

ACKNOWLEDGMENTS

The editors wish to offer sincere thanks to all the contributors for the expertise they provided during the writing and compilation of this book. Thanks also to Jonathan Rose at John Wiley & Sons, Inc. for approaching us about the project, to the artist Glenn Vilppu for the drawing of Phil Marder, and Ira Schieren for the cover design.

CONTRIBUTORS

Fatima Abbasi, CBER, FDA, NIH, Bethesda, MD, USA

James Andahazy, Partec North America, Swedesboro, NJ, USA

C. Bruce Bagwell, Verity Software House, Topsham, ME, USA

Sean C. Bendall, Stanford University, Stanford, CA, USA

Thomas N. Denny, Duke University Medical Center, Durham, NC, USA

Roy Edward, Biostatus Ltd., Shepshed, Leicester, UK

Bruce S Edwards, University of New Mexico Center for Molecular Discovery, Albuquerque, NM, USA

Rachel J. Errington, Cardiff University, Cardiff, UK

John Ferbas, Amgen Inc., Thousand Oaks, CA, USA

Matthew B. Hale, Stanford University, Stanford, CA, USA

Kristi R. Harkins, Harkins Strategic Consulting, LLC, Madrid, IA, USA

David W. Hedley, Princess Margaret Hospital/Ontario Cancer Institute, Toronto, ON, Canada

Carla G Hill, ICON Central Laboratories, Farmingdale, NY, USA

Jonathan M. Irish, Stanford University, Stanford, CA, USA

Siddha Kasar, NJMS/UMDNJ, Newark, NJ, USA

Peter O. Krutzik, Stanford University, Stanford, CA, USA

Virginia Litwin, Covance Central Laboratory Services, Indianapolis, IN, USA

Byron H. Long, Doylestown, PA, USA

Raul Louzao, Duke University Medical Center, Durham, NC, USA

Philip Marder, Redram Consulting, LLC, Indianapolis, IN, USA

Gerald E. Marti, CBER, FDA, NIH, Bethesda, MD, USA

David McFartand, iCyt, a Sony Company, Champaign, IL, USA

Garry P. Nolan, Stanford University, Stanford, CA, USA

Denise M. O'Hara, Pfizer (formerly Wyeth), Andover, MA, USA

Elizabeth Raveche, NJMS/UMDNJ, Newark, NJ, USA

Carmen Raventos-Suarez, Bristol-Myers Squibb Company, Princeton, NJ, USA

Manjula P. Reddy, Ortho Biotech, Unit of Centocor R&D, Johnson and Johnson, Radnor, PA, USA

Susan M. Richards, Genzyme, Framingham, MA, USA

Erica Salerno, NJMS/UMDNJ, Newark, NJ, USA

Larry A. Sklar, University of New Mexico, Albuquerque, NM, USA

Paul J. Smith, Cardiff University, Cardiff, UK

Michelle J. Schroeder, Amgen Inc., Thousand Oaks, CA, USA

Valerie Theobald, Genzyme, Framingham, MA, USA

Ole Vesterqvist, Covance Central Laboratory Services Inc., Indianapolis, IN, USA

Brooke Walker, Duke University Medical Center, Durham, NC, USA

John Wong, Duke University Medical Center, Durham, NC, USA

Dianna Y. Wu, Bristol-Myers Squibb Company, Pennington, NJ, USA

Yuanxin Xu, Genzyme, Framingham, MA, USA

Yao Yuan, NJMS/UMDNJ, Newark, NJ, USA

PART I

INTRODUCTION

PHILIP MARDER AND VIRGINIA LITWIN

The discovery and development of novel therapeutic compounds is a lengthy, difficult, and expensive process with recent estimates of more than 1.2 billion dollars required for each new drug brought to market. As a result, the standard processes of the pharmaceutical industry are being reevaluated and modified in order to increase efficiencies in the drug development process. One approach in process transformation is to promote more informed decision making by incorporating advanced technologies such as flow cytometry.

A wide variety of flow cytometric methods are employed during various stages of the drug development life cycle. This book explores many of the benefits and complexities associated with this unique application of the technology. Part I is intended to provide the reader with essential background information regarding both flow cytometry (Chapters 1 and 2) and drug development (Chapter 3).

Flow Cytometry in Drug Discovery and Development, Edited by Virginia M. Litwin and Philip Marder

1

INTRODUCTION TO FLOW CYTOMETRY

ELIZABETH RAVECHE, FATIMA ABBASI, YAO YUAN, ERICA SALERNO, SIDDHA KASAR, AND GERALD E. MARTI

1.1 INTRODUCTION

This chapter presents, in basic terms, the concepts and principles of flow cytometry. Numerous books and articles describing flow cytometers and their use in a clinical and biomedical research setting have been published [1–7]. In this chapter, flow cytometers will be discussed from their infancy arriving at the current instrumentation that allows for detection of numerous features of individual cells or particles, including determination of size and granularity, surface marker expression, DNA content, intracellular protein expression, and function. The key to flow cytometers is that the analysis is done on cells in suspension [8–10]. The analysis of individual cells (or particles) rather than the whole population allows for detection of multiple properties measured on the same cell. The detection is rapid (as fast as the cell in the fluid sheath passes through the laser beam). In addition to analysis of individual cells, some types of flow cytometers can physically sort cells based on signals associated with the parameters being detected. The term fluorescence-activated cell sorter or FACS has been adopted to refer to this type of analysis [11]. Flow cytometry is a very useful tool for both clinical diagnosis and scientific research. The history of flow cytometers has been the subject of numerous reviews [12–20]. The first flow cytometers were introduced in the mid-1970s and first used for DNA analysis and leukemia immunophenotyping [7, 21–25]. A further impetus to bring flow cytometers to the forefront of clinical labs came in the early 1980s with the discovery that individuals infected with the HIV virus developed AIDS, which could be monitored

Flow Cytometry in Drug Discovery and Development, Edited by Virginia Litwin and Philip Marder by enumerating the number of $CD4^+$ T cells by flow cytometric analysis [26–30]. Currently, there are emerging areas with flow cytometric applications including the enumeration of $CD34^+$ hematopoietic stem cells [29, 31, 32], detection of circulating metastatic tumor cells [33–37], determination of antigen-specific T cells [38–40], and identification of pathogens [41–45], to list a few. Combination of sorting with molecular analysis represents an important use of the sorting aspects of flow cytometers. There are over 100,000 flow cytometers in use and the employment of this instrument in clinical diagnostics has increased dramatically, particularly with the increase in FDA-approved fluorochrome reagents for *in vitro* diagnostics (fluorochrome-conjugated antibodies). However, in third-world countries, access to clinical flow cytometers is not optimal [46, 47]. The use of flow cytometers and the impact of this instrument on biomedical and clinical studies can be appreciated by looking at the increase in publications in which the word "flow cytometry" appeared in the abstract or title with time (Figure 1.1).

Improvements in instrumentation and computer-assisted analysis have made the flow cytometer a critical instrument in biomedical research, clinical diagnostics, and drug discovery. Herzenberg was honored for his work in flow cytometry by the American Association for Clinical Chemistry with the Ullman Award in 2002 and some of the history described in this chapter comes from his lecture and the accompanying article [19]. The original description of the first flow cytometer was provided in *Scientific American* [48]. This instrument consisted of one laser and two light detectors, one for forward scatter to measure cell size and the other for fluorescence. This meant that one was restricted to measuring a single marker. When one of the authors of this article used that prototype instrument, the LASL, we were measuring the DNA content of individual cells. This was one of the first uses of these

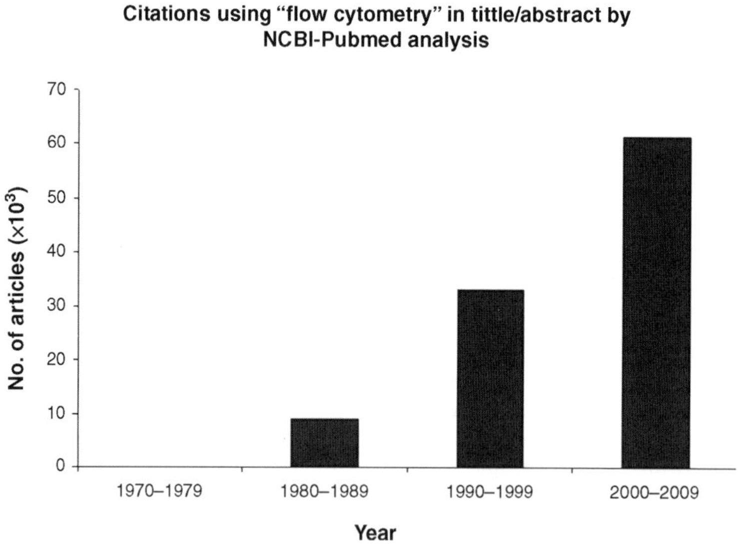

FIGURE 1.1 A bar graph showing the number of publications having "flow cytometry" in their title/abstract since 1970 to present. There is almost a 150% increase since 1980–1989.

early flow cytometers since reagents were available that not only bound specifically to DNA (e.g., ethidium bromide developed by Dittrich and Gohde in 1969 [49]) but also emitted fluorescence when excited with a laser. Much of the essentials of the modern-day FACS are the same as those in the early flow cytometers. However, these early flow cytometers were cumbersome and required an on-site engineer. The laser was water cooled and alignment issues were critical. In addition, no computer was attached to these early flow cytometers, nor were programs available for data analysis [50]. At one point, we took Polaroid pictures of oscilloscopes and sent data to a DEC10 supercomputer and wrote our own programs for cell cycle analysis.

Although the development of FACS depended on many advances in various disciplines including dye chemistry, electronics, and computers, one important breakthrough that was critical for the development of flow cytometers was the principle of measuring cells or particles in liquid suspension. Advances in the flow principle began in 1940 with Crosland-Taylor using the flow principle and light scatter to measure blood cells [51]. The breakthrough technology was first developed by Coulter and the Coulter principle describes changes in the electrical conductivity of a small saline-filled orifice as a cell passes through it. In 1953, Wallace Coulter and his brother Joe obtained a U.S. patent for the Coulter counter that automated counting of particles, particularly cells in the blood [52]. The use of a liquid stream (or a sheath) to which a sample is introduced allows individual cells to be distributed in the sheath that then passes through a nozzle (detecting electrical conductivity changes) to generate a trigger, which indicates the presence of a signal that exceeds the threshold level.

Many of the applications for FACS analysis involve the identification of membrane markers via the use of fluorochrome-tagged antibodies, which recognize these markers. Many of these membrane markers are surface proteins or surface antigens, which help to define the cell. These antigens are used to classify the cells and are often assigned a cluster of differentiation number or a CD number. Antibodies (which are normally produced by B lymphocytes) can be made that specifically bind to these CD molecules. There are more than 200 CD molecules that have been identified and specific antibodies have been produced that recognized these CD markers [53–55]. In addition, many of these antibodies are commercially available as labeled antibodies with different fluorochromes.

1.2 BASIC PRINCIPLES OF HOW A FLOW CYTOMETER WORKS

The basic components of a flow cytometer (Figure 1.2) consist of (1) a flow cell that forces single cells into the middle of a fluidic sheath, (2) a laser source of light, (3) optical components to focus light of different wavelengths (colors) onto a detector, (4) a photomultiplier to amplify the signal, and (5) a computer.

In a basic flow cytometer, the sample (containing the cells tagged with fluorochromes in a liquid) is drawn up and pumped into the flow cell through tubing. The cells flow through the flow chamber rapidly and singly and are passed through one or more laser light beams. As the laser beam hits the cells, the light beam is scattered in a forward direction and a side direction. Fluorescence emission can also be detected.

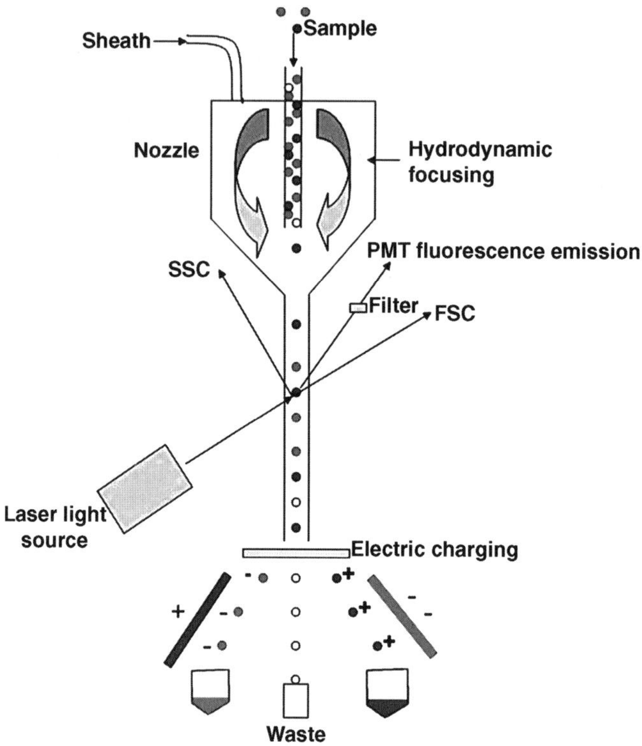

FIGURE 1.2 Diagrammatic representation of a basic flow cytometer. The fluorescently labeled cells are hydrodynamically focused into a single file in the flow cell. Individual cells are excited by the laser light source and the fluorescence emissions, FSC, and SSC are detected. The cells can then be given a particular charge based on their fluorescence profile and deflected toward the oppositely charged plates. In the figure, light grey cells and dark grey cells are given negative and positive charges, respectively, and are thus deflected toward two different tubes.

Scatter or fluorescence is captured, filtered (based on the wavelength), and directed to the appropriate photodetectors for conversion to electronic signals. The electronics in the flow cytometer amplify the signal and convert the analog data to digital data, which can then be analyzed by computer software programs.

1.3 FLUIDICS

1.3.1 Flow Cells

In order to perform flow cytometric analysis, the sample must be in a suspension and the cell in the sample stream must be centered in the laminar flow [49]. Hydrodynamic focusing induces cells to orient with their long axis parallel to the flow. The end result is that the introduced sample passes by the laser with each cell oriented in the center of the sample stream in a particular manner in three dimensions.

1.4 OPTICS

Flow cytometers depend on the laws of optics, such as reflection, refraction, and other principles, which are not new but based on works established centuries ago [56]. Optics are present on both the excitation and the emission side. The excitation optics encompass the lasers and the lenses that focus the laser beam. The emission optics are involved in collecting the emission following excitation. These involve lenses to collect emitted light and mirrors and filters to route specified wavelengths of the collected light to designated optical detectors. Light coming out of a laser may be considered a beam but fluorescence must be considered as a photon.

1.4.1 Light Scatter

Due to differences between the refractive indices of cells and the surrounding sheath fluid, light impinging upon the cells is scattered. The forward light scatter (FSC) provides empirical information on cell size. Light scattered in an orthogonal direction or side scatter (SSC), which is collected by a different detector, provides information about granularity.

1.4.2 Types of Lasers

Laser stands for light amplification by stimulated emission of radiation. Gas lasers have mirrors at each end of a cylinder or plasma tube filled with an inert gas. The gas is ionized to a higher energy state by a high-voltage electric current. When these excited atoms return to the ground state, they give off photons of a characteristic wavelength. The photons can be reflected by the mirrors and the excitation of the atoms in the plasma can be amplified but the wavelengths of the emission still are the characteristic wavelengths for that gas [57]. In the front of the laser there is a small optic that allows the transmitted light to form a laser beam of desired output wavelengths. The light from lasers is a stimulated emission and it has uniform characteristics. For current stream-in-air instrumentation, it is desirable to have at least 50 mW of power for each laser line in use, since the fluorescence signal (and thus sensitivity) increases with laser power. Cytometers use multiple lasers that are positioned spatially such that there is a time delay for each laser beam intercept with the cell. Newer solid-state diode lasers [58–60] are becoming prevalent and these are significantly cheaper than the older gas ion lasers. Diode lasers are pumped by input of electric current. A partial list of different lasers is presented in Table 1.1.

The most common lasers for flow cytometers are the argon ion lasers that run at 488 nm. The lasing medium in an ion laser is plasma. A high-voltage pulse is used to ionize the gas to start the plasma. Ion lasers require a high current to maintain the plasma discharge. In addition to the 488 nm emission, argon ion lasers also emit at 515 nm (green) and 457 nm (violet-blue). Other emissions can be obtained using specially coated mirrors. The new low-power, air-cooled argon laser gives out 25 mW at 488 nm. To obtain other lines of emission, large lasers capable of giving 100 mW in UV must be used.

TABLE 1.1 Partial List of Laser with Their Excitation Wavelength Line and the Fluorochromes which Can be Detected

Laser	Excitation Line (nm)	Fluorochrome
UV	355	Hoescht 33342, 33250
He–Cd	325	DAPI, ELF-97, AMCA (AlexaFluor 350), INDO-1
Mercury lamp		
Violet	405	Pacific Blue, CasB
Krypton ion	435	CasY, AlexaFluor 405 (AF405)
Blue (argon)	488	PE-TR, 6FP, FITC, PE, AF488, PE-Cy7, PerCP, PE-Cy5, SYTO 9, PerCP-Cy5.5
Red (solid state)	640	APC, APC-Cy7, SYTO 59–61
He–Ne	633	AF647, APC-Cy7
Red diode	635	APC-Cy5.5, AF700
Yellow/green	561	PE-Texas Red, PerCP, PE

Krypton lasers can give out strong blue-green lines and UV and violet lines. Krypton lasers need to be water cooled and optimized and the alignment is very difficult. Another type of laser is a dye laser and the lasing medium in a dye laser is a fluorescent dye. The selection of dye depends on the wavelength at which the operation is desired. Helium–neon (He–Ne) lasers are also small, air cooled, and stable. The most common lasers emit at 633 nm and have power outputs ranging from 1 to 50 mW. He–Ne lasers are available at 633, 543, 594, and 611 nm. Helium–cadmium (He–Cd) lasers emit 5–200 mW in blue (441 nm) and 1–50 mW in UV (325 nm). They plug into the wall and do not require water cooling.

1.4.3 Filters for Emission

All signals that are emitted from fluorochromes that are excited as the cells to which they are bound are interrogated by the laser beams are routed to detectors via a system of mirrors and optical filters. In addition, beam splitters direct light of different wavelengths in different directions. The most commonly used filters are short-pass filters (which transmit wavelengths of light equal to or shorter than the specified wavelength), long-pass filters (which transmit wavelengths of light equal to or larger than the specified wavelength), and band-pass filters (which allow a narrow range of wavelengths to reach the detector). An example of these types of filters is presented in Figure 1.3. Because each fluorochrome has an emission spectrum, the choice of filters optimizes detection of the specific fluorochrome by one detector or photomultiplier tube (PMT).

Detection of fluorochromes requires selection of appropriate filters that are placed before each detector or PMT. The type of filter selected must collect as much emitted light from the primary fluorochrome for high sensitivity, but as little as possible from other fluorochromes to reduce the compensation required. A partial list of filters is presented in Table 1.2.

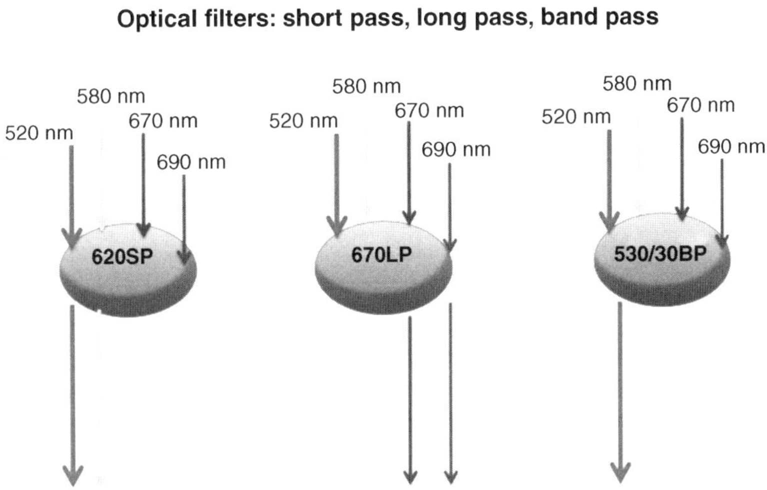

FIGURE 1.3 An example of fluorescence emission of various wavelengths (top) as it passes through different types of optical filters (bottom).

TABLE 1.2 Partial List of Filters Typically Employed with Various Fluorochromes

Fluorochrome	Filter
Pacific Blue, BD Horizon V450, CasB	440/40
AmCyan	525/40
FITC, AlexaFluor 488 (AF488)	530/30
CasY, AF430	545/90
PE	585/40
PE-Texas Red, AF595	625/40
APC, AF647	660/20
PE-Cy5, PerCP-Cy5.5, PerCP	695/40
APC-Cy5.5	705/50
AF700	720/45
PE-Cy7, BD-APC H7	780/60

1.5 TYPES AND CHOICE OF FLUOROCHROMES

A fluorochrome is a fluorescent marker that emits a particular wavelength when a laser light hits it. Fluorescence occurs when a molecule, which is excited by light from a laser at one wavelength, loses its energy and emits light of a longer wavelength. The emitted wavelength is what is detected. The excited and emitted light are of different wavelengths. The fluorescence intensity that is emitted is proportional to the quantity of binding sites for the fluorescent compound on the cell. Therefore, the more the fluorescence that is emitted the more the binding sites on the cell. For instance, for an antibody tagged with FITC (fluorescein isothiocyanate, which is excited by a 488 nm

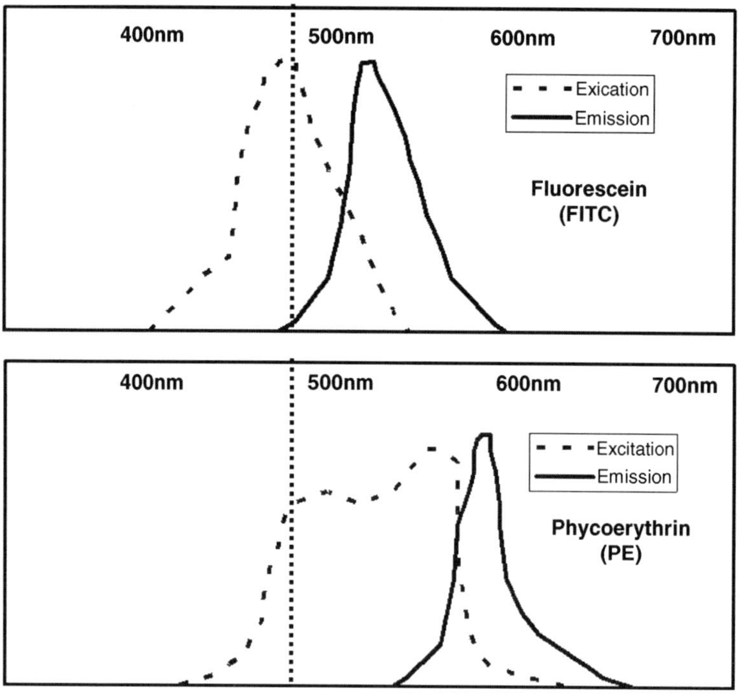

FIGURE 1.4 Excitation and emission spectra of FITC and phycoerythrin (PE). Fluorescent molecules absorb light of a characteristic wavelength and emit light of a longer wavelength. FITC and PE that are commonly used for flow cytometry absorb at 488 and 488–560 nm, respectively, but emit at 520 and 590 nm, respectively. Thus, they can be excited by the same laser line and used together in the same tube [10].

argon laser but emits in the 520 nm (green) range) that recognizes and binds to CD4, the more the 520 nm emission the more the CD4 on the cell (Figure 1.4).

The fluorochrome label for a reagent depends on instrument configuration (type and number of lasers and type of optical filters and detectors), which determines if a given instrument can excite a given fluorochrome and detect the emission. While it is not possible to uniformly state the best fluorochrome combination, there are a few guidelines that can help in this choice. The first issue is to determine what is the reagent brightness, which takes into account the resolvable signal associated with the presence of the marker being detected by comparing a negative and a positive sample. The negative population emission is the background emission. Background is signal (emission) due to electronic noise (dark current), cell autofluorescence, nonspecific staining, and background emission that is a spillover from another fluorochrome [61, 62]. The rule of thumb is to use the brightest reagents possible [63, 64]. There is a caveat to this statement. The spillover problems increase as the number of colors to be resolved (different emissions) increases. Compensation can help prevent the spillover contribution, but as a rule of thumb, one should use fluorochromes whose emissions have the least amount of spectral overlap [65, 66]. In addition, logically the markers with the least amount of expression on a given population should be detected with a reagent that is labeled with the brightest fluorochrome. However, this weakly expressed antigen should be stained with a fluorochrome-tagged reagent that does not have spillover issues with another fluorochrome reagent recognizing a cell marker that is highly expressed [64]. A final word of caution is to take into account that a fluorochrome as a single reagent may give different results when employed in a multicolor reagent cocktail and this is a fidelity issue. To determine if this is a problem, one can compare the antibody–fluorochrome conjugate by itself and compare the results with the results obtained for this reagent when it is in the multicolor reagent cocktail. One should try and use reagent combinations that have good fidelity when used in a multicolor reagent cocktail.

1.6 COMPENSATION

Due to overlap in the emission spectra of different dyes, it is often not possible to choose emission filters that uniquely measure only one of the dyes in a multicolor experiment. Due to this spectral overlap, one fluorochrome can contribute a signal to several detectors; therefore, the contribution in detectors not assigned to that fluorochrome must be removed from the total signal in those detectors. Compensation is an artificial means of eliminating spectral overlap between two different fluorochromes by mathematical means and is not just a subtraction process [65, 66]. Compensation between detectors can be performed either by hardware, after signal detection but before logarithmic conversion and/or digitization, or by uncompensated data that are analyzed post-collection by software (Figure 1.5).

1.7 THRESHOLD AND GATES

An electronic threshold is defined as a gate for acquiring signals. Only events with intensity greater than the threshold will be processed and analyzed. Thresholding is most often used to eliminate debris. A gate is a boundary that can be used to identify subpopulations and this limits the number of events that are analyzed (note that these events are acquired in the listmode data file but not analyzed). Gates are often used to identify the lymphoid cells for analysis.

1.8 ANALYSIS

The electric pulses that are detected by the PMTs are amplified (log amplification is most often used to measure fluorescence). These amplified signals are converted from analog to digital. Data can be stored as a listmode file, which consists of a complete listing of all events and parameters that were measured [67]. One can take a listmode file and subject the data to analysis such as regions and gating but one cannot adjust amplification or fluorescence [68].

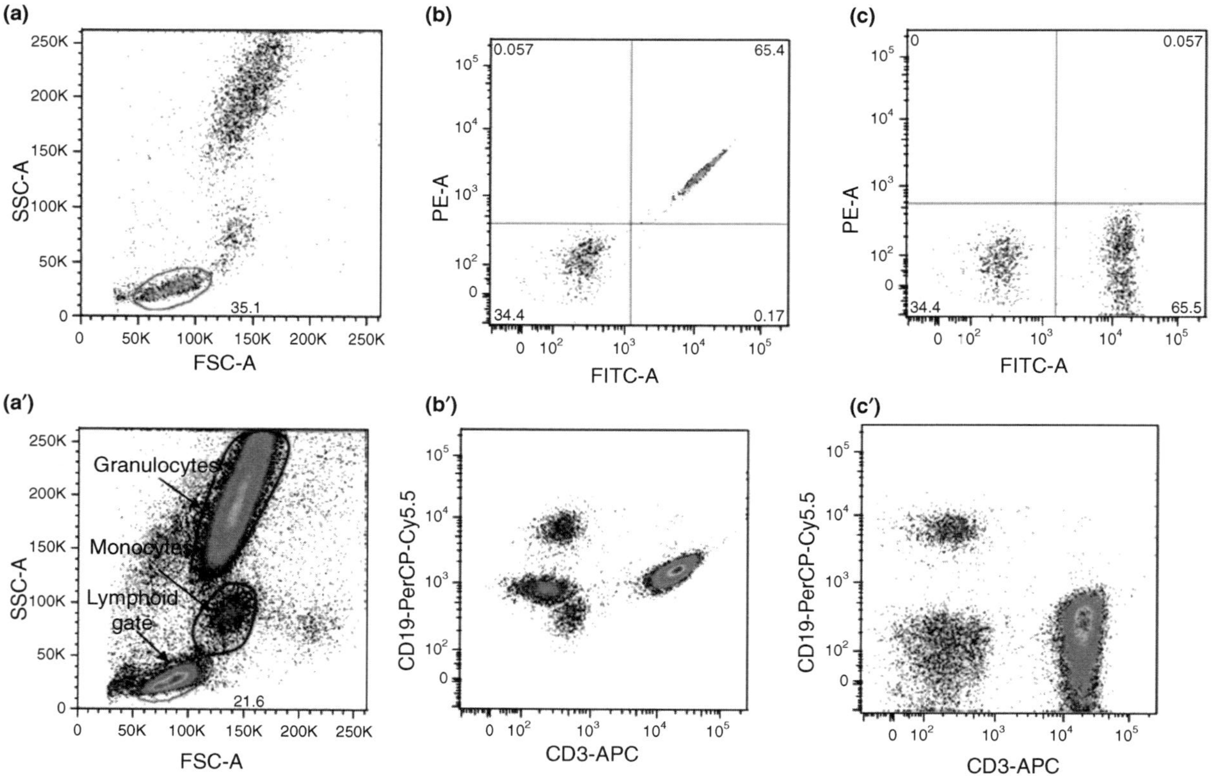

FIGURE 1.5 Compensation controls (a–c) human PBMCs stained with a single FITC-labeled antibody. (a) Cells are gated on the lymphoid gate based on forward scatter (x-axis) and low side scatter (y-axis). (b) Uncompensated data demonstrating that the FITC signal is "spilling" into the PE channel with 65.4% of the cells demonstrating dual positivity incorrectly. (c) Data after compensation with only single positive cells expressing FITC and no PE. Note that the PE mean for the FITC negative and FITC positive cells is nearly identical. (a′–c′) Two-color data of human PBMCs stained with anti-CD19 labeled with PerCP-Cy5.5 (B-cell marker) and anti-CD3 labeled with APC (T-cell marker). There should be no dual positive cells. Data were collected as listmode and compensation was added after data collection: (a′) lymphoid gate; (b′) uncompensated data; (c′) data after compensation. (See the color version of the figure in the Color Plates section.)

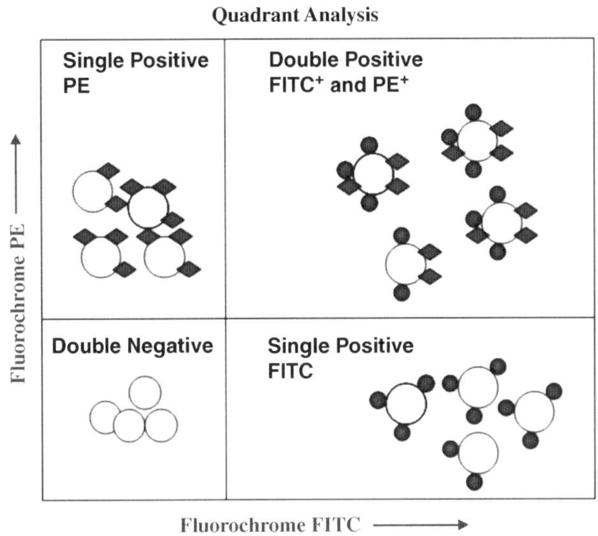

FIGURE 1.6 Idealized two-parameter quadrant analysis. A population of cells is stained for two markers labeled with PE indicated with a diamond (y-axis) and FITC indicated with a solid small circle (x-axis). Lower left quadrant: cells lacking both the markers, and hence double negative. Lower right quadrant: cells FITC and shown as cells with solid circles. Upper left quadrant: cells PE and shown as cells with diamonds. Upper right quadrant: cells expressing both markers, also called double positive, and shown as individual cells (diamonds and solid circles) together.

1.8.1 Flow Histogram

For single-color analysis, the events can then be plotted as a single parameter such as a histogram, in which the x-axis is the measurement and the y-axis is the number of events. Usually the x-axis corresponds to channels (typically, 1024 channels); the brighter the specific fluorescence the higher the channel number. A new "Logicle" display method (also known as biexponential method) when analyzing flow data enables the close to zero signals to be shown on the plot graph that combines both the logarithmic and linear scales, providing a more complete way of interpretation of data [11, 69, 70]. Multicolor flow analysis is often displayed as two-color analysis. In Figure 1.6, an idealized phenotype of cells in two-color analysis is shown.

1.8.2 Thresholding and Doublet Discrimination

The flow cytometer parameters can be set such that only events whose intensity is greater than a particular threshold value are recorded. This is called thresholding and it can be used to eliminate debris (Figure 1.7a), that is, cells having a very low FSC and SSC. Cells of interest can be gated based on fluorescence parameters; for example, expression of CD45 with low side scatter predominantly identifies lymphoid cells

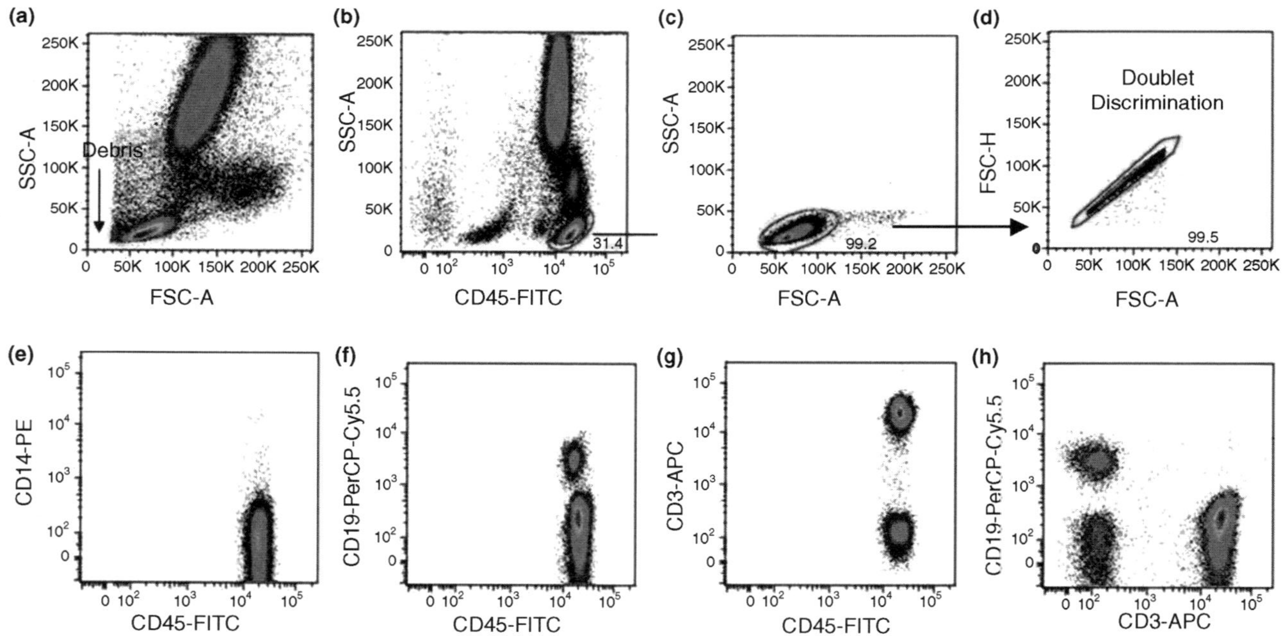

FIGURE 1.7 Typical gating for lymphoid cells. (a) The different populations of cells present in the sample are visualized based on their size (FSC) and complexity (SSC). Thresholding was performed to remove debris (arrow). (b) The cells are stained with a FITC-labeled antibody that recognizes CD45, a hematopoietic lineage marker (referred to as CD45-FITC). The circled population indicates lymphoid gate (high FSC but very low SSC). (c) CD45-gated lymphoid cells are reevaluated to exclude large $CD45^+$ cells. (d) Doublets that are distinguished from singlets on the basis of FSC-H/FSC-A ratio are excluded from the area of interest indicated with the circle. Detection of macrophages ($CD45^+CD14^+$) (e), B cells ($CD45^+CD19^+$) (f), and T cells ($CD45^+CD3^+$) (g). (h) Two subpopulations, B and T cells, can be easily visualized on the same dot plot.

(Figure 1.7b and c). Although hydrodynamic focusing streams the cells into a single file, occasionally two cells stick together. A doublet made of two single positive cells (each one positive for a different fluorochrome) can be erroneously recorded as a double positive cell. Hence, doublet discrimination is crucial. Doublets will have a higher FSC height/FSC area ratio as compared to singlets. One can set a gate around the events in which there is a linear relationship between FSC-H and FSC-A (Figure 1.7d) and these cells can then be analyzed for different markers (Figure 1.7e–h).

1.8.3 Two-Color Dot Plot Versus Contour Plot

The fluorescence of two different markers can be represented in 2D using the two-color dot plot (Figure 1.8a–d) or a contour plot (Figure 1.8b′–d′). There are advantages to data displayed in either mode and data are amenable to either type of analysis. In Figure 1.8, identical data are displayed as a dot plot and a contour plot. In a dot plot, each dot represents one or more events (events are usually cells that have passed the criteria of thresholding and gating) that are determined by the user. The density of events can be color coded (e.g., red implying highest density). The events are gated on the lymphoid gate on the basis of FSC and SSC (Figure 1.8a) with a second gate to include only lymphoid cells that are CD3 positive (Figure 1.8c′). There exists a small population of cells positive for both CD4 and CD8, which is visible in the dot plot (Figure 1.8d). Contour plots show the same data in which rings represent a defined percentage of total events with a particular combination of fluorescence intensities (Figure 1.8b′–d′). This type of data display removes the outliers allowing one to clearly see subpopulations of cells. In addition, one can smoothen the contour plots making it even more difficult to visualize the rare population (Figure 1.8d′, $CD4^+/CD8^+$ cells). Using contour plots as a formatting option, other display options include linear density and log density contour plots in which the contour lines are defined as a percentage of the maximum numbers. There is an additional option in which one can combine a contour plot and a dot plot in which dots are displayed below the lowest contour line allowing the observation of the rare population in a contour plot. Despite different display options, the data quadrant statistics would be identical in both the dot plot and the contour plot.

1.9 SORTING

Cells (or particles) of interest (expression of desired markers) can be purified or sorted [71]. In most flow cytometers equipped with sorting capabilities, the liquid sheath stream is regularly broken into droplets by the vibration of the piezoelectric crystal attached to the flow chamber. A cell passing through the laser meeting the selection criteria based on the fluorescence pattern is electrically charged in the droplet. These droplets containing the charged cells are then deflected and collected into awaiting tubes/wells or onto slides. Depending on the further downstream applications of these sorted cells, they may have to be collected in appropriate buffers or under sterile conditions. The temperature of the sheath fluid and sample collection tubes may need to be controlled. In the example shown, mouse spleen cells to be sorted are shown on the

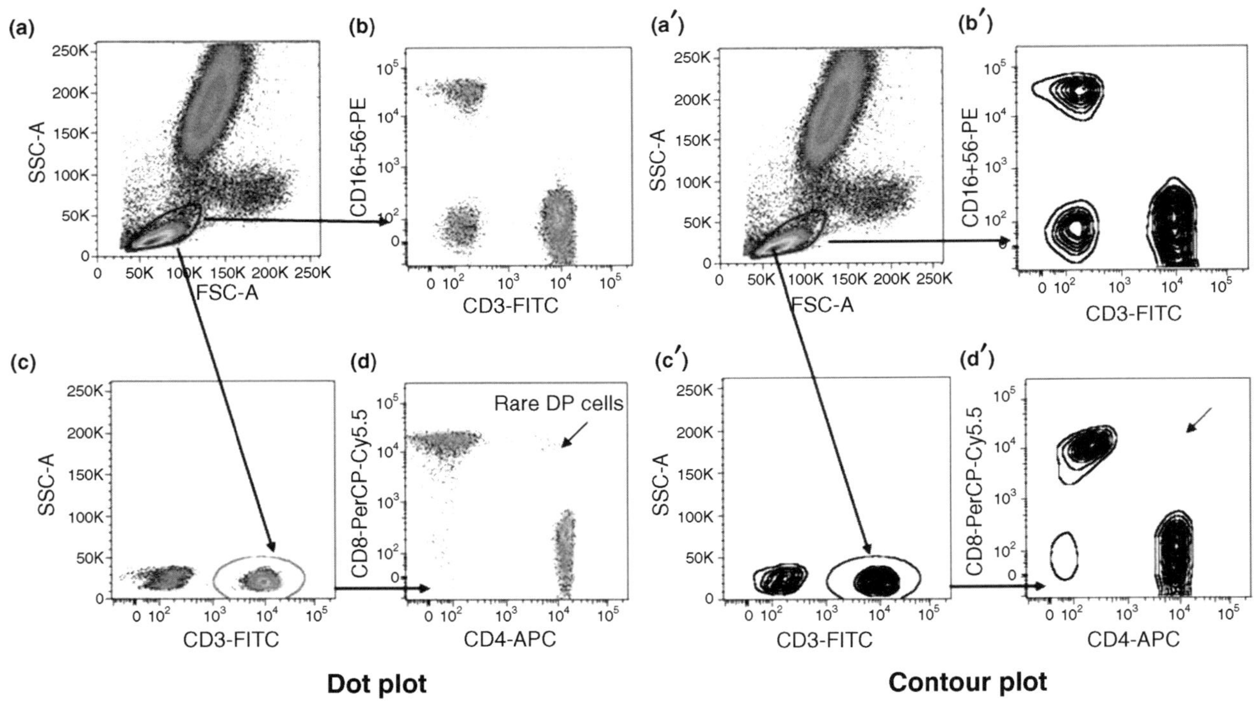

FIGURE 1.8 Dot plot versus contour plot of the same data. Cells were stained with CD3-FITC, CD16 and CD56-PE, CD4-APC, and CD8-PerCP-Cy5.5. Plots (b)–(d) and (b′)–(d′) show similar results by both dot plot and contour plot. However, the rare population of double positive cells can be visualized in the dot plot (d) but not in the contour plot (d′). For details refer to the text.

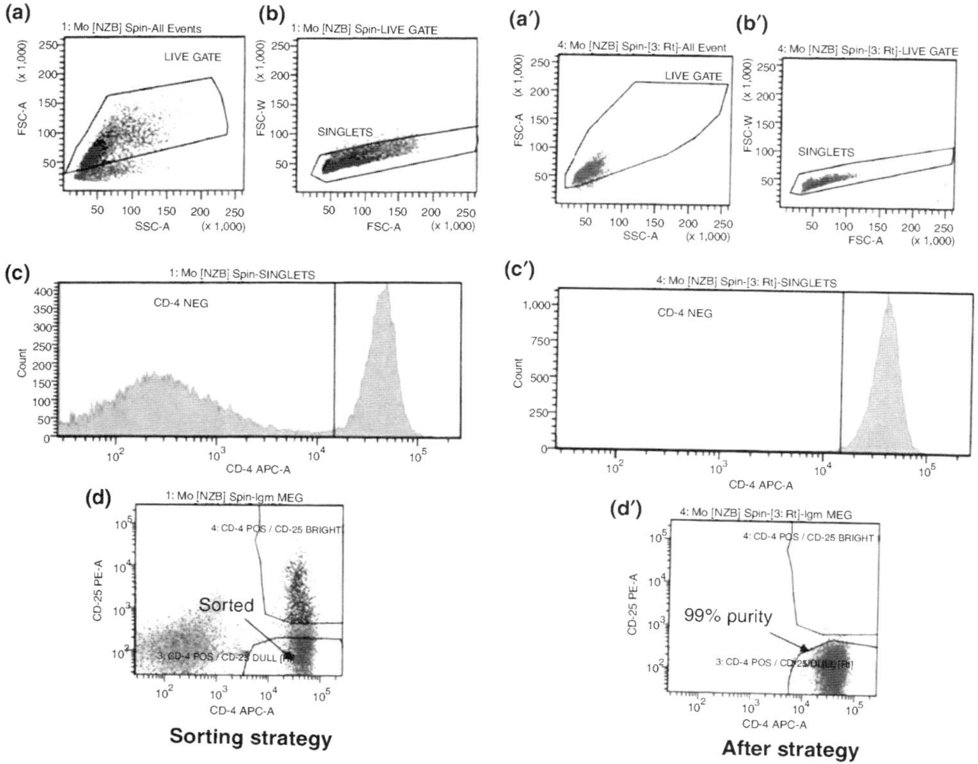

FIGURE 1.9 Sorting of CD4$^+$ cells from cells stained with CD4-APC and CD25-PE. (a) After thresholding, the live cells were gated. (b) Doublet discrimination was performed on the live cells. (c) Single-color histogram showing the CD staining before sorting (left side). (d) Two-color dot plot representing the different subpopulations present after thresholding and gating but before sorting. The indicated subpopulation was sorted. (a′–d′) The sorted population was reacquired to assess the efficiency of sorting. The FSC and SSC of the sorted cells (a′) were limited to the cells present in the gate region of (a). (b′) The sorted cells were predominantly singlets. (c′) The single-color histogram of CD4 showed predominantly CD4$^+$ population after sorting. (d′) Two-color dot plot representing the different subpopulations present after sorting. The quadrant statistics indicate that the sorted cells are a 99% pure population of CD4$^+$CD25$^-$ cells.

left (Figure 1.9a–d) and analysis of the cells following sorting is shown on the right (Figure 1.9a′–d′). The live cell gate (Figure 1.9a and a′) indicates that following sorting the cells that were gated on CD4 and the live gate resulted in 99% of the gated cells meeting the gating criteria (live, singlets, and $CD4^+$).

1.10 CONCLUSION

Since their introduction in the 1970s, the design and applications of flow cytometers have undergone tremendous change. Current flow cytometers are rapid, use multiple lasers (five lasers on several instruments), and can detect more than 20 different fluorochrome tags. Some of the novel applications include fluorescence *in situ* hybridization (FISH) using flow cytometer and Amnis ImageStream, which is a blend of flow cytometry and microscopy and allows the visualization of single cells. The field of flow cytometry holds great promise for research and clinical diagnostics.

REFERENCES

1. Shapiro HM. *Practical Flow Cytometry*, 4th ed., Wiley-Liss, 2003, 736 pp
2. Givan AL. *Flow Cytometry: First Principles*, 2nd ed., Wiley-Liss, 2001, 296 pp.
3. Valeriote FA. *Basic and Clinical Applications of Flow Cytometry*, 1st ed., Springer, 1995, 336 pp.
4. Owens MA. *Flow Cytometry Principles for Clinical Laboratory Practice*, 1st ed., Wiley-Liss, 1994, 228 pp.
5. Matsui Y, Shapiro HM, Matsuo Y, Ohkubo K, Matsumoto K, Yokoyama M. Clinical application of flow cytometry. *Nippon Rinsho* 1983;41(5):1107–1119.
6. Ault KA. Clinical applications of fluorescence-activated cell sorting techniques. *Diagn. Immunol.* 1983;1(1):2–10.
7. Keren DF. *Flow Cytometry in Clinical Diagnosis*, 3rd ed., American Society Clinical Pathology, 2001, 739 pp.
8. Omann GM, Coppersmith W, Finney DA, Sklar LA. A convenient on-line device for reagent addition, sample mixing, and temperature control of cell suspensions in flow cytometry. *Cytometry* 1985;6(1):69–73.
9. Hawley RG. *Flow Cytometry Protocols*, 2nd ed., Humana Press, 2004, 448 pp.
10. Robinson JP. *Handbook of Flow Cytometry Methods*, Wiley-Liss, 2008, 260 pp.
11. Tung JW, Heydari K, Tirouvanziam R, Sahaf B, Parks DR, Herzenberg LA. Modern flow cytometry: a practical approach. *Clin. Lab. Med.* 2007;27(3):453–468.
12. Shapiro HM. Multistation multiparameter flow cytometry: a critical review and rationale. *Cytometry* 1983;3(4):227–243.
13. Fulwyler MJ. Flow cytometry and cell sorting. *Blood Cells* 1980;6(2):173–184.
14. Steen HB. Light scattering measurement in an arc lamp-based flow cytometer. *Cytometry* 1990;11(2):223–230.
15. Noguchi Y, Sugishita T. A historical review of flow cytometers and characteristics of commercial instruments. *Nippon Rinsho* 1992;50(10):2299–2306.

16. Darzynkiewicz Z, Bedner E, Li X, Gorczyca W, Melamed MR. Laser-scanning cytometry: a new instrumentation with many applications. *Exp. Cell Res.* 1999;249 (1):1–12.

17. Zuba-Surma EK, Kucia M, Abdel-Latif A, Lillard JW, Jr., Ratajczak MZ. The ImageStream system: a key step to a new era in imaging. *Folia Histochem. Cytobiol.* 2007;45(4):279–290.

18. Robinson JP. Comparative overview of flow and image cytometry. *Curr. Protoc. Cytom.* 2005; Chapter 12:Unit 12.1.

19. Herzenberg LA, Parks D, Sahaf B, Perez O, Roederer M. The history and future of the fluorescence activated cell sorter and flow cytometry: a view from Stanford. *Clin. Chem.* 2002;48(10):1819–1827.

20. Shapiro HM. Trends and developments in flow cytometry instrumentation. *Ann. N.Y. Acad. Sci.* 1993;677: 155–166.

21. Loken MR, Stall AM. Flow cytometry as an analytical and preparative tool in immunology. *J. Immunol. Methods* 1982;50(3):R85–R112.

22. Dean PN, Jett JH. Mathematical analysis of DNA distributions derived from flow microfluorometry. *J. Cell Biol.* 1974;60(2):523–527.

23. Krishan A. Rapid flow cytofluorometric analysis of mammalian cell cycle by propidium iodide staining. *J. Cell Biol.* 1975;66(1):188–193.

24. Crissman HA, Tobey RA. Cell-cycle analysis in 20 minutes. *Science* 1974;184 (143):1297–1298.

25. Shapiro BE, Felberg NT, Donoso LA, Augsburger JJ, Shields JA, Gamel J. Flow cytometry of uveal melanomas. *Cancer Biochem. Biophys.* 1986;8(3):235–238.

26. Schnizlein-Bick CT, Spritzler J, Wilkening CL, Nicholson JK, O'Gorman MR. Evaluation of TruCount absolute-count tubes for determining CD4 and CD8 cell numbers in human immunodeficiency virus-positive adults. Site Investigators and The NIAID DAIDS New Technologies Evaluation Group. *Clin. Diagn. Lab. Immunol.* 2000;7 (3):336–343.

27. Mellors JW, Munoz A, Giorgi JV, Margolick JB, Tassoni CJ, Gupta P, Kingsley LA, Todd JA, Saah AJ, Detels R, Phair JP, Rinaldo CR, Jr., Plasma viral load and $CD4^+$ lymphocytes as prognostic markers of HIV-1 infection. *Ann. Intern. Med.* 1997;126(12):946–954.

28. HPPMC Study. Predictive value of absolute CD4 cell count for disease progression in untreated HIV-1-infected children. *AIDS* 2006;20(9):1289–1294.

29. Barnett D, Granger V, Whitby L, Storie I, Reilly JT. Absolute $CD4^+$ T-lymphocyte and $CD34^+$ stem cell counts by single-platform flow cytometry: the way forward. *Br. J. Haematol.* 1999;106(4):1059–1062.

30. Pinkerton PH, Rachlis A, Reis MD, Robinson JB. Flow cytometry in the management of infection with human immunodeficiency virus. *Transfus. Sci.* 1995;16(4):321–332.

31. Chin-Yee I, Anderson L, Keeney M, Sutherland DR. Quality assurance of stem cell enumeration by flow cytometry. Canadian QASE Study Group. *Cytometry* 1997;30(6):296–303.

32. Fritsch G, Buchinger P, Printz D. Use of flow cytometric CD34 analysis to quantify hematopoietic progenitor cells. *Leuk. Lymphoma* 1993;10(6):443–451.

33. Ling J, Wiederkehr U, Cabiness S, Shroyer KR, Robinson JP. Application of flow cytometry for biomarker-based cervical cancer cells detection. *Diagn. Cytopathol.* 2008;36(2):76–84.

34. Bartal AH, Gazitt Y, Zidan G, Vermeulen B, Robinson E. Clinical and flow cytometry characteristics of malignant pleural effusions in patients after intracavitary administration of methylprednisolone acetate. *Cancer* 1991;67(12):3136–3140.

35. Finn WG, Peterson LC, James C, Goolsby CL. Enhanced detection of malignant lymphoma in cerebrospinal fluid by multiparameter flow cytometry. *Am. J. Clin. Pathol.* 1998;110(3):341–346.

36. Galanzha EI, Shashkov EV, Spring PM, Suen JY, Zharov VP. *In vivo*, noninvasive, label-free detection and eradication of circulating metastatic melanoma cells using two-color photoacoustic flow cytometry with a diode laser. *Cancer Res.* 2009;69(20):7926–7934.

37. Georgakoudi I, Solban N, Novak J, Rice WL, Wei X, Hasan T, Lin CP. *In vivo* flow cytometry: a new method for enumerating circulating cancer cells. *Cancer Res.* 2004;64 (15):5044–5047.

38. Maecker HT, Maino VC. Analyzing T-cell responses to cytomegalovirus by cytokine flow cytometry. *Hum. Immunol.* 2004;65(5):493–499.

39. Maecker HT, Auffermann-Gretzinger S, Nomura LE, Liso A, Czerwinski DK, Levy R. Detection of CD4 T-cell responses to a tumor vaccine by cytokine flow cytometry. *Clin. Cancer Res.* 2001;7(3 Suppl.):902s–908s.

40. Maecker HT. Multiparameter flow cytometry monitoring of T cell responses. *Methods Mol. Biol.* 2009;485: 375–391.

41. Hercher M, Mueller W, Shapiro HM. Detection and discrimination of individual viruses by flow cytometry. *J. Histochem. Cytochem.* 1979;27(1). 350–352.

42. Johnston SL, Dalal A, Mason S, Wilson JW, Robinson BS, Holgate ST. Detection of respiratory syncytial virus infection in nasal aspirate samples by flow cytometry. *Clin. Diagn. Virol.* 1994;2(4–5):279–289.

43. Shapiro HM. Flow cytometry of bacterial membrane potential and permeability. *Methods Mol. Med.* 2008;142: 175–186.

44. Shapiro HM, Nebe-von-Caron G. Multiparameter flow cytometry of bacteria. *Methods Mol. Biol.* 2004;263: 33–44.

45. Shapiro HM. Multiparameter flow cytometry of bacteria: implications for diagnostics and therapeutics. *Cytometry* 2001;43(3):223–226.

46. Janossy G, Mandy F, O'Gorman MR. Diagnostics in the shadow of HIV epidemics. *Cytometry B Clin. Cytom.* 2008;74 (Suppl. 1): S1–S3.

47. Glencross DK, Aggett HM, Stevens WS, Mandy F. African regional external quality assessment for CD4 T-cell enumeration: development, outcomes, and performance of laboratories. *Cytometry B Clin. Cytom.* 2008;74 (Suppl. 1): S69–S79.

48. Herzenberg LA, Sweet RG. Fluorescence-activated cell sorting. *Sci. Am.* 1976;234(3):108–117.

49. Dittrich W, Gohde W. Impulse fluorometry of single cells in suspension. *Z. Naturforsch. B* 1969;24(3):360–361.

50. Shapiro HM. Technical developments in flow cytometry. *Hum. Pathol.* 1986;17(7):649–651.

51. Crosland-Taylor PJ. A device for counting small particles suspended in a fluid through a tube. *Nature* 1953;171(4340):37–38.

52. Coulter WH. High speed automatic blood cell counter and cell size analyzer. In *Proceedings of the 12th National Electronics Conference*, 1956, pp. 1034–1042.

53. Zola H, Swart B, Banham A, Barry S, Beare A, Bensussan A, Boumsell L, Buckley C, Buhring HJ, Clark G, Engel P, Fox D, Jin BQ, Macardle PJ, Malavasi F, Mason D,

Stockinger H, Yang X. CD molecules 2006: human cell differentiation molecules. *J. Immunol. Methods* 2007;319(1–2):1–5.

54. Fiebig H, Behn I, Gruhn R, Typlt H, Kupper H, Ambrosius H. Characterization of a series of monoclonal antibodies against human T cells. *Allerg. Immunol. (Leipz.)* 1984;30 (4):242–250.
55. Herzenberg LA, De Rosa SC. Monoclonal antibodies and the FACS: complementary tools for immunobiology and medicine. *Immunol. Today* 2000;21(8):383–390.
56. Watson JV. The early fluidic and optical physics of cytometry. *Cytometry* 1999;38 (1):2–14; discussion 11.
57. Shapiro HM. Lasers for flow cytometry. *Curr. Protoc. Cytom.* 2004; Chapter 1:Unit 1.9.
58. Telford WG. Analysis of UV-excited fluorochromes by flow cytometry using near-ultraviolet laser diodes. *Cytometry A* 2004;61(1):9–17.
59. Telford WG. Small lasers in flow cytometry. *Methods Mol. Biol.* 2004;263: 399–418.
60. Shapiro HM, Perlmutter NG. Violet laser diodes as light sources for cytometry. *Cytometry* 2001;44(2):133–136.
61. Maecker HT, Trotter J. Flow cytometry controls, instrument setup, and the determination of positivity. *Cytometry A* 2006;69(9):1037–1042.
62. Alberti S, Parks DR, Herzenberg LA. A single laser method for subtraction of cell autofluorescence in flow cytometry. *Cytometry* 1987;8(2):114–119.
63. Roederer M, Kantor AB, Parks DR, Herzenberg LA. Cy7PE and Cy7APC: bright new probes for immunofluorescence. *Cytometry* 1996;24(3):191–197.
64. Maecker HT, Frey T, Nomura LE, Trotter J. *Selecting fluorochrome conjugates for maximum sensitivity. Cytometry A* 2004;62(2):169–173.
65. Roederer M. Compensation in flow cytometry. *Curr. Protoc. Cytom.* 2002; Chapter 1:Unit 1.14.
66. Stewart CC, Stewart SJ. Four color compensation. *Cytometry* 1999;38(4):161–175.
67. Robinson JP, Durack G, Kelley S. An innovation in flow cytometry data collection and analysis producing a correlated multiple sample analysis in a single file. *Cytometry* 1991;12(1):82–90.
68. Herzenberg LA, Tung J, Moore WA, Parks DR. Interpreting flow cytometry data: a guide for the perplexed. *Nat. Immunol.* 2006;7(7):681–685.
69. Parks DR, Roederer M, Moore WA. A new "Logicle" display method avoids deceptive effects of logarithmic scaling for low signals and compensated data. *Cytometry A* 2006;69 (6):541–551.
70. Tung JW, Parks DR, Moore WA, Herzenberg LA. New approaches to fluorescence compensation and visualization of FACS data. *Clin. Immunol.* 2004;110(3):277–283.
71. Ibrahim SF, van den Engh G. Flow cytometry and cell sorting. *Adv. Biochem. Eng. Biotechnol.* 2007;106: 19–39.

2

RECENT ADVANCES IN FLOW CYTOMETRY: PLATFORMS, TOOLS, AND CHALLENGES FOR DATA ANALYSIS

PAUL J. SMITH, ROY EDWARD, AND RACHEL J. ERRINGTON

2.1 INTRODUCTION AND OVERVIEW

An expanding range of both new and refined technologies is being applied at all stages of modern drug discovery by the pharmaceutical industries. There is also a growing involvement of the academic sector in the early phases of target identification, elucidation of mechanisms of action, invention of discovery technologies, and creation of molecular probes and reporter systems. In parallel, there is a drive to advance the means of extraction of knowledge from acquired data through bioinformatics. Flow cytometry lends itself to exploitation within this discovery and development pathway due to its unparalled capacity to analyze heterogeneous cellular systems, provide multiparameter functional information at the single-cell level, offer flexibility for different reporter systems, and acquire data in standard structures. The working context for flow cytometry in discovery and development essentially reflects the emphasis and expectation placed upon cell-based analyses at different stages within the pathway. Fundamentally, the cell-based assay provides an opportunity to examine the target within a minimally relevant environment and to perceive complex outcomes such as molecular translocation events as part of a downstream pharmacodynamic response that may have a stochastic element to the timing of expression or be driven by other permissive events such as cell cycle progression.

Flow Cytometry in Drug Discovery and Development, Edited by Virginia Litwin and Philip Marder

The high impact of flow cytometry on clinical practice and research, together with the availability of instrumentation, has fostered the view that flow cytometry is a mature technology. This belies the continued appearance of advances in analysis capacity, improvements in sorting, integration of sample handling solutions, and enhanced event detection and resolution. Such developments carry burdens for platform evaluation and cross-platform comparisons particularly when data from trials depend on multicenter analyses. For example, in the discrete area of hematology analyzers, the introduction of new platforms (e.g., LH 750, Beckman Coulter; Advia 120, Bayer Diagnostics; XE 2100, Sysmex; Excell 2280, Drew Scientific Inc.) required cross-platform evaluations for analyte sample stability [1] or operation within a given workload environment [2].

An emerging feature of the advance in the flow cytometry platform is the expansion of multiparameter capacity and the provision of supporting reagent technologies as addressed elsewhere in this book. The success of the technology has also primed awareness in microfluidics and the search for new modalities, beyond light scatter and fluorescence, for both cell sensing and analysis. For the drug discovery sector, innovation in cytometry presents issues for platform selection when coping with biological/chemical diversity, the need for improved robotic handling, sample throughput, and choice within an expanding range of screening applications that require validation on a given platform [3]. This is a particular challenge when faced with new and highly informative model systems, such as zebrafish (*Danio rerio*), for discovery and drug development in the screening of lead compounds, target identification, target validation, and physiology-based assays including toxicity testing [4]. High-throughput multifactorial analysis has previously promised an enhanced efficiency with which novel bioresponse-modifying drugs may be identified and characterized [5]. Thus, in the drug discovery and development (DD&D) area, the demand remains for higher throughput performance and "high-content" screening applications [6, 7].

Multiparameter phenotypic profiling technologies for small molecules can improve success rates of lead selection and optimization in the drug discovery pathway [8]. In parallel, there is a recognition of the value of "cell-centric technologies" particularly in oncology [9]. Recent advances in gene and small molecule delivery modalities and automated systems for cell manipulation have supported such approaches [9]. Flow cytometry stands alongside a range of specialized techniques for performing cell-based assays in drug discovery and even has an expanding role in profiling toxicology for ADME/Tox (absorption, distribution, metabolism, excretion, and toxicology) evaluations. With its unique capacity to assess the dimensions of heterogeneity in complex cellular systems, it continually finds new areas of application.

Flow cytometry has a distinguished track record in providing elegant solutions in the screening for antivirals, given the convenient cell surface presentation of key target molecules. For example, hepatitis C virus (HCV) presents a significant health challenge with an estimated 170 million individuals affected and is a challenge for compound screening [10]. Attachment of the virus to the cell surface and viral entry are critical for the initiation of infection, and an interaction between the envelope glycoprotein E2 and the human tetraspanin protein CD81 is a key event for inhibitor targeting. A flowcytometric assay using human hepatoma cells and fluorescently labeled anti-CD81 antibody (JS81) has been used to profile the potencies of the compounds inhibiting the E2–CD8 interaction [ref. 11]. However, new foci of interest in flow cytometry are also generated by the introduction of new concepts in biology and the clinical sciences, such as the role of programmed cell death in disease processes and the hierarchical nature of tumor cell populations imposed by cells with stem-like qualities. Exploration of such concepts in screening activities has been aided by RNA interference (RNAi) technology in areas including oncogene addiction/oncogenic shock, cancer stem cell biology, lineage dependency, and the epithelial–-mesenchymal transition [12].

The continued application of the principles of cytometry is apparent in the development of cytometric bead array technologies, now well established, but arising from the historical association between microspheres and flow cytometry [13]. Bead arrays have provided a method of quantifying soluble analytes using earlier systems such as the Luminex FlowMetrix platform [13]. The multiplexed assay format now has applications in various areas of biology including protein expression profiling, cancer markers, cardiac markers, cellular signaling, cytokines, chemokines, growth factors, endocrinology, isotyping, matrix metalloproteinases, metabolic markers, neurobiology, and transcription factors/nuclear receptors [14]. Critical to bead technology is quality control, and staining protocols are available for using the amine-reactive CBQCA dye to measure the amount of protein, without binding partners, on the surface of a microsphere [15]. The high-speed spectral analysis of individual particles in flow is set to enable new applications [16]. The measurement of fluorescence and Raman spectra of individual particles at high speeds has been performed on a modified Union Biometrica COPAS Plus instrument allowing red excitation and optical fiber-based light collection and spectral analysis using a spectrograph and a CCD array detector [16].

The application of the principles of cytometry is apparent in the development of microscale devices and *in vivo* cytometry. An example of the latter relates to the problem of cell assessment in lymphatics. Galanzha et al. have described how cells in living animals can be counted by laser generation of photoacoustic signals in individual cells, exploiting the principles of flow cytometry through a natural "cell-focusing" phenomenon found in lymph vessels [17]. Thus, there is every expectation that the involvement of flow cytometry in drug discovery will expand and that it will be indispensable in new areas such as cell-based therapeutics from discovery to quality control for regulatory compliance. Critically for drug discovery applications there has been a drive to make advanced instrumentation user-friendly and to remove the burden of expert knowledge required to extract reliable and robust data. However, in parallel, flow cytometry has widened its attractiveness for the development of novel reagents, requiring increased support from technology providers to navigate new applications to appropriate platforms.

In addressing cellular heterogeneity, flow cytometry and single-cell analyses in general start to meet the aspiration to acquire systems level insights from the molecular to the whole organism level in describing disease processes and revealing opportunities for therapeutic intervention. As a result, there is an increasing recognition of the need to progress the "flow of cytometry into systems biology" [18], a common aim being to bridge the molecular–cellular systems gap [19–21]. Here, both instrumentation and knowledge from data integration issues present significant challenges, but if solved, will move *systems cytometry* to an interface with wider and more complex systems biology frameworks.

The aim of this chapter is to emphasize the context for flow cytometry in DD&D, recent advances in the instrumentation offered, new cytometric technologies, a selective view of some new reagents, and the data handling challenges associated with cytometry platforms. Throughout we have attempted to reference exemplar applications that particularly apply to DD&D.

2.2 CURRENT INSTRUMENTATION: ADVANCES IN FLEXIBILITY, PERFORMANCE, AND THROUGHPUT

A major characteristic of flow cytometry is its inherent versatility as a platform, allowing refinement for the demands of different stages of drug discovery. Parallel advances have occurred in the range of lasers available [22], providing new excitation options for fluors and reducing cost and operation overheads. There has been a trend over the past decade for flow cytometer manufacturers to move toward digital electronics with the principal advantages of accurate logarithmic conversions, convenient compensation matrices, and event triggering on any parameter. Designs attempt to reduce the disadvantages of digital electronics in the possible loss of resolution at the low end of the scale, aliasing, and digital pulse processing. The applications of flow cytometry are also being extended beyond cell profiling into engineered particle technologies and the analysis of molecular level event detection [23]. Flow-based approaches are seen as complementary to imaging methods [24] or even combined, as in imaging and laser scanning cytometers [25, 26]. Current instrumentation therefore offers DD&D a range of options for analysis and cell sorting (Table 2.1). Instruments often share similar performance characteristics for many applications in DD&D but offer different routes for integration and ease of access. Here we review selected features of some current platforms to provide a sense of the range platforms offered.

Significant problems for cell sorting uptake are the operational burden of the instrumentation setup and the reliability of high-speed preparative performance. There has been a move toward making cell sorting easier for biologists across a range of applications. Becton Dickinson Biosciences recently introduced their BD FACSAria™ II flow cytometry system based on a true fixed alignment flow cell that is gel coupled to the collection optics addressing the need to minimize start-up time. BD's recent acquisition of Cytopeia also brings the cell sorting BD Influx™ into wider availability, offering an open, configurable cell sorting platform with particular performance characteristics advantageous for small particle detection. BD Flow Cytometry Analyzers include BD FACSArray™, BD FACSCalibur™, BD FACSCanto™ II, BD LSR™ II, and BD LSRFortessa™. Instrument innovations

TABLE 2.1 Examples of Current Flow Cytometry Platforms and Key Features

Supplier	Instrument	Main Purpose	Excitation Sources	Maximum Detection Channels (Fluors/Scatter)	Key Features	Main Application Area
Accuri Cytometers, Inc.	C6 Flow Cytometer	Cell analyzer	2 (488, 640 nm diode lasers)	4 + FSC + SSC	Benchtop; 6 log dynamic range; low-pressure fluidics	Basic and clinical research
Amnis Corp.	ImageStreamX	Cell analyzer	Up to 5 (405, 488, 560, 592, 658 nm)	10 + bright field + dark field (with second camera)	Imaging cytometer; dots are also images; phenotype + morphology	Basic and clinical research
Axetris Division, Leister Process Technologies	Axetris Impedance Flow Cytometer	Cell analyzer	None, impedance measurement	Cell size; membrane capacitance; cytoplasmic conductivity	Biomarker independent; microfluidic chip system; reduced sample processing	Industrial and basic research
BD Biosciences	BD Influx™	Cell sorter	Up to 5 (from 405, 457, 488, 532, 561, 640 nm)	3/4-plex PMT modules per laser line	Fully aseptic sorting; multiparameter sorting; flexible platform	Clinical research
BD Biosciences	BD FACSAria™ II	Cell sorter	Up to 4 (from 375, 405, 488, 633 nm)	13 + FSC + SSC	Aseptic sorting; optimized for stem cells; fixed alignments	Basic and clinical research
BD Biosciences	BD FACSCanto™ II	Cell analyzer	Up to 3 (from 405, 488, 633 nm)	8 + FSC + SSC	Approved for CIVD; carousel and HTS option; fixed alignment	Clinical use
BD Biosciences	BD FACSCount™ Instrumentation System	Cell analyzer	1	2 + FSC + SSC	Approved for CIVD; dedicated for HIV/AIDS; minimal user intervention	Clinical use (HIV/AIDS)
BD Biosciences	BD LSR™ II	Cell analyzer	4–7 (special order, from 11 wavelengths)	18 + FSC + SSC	Fully flexible platform; upgradeable; optimizes excitation/emission	Basic and clinical research
BD Biosciences	BD LSRFortessa™	Cell analyzer	5 (from 11 wavelengths)	18 + FSC + SSC	Reduced footprint (cf. LSR II); upgradeable; optimizes excitation/emission	Basic and clinical research

(Continued)

TABLE 2.1 (*Continued*)

Supplier	Instrument	Main Purpose	Excitation Sources	Maximum Detection Channels (Fluors/Scatter)	Key Features	Main Application Area
Beckman Coulter	Cell Lab Quanta™ SC (with plate loader option)	Cell analyzer	2 (488 nm + arc lamp)	3 + Coulter volume + SSC	Absolute cell counts; HTS with sample prep; sample track and 21 CFR Pt 11	Clinical research and drug discovery
Beckman Coulter	Coulter® EPICS XL™ and XL-MCL™	Cell analyzer	1 (488 nm)	4 + FSC + SSC	Digital processing; HTS carousel option; single laser simplicity	Clinical
Beckman Coulter	Cytomics FC 500	Cell analyzer	2 (488, 635 nm)	5 + FSC + SSC	Five-color detection off blue line; digital compensation; plate-based HTS option	Clinical
Beckman Coulter	Gallios™	Cell analyzer	2 (blue and red) or 3 (+violet)	6 (5 + 1), 8 (5 + 3), or 10 (5 + 3 + 2) + FSC + SSC	Optimized with reagents; Boulevard light path; easy change filters	Basic and clinical research
Beckman Coulter	MoFlo™ XDP	Cell sorter	3 (from 355, 405, 488, 635 nm)	12 + FSC + SSC	Multiple sort gates; four-way sorting; 5 log dynamic range	Basic and clinical research
Beckman Coulter	CyAn ADP	Cell analyzer	2 (488, 642 nm) or 3 (+405 nm) laser versions	7 (5 + 2) or 9 (5 + 2 + 2) + FSC + SSC	Benchtop footprint; automated compensation; walk-up operation	Basic and clinical research
Beckman Coulter	Navios	Cell analyzer	Awaiting data	Awaiting data	Awaiting data	Awaiting data
iCyt/Sony Group	Reflection	Cell sorter	7 (from 355, 405, 488, 532, 561, 592, 640 nm)	Multiple PMTs and FSC/SSC settings	Multiple sort streams; index or proportional sorting; highly flexible platform	Basic and clinical research
Millipore	Guava PCA	Cell analyzer	1 (green)	2 + FSC	Compact benchtop; absolute counting; microcapillary flow cell	Basic and clinical research
Millipore	Guava PCA-96 Base System	Cell analyzer	1 (green)	2 + FSC	Compact benchtop; automated MTP sampling; microcapillary flow cell	Basic and clinical research

Millipore	Guava EasyCyte™ Mini	Cell analyzer	1 (blue)	3 + FSC + optional SSC	Compact benchtop; absolute counting; microcapillary flow cell	Basic and clinical research
Millipore	Guava easyCyte 8HT	Cell analyzer	2 (red, blue)	6 + FSC + SSC	Compact benchtop; automated MTP sampling; microcapillary flow cell	Basic and clinical research
Partec GmbH	CyFlow® SL	Cell analyzer	1 (488 nm, with option for others)	3 + FSC + SSC	Mobile, can run on 12 V DC; volumetric counting; submicron detection	Basic and clinical research
Partec GmbH	CyFlow® ML	Cell analyzer	5 (from 375, 405, 488, 638, 532, 561 nm, UV lamp/LED)	13 + FSC (2) + SSC	Compact benchtop; volumetric counting; submicron detection	Basic, industrial, and clinical research
Partec GmbH	CyFlow® space	Cell analyzer (sorter upgrade)	3 (from 375, 405, 488, 635 nm)	7 + FSC + SSC	Compact benchtop; volumetric counting; submicron detection	Basic, industrial, and clinical research
Partec GmbH	CyFlow® Ploidy Analyser	Cell analyzer	2 (from UV LED and 532 nm)	2 + SSC	Compact with integrated GUI; genome sizing/ploidy; plant, animal, microbes	Agriculture, aquaculture, and basic research
Union Biometrica	COPAS™ Large Particle Flow Cytometry	Cell and large particle sorter	488/514 nm multiline for fluors and 670 nm for size/density	2 + size + optical density	250 2000 µm flow cell options; sorts clusters, organisms; liquid handling options	Basic and clinical research
Miltenyi Biotec GmbH	MACSQuant	Cell analyzer	3 (405, 488, 635 nm)	7 + FSC + SSC	Compact benchtop design; couples to magbead column; integrated fluidics and GUI	Basic and clinical research
Stratedigm, Inc.	S1000 Flow	Cell analyzer	4 (from 372, 405, 488, 532, 640 nm)	8 + FSC + SSC	Benchtop design; temperature-controlled HTS loader; user-replaceable modules	Basic and clinical research

continue to permit researchers to explore multicolor flow cytometry with new options such as the BD Horizon™ V450 violet laser reagents aiding that process.

Beckman Coulter's CyAn ADP flow cytometer is a two-option platform incorporating fast electronics (70,000 events/s) with a software package to efficiently handle the demands of rare event analysis especially where large data files may have to be generated. The CyAn ADP 7 Color has nine parameters being equipped with 488 and 642 nm solid-state lasers with capacity for multiple applications including signal transduction monitoring. CyAn ADP 9 Color adds a 405 nm solid-state laser and two fluorescent channels permitting stem cell/SP analysis, UV-excited fluor analysis, and CFP-YFP FRET detection. The MoFlo™ platform provides high-speed sorting capacity and when configured with a 351 nm UV laser and fluorescent emission detection both in the blue region, using a 450/20 band-pass filter, and in the red region, using a 675 eFLP filter, is suitable for sorting rare side populations in stem cell isolation. The new Gallios™ flow cytometer has high sensitivity and high-speed data collection with convenient filter change options with three main configurations providing 6–10 color, 2–3 laser options and new log/lin transform for visualization of negative data and compensation optimization.

Partec offers a flexible range of multilaser and multiparameter instrumentation on their high-end flow cytometers (CyFlow® ML and the CyFlow® space flow cytometer) that complement their range of specialized platforms for cell counting and DNA cell cycle analysis (CyFlow® CCA Cell Counter Analyser), HIV monitoring (CyFlow® Counter), and plant ploidy analysis (CyFlow® Ploidy Analyser). The CyFlow® SL, for example, has been used for quality control in liquid-based cytology [27]. The Partec Particle and Cell Sorter PPCS is a closed sorting system add-on for Partec CyFlow® space, Partec PAS, and Partec PA-II platforms.

Specialized applications often require the high-speed preparations of cell subsets or the collection of rare cells from heterogeneous mixtures. Rare event selection (five-decade multichannel digital acquisition system with 32-bit data processing improving resolution) and high-speed sorting (four-way sort capability) are provided by Beckman Coulter's MoFlo™ XDP (analyzes 100,000 events/s and sorts 70,000 events/s). The iCyt Reflection System achieves high-speed parallel sorting with multiple Highly Automated Parallel Sort (HAPS) modules, providing collection rates past conventional limits with platforms of the same footprint. An attractive feature of the iCyt Reflection System is the mounting of the HAPS inside a level III biological safety cabinet to protect samples and operators from biohazard contamination. The iCyt Synergy sorter/analyzer platform supports two high performance cell sorter modules, two automated cell analyzer modules, or a sorter/analyzer combination with seven standard laser options.

Accuri Cytometers' C6 Flow Cytometer is a small footprint platform incorporating fluorescence and two scatter detectors with a sheath-focused core for event rates of up to 10,000 events/s and sample concentrations over 5×10^7 cells/mL. The C6 is equipped with a blue and a red laser, two scatter detectors, and four fluorescence detectors with interference filters optimized for the detection of FITC, PE, PerCP-Cy5.5, and APC. The C6 measures the volume drawn and hence cell concentration. Recent evaluations have shown that the system permits the introduction of an agent without sample removal, facilitating the tracking of rapidly changing events such as calcium release. The convenience of the unusually large dynamic range of the C6 system has been exploited in the simultaneous analysis of plant species whose nuclear content DNA values (*C*-values) span at least two orders of magnitude and reflect almost the entire described angiosperms [28]. The incorporation of a fully digital data collection system facilitates post-collection analysis and highlights the potential value for data mining in screening applications.

The Amnis ImageStreamX combines the advantages of flow cytometry in sample handling and fluorescence sensitivity with the imaging of cells directly in suspension at the resolution of a 60× microscope. The EDF (extended depth of field) capacity increases image quality and ensures that the whole cell stays in focus [29]. This platform permits phenotypic and functional studies with data extraction from 12 images per cell. Analysis performance is in the range of 1000 cells/s with a software package to quantify predefined fluorescence and morphologic parameters for multiple applications [26, 29–36]. The instrument adds a complementary technology for coping with the problem of rare event analysis. For example, the mobilization of nonhematopoietic very small embryonic-like stem cells (VSELs) in acute myocardial infarction has been studied by VSELs isolated using FACSAria™, and expression of pluripotent markers, early cardiac and endothelial markers, and chemokine receptor CXCR4 then confirmed using an ImageStream system [36]. Given that flow cytometry is restricted when attempting to address morphological or subcellular spatial details, image cytometry has the combined advantages of both, for example, in the multispectral imaging of stages of erythroid maturation by measuring changes in Ter119 mean intensity and area, DNA (DRAQ5 stain) mean intensity and area, and RNA content (thiazole orange stain) [34]. The ImageStream has the capacity to approach difficult forms of in-flow analysis such as receptor internalization and nuclear translocation, providing a route to highly multiplexed assays, compound screening, and profiling [37]. The imaging cytometer also highlights the potential for the definition of new complex analytical parameters that can be exploited to identify a change in a cell state of interest. Of course, this still applies for the reexamination of conventional parameters, such as light scatter, as recently demonstrated in the tracking of changes in prostate cancer cellular granularity, monitored on the HyperCyt high-throughput flow cytometer system, to screen for small molecule inducers of intracellular granularity revealing a candidate class of aryl-oxazoles that increased intracellular granularity in both androgen-sensitive and androgen-independent cell lines [38].

Multiple approaches to increasing throughput are now available with robotics and automated sample processing interfaces being important components in workflows. Solutions are available for many instrument platforms. For example, the Hudson Flow Cytometry Workcell from Hudson Robotics, Inc. enables automated sample acquisition system in an integrated platform combining the Guava® EasyCyte™ Plus Flow Cytometer with Hudson's PlateCrane EX® microplate robot arm and microvolume sampling, multiplex analysis, and up to six detection parameters. The allied software package facilitates data tracking and scheduling. Millipore's Guava instrument platform provides a microcapillary flow cytometry operating without sheath fluid, reducing sample volumes, waste, and maintenance requirements. The new easyCyte

8HT system provides extended cell analysis capabilities using blue (488 nm) and red (635 nm) excitation lasers and eight detection parameters, including six fluorescent colors plus forward and side scatter for size and morphology determination. High-throughput analysis is made possible with a robotic sample tray that can handle a 96-well microplate.

The HyperCyt® system, comprising the HyperCyt Autosampler and HyperView® Data Analysis Software, enables existing flow cytometers to analyze samples from 96- or 384-well microplates, with extremely fast rates. Earlier evaluations showed real promise for enhanced library throughput using a HyperCyt configured platform, operating with air bubble separated compounds in one sample line and a continuous stream of cells in another being mixed in-line for serial flow cytometric cell response analyses [39]. High-throughput flow cytometry (HTFC) using a HyperCyt system has been used to detect potential anti-inflammatory compounds that block ligand binding to the formyl peptide receptor (FPR) family of G-protein-coupled receptors (GPCRs) associated with the localization and activation of tissue-damaging leukocytes at sites of chronic inflammation [40] In HTS mode, samples were routinely processed at 1.5 s/well (approximately 2500 cells analyzed per sample), allowing a 96-well plate to be processed in less than 2.5 min [41]. The S1000 flow cytometry platform from Stratedigm, Inc. incorporates an architecture that allows up to four lasers to be mounted on a single plate with solid-state options from 372 to 640 nm. Operationally, the instrument has a small dead volume of <8 μL, a minimum sample size of <20 μL, and a temperature-controlled 168-tube loader with a bar code reader.

Flow cytometry combined with HCS has the potential to improve biomarker identification. However, a combination of robotic fluid handling, high-performance flow cytometric instrumentation, and linked bioinformatics software reveals a limiting step in the processing and analysis of large sample sets in a short period of time: the degree of automation of the data processing steps [42]. This issue was recently explored for the extraction of a "cellular signature definition" for acute graft versus host disease [42].

2.3 ADVANCES IN OPERATING PRINCIPLES: OBJECT MANIPULATION, SEPARATION, ANALYSIS, AND LIGHT SOURCES

The requirement for cells to be in suspension form for flow cytometric analysis offers convenient opportunities to modify in-line manipulation of samples and objects of interest and to integrate analysis of principle of interest in drug development. A typical example is the extension to flow cytometry of electronic volume measurements, based on the robust sizing and counting concept introduced by Wallace Coulter (W.H. Coulter, Means for counting particles suspended in a fluid, U.S. Patent 2,656,508, priority August 27, 1949). Combination of the technologies allows precise determination of changes in cell size beyond the resolution provided by the surrogate light scattering parameters acquired using flow cytometry and provided a basis for initial cell sorting development [43]. The Cell Lab Quanta™ SC flow cytometer integrates the power of electronic sizing with flow cytometry and is highly informative for tracking changes in cell volume during apoptotic commitment [44], with earlier work showing the application to electronic nuclear volume (ENV) and DNA content using a NPE Quanta [45, 46]. Again, the fact that cells are in suspension for flow applications provides convenient methods for the evaluation of novel, fluid-based approaches in drug delivery research. For example, ultrasound-directed drug delivery using microbubbles is a promising method [47] and flow cytometry has been used to investigate the effect of ultrasound exposure parameters on the "sonoporation" [48]. Here we selectively review some interesting innovations applicable to fluid-based platforms.

In flow cytometry, object manipulation technologies determine the method and performance envelope of cell presentation and separation. It is important to appreciate that other solutions may be available when preparative cell separation is an intrinsic requirement of a diagnostic, biotechnological, or biomedical application. Matrix-based separation methods such as affinity chromatography have to overcome the problems of the availability of compatible matrices with convenient pore sizes and the means of elution of cells from the affinity capture surface surfaces. In this area, a novel solution is the use of supermacroporous monolithic high-elasticity cryogels with highly interconnected large (10–100 μm) pores permitting cell detachment through mechanical compression [49]. There continue to be advances in cell separation using micro- and nanoscale technologies, offering advantages over conventional, macro-scale separation systems in terms of sample volumes, low cost, portability, and potential for integration with other analytical techniques [50]. A recent review [51] has discussed various cell separation principles based on size, magnetic attraction, fluorescence, and adhesion to surfaces in addition to new emerging technologies such as affinity capture planar flow cell arrays, dielectrophoresis, field flow methods, and column separation devices. Such approaches stand alongside FACS for applications in medicine and biotechnology [51, 52].

2.3.1 Pneumatic Delivery

There is an expansion in the size range within which flow cytometry can now operate as interest arises in more diverse and complex model systems. Here one issue is the problem of hydrodynamic focusing becoming inherently limiting due to the potential impact of high sheath pressures on object integrity and downstream cell viability. The COPAS™ instruments of Union Biometrica provide analysis and low-stress (2–5 psi) pneumatic sorting of biological objects up to 1500 μm including small multicellular animals (e.g., *Caenorhabditis elegans*, and the eggs and larvae of *Drosophila*, *Hydra*, *Xenopus*, mosquitoes, *Medaka*, and zebrafish *D. rerio*). A drug compound library might be tested by arraying the library in the wells of a microtiter plate, and then dispensing a model system such as nematodes into each well with subsequent aspiration of well content for reanalysis. Thus, of particular interest to the drug discovery and development sector is the potential for the analysis and isolation of large cells and cell clusters (e.g., adipocytes, duct cells (kidney, breast, pancreas), hepatocytes, pancreatic islet cells, imaginal discs, stem cells, and clusters). For automated detection and sorting of live *D. rerio* embryos from dead eggs, the COPAS XL instrument (operating on size and optical parameters) has been used on

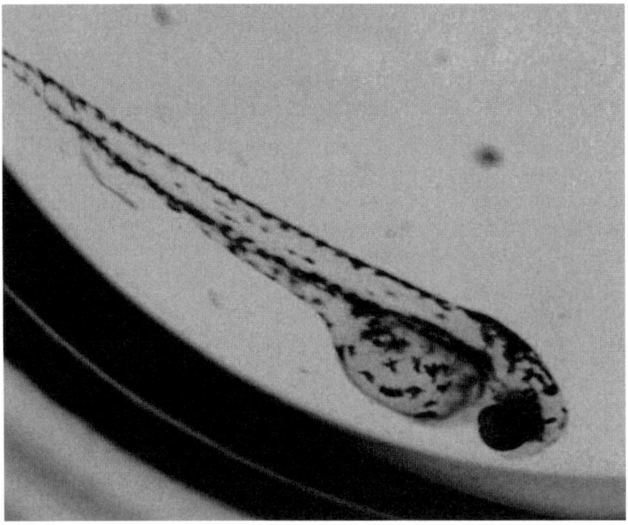

FIGURE 2.1 Live *D. rerio* (zebrafish) embryo immobilized in CyGEL Sustain™ (image courtesy of Dr. Yolanda Alvarez, UCD School of Biomolecular and Biomedical Sciences, Conway Institute, University College, Dublin). (See the color version of the figure in the Color Plates section.)

zebrafish embryos with high viability (94.7%). Complementary technologies have been developed to aid downstream analysis of *D. rerio* embryos in functional assays that require immobilization and kinetic imaging. For example, an application-specific formulation of a thermoreversible hydrogel (CyGel, Biostatus Ltd, UK) has been used to permit dechorionated embryo analysis by conventional imaging by thermally driven immobilization and post-analysis recovery. Figure 2.1 shows CyGEL Sustain™ immobilization of live anesthetized zebrafish embryos facilitating other techniques such as tissue-specific microinjections, microsurgery, and HCS applications.

2.3.2 Ultrasonic Object Focusing

Although the creation of inexpensive, small flow cytometers may be of critical importance for disease diagnosis in resource-poor areas, the early stages of the discovery process are less dependent on such innovation. However, it is likely that there will be increased cell analysis applications in therapeutic monitoring and selection that will demand increased instrument portability and reduced cost. Alternatives to hydrodynamic focusing, primarily to dispense with the burden of sheath fluid, are being sought with a recent demonstration that acoustic focusing has no effect on cellular viability in research into volume sensing [53, 54]. In early 2010, Life Technologies Corporation announced the release of the Attune™ Acoustic Focusing Cytometer, a first of its kind cytometer system that uses acoustic waves to control the movement of cells during analysis. Potential advantages include longer transit times, higher throughput, and potentially a wider variety of assay types than can be performed on hydrodynamically focused cytometer systems.

2.3.3 Cellular Dielectric Properties

The specific membrane capacitance and conductance of mammalian cells reflect the surface morphological complexities and barrier functions of the cell membrane, providing an opportunity to use such dielectric properties to identify cell physiological and pathological changes. Rapid and noninvasive cell characterization based on electric and dielectric properties is of particular interest in novel approaches to disease diagnosis and monitoring, basic research, and cell engineering. Dielectrophoresis (DEP) is the motion of neutral but polarizable particles subjected to nonuniform electric fields. The surface properties of submicron particles dominate their dielectrophoretic behavior as demonstrated, for example, by tobacco mosaic virus and herpes simplex virus that can be manipulated and spatially separated in a microelectrode array [55]. Dielectric properties can also be exploited for the separation and controlled manipulation of micro- or nanostructures of living organisms. Recent studies have advanced the application of DEP and chip-based devices for human cells [56, 57]. The results point toward promising new possibilities for early detection of malignancies in whole blood mixture using integrated DEP devices.

Moving dielectrophoresis (mDEP) can be generated by sequentially energizing a single electrode or an array of electrodes to form an electric field that moves cells continuously along a microchannel [58]. DEP provides an increased measurement precision and sensitivity in the detection of cells with different dielectric properties, without any need for the use of labeling technology. It also allows for the use of much reduced sample volumes when compared to traditional methods of cell analysis, such as flow cytometry, potentially providing an efficient alternative for clinical use. DEP differences could represent genetically determined differences (e.g., gene mutation or deletion) and epigenetic effects (e.g., CpG methylation effects) of phenotypic plasticity arising from the structural modification of cellular macromolecules such as cell surface glycoproteins. There is a role for DEP-based devices in cell-based analyses [59] and DEP has been used to differentiate drug accumulation phenotypes in breast cancer cell lines using cytoplasmic conductivity [60], opening the areas of drug discovery in which DEP could be linked with functional determinations of drug micropharmacokinetics [61].

Negative dielectrophoretic force (nDEP force)-assisted positioning and electrorotation (ROT) measurements have been used to provide a quantitative analysis of the toxic damage to cells with potential for the testing and development of new pharmaceuticals [62, 63]. Such an approach provides a "holistic" view of the cellular response and therefore is not dependent on the identification of a marker molecule as a signature of change. Different cell properties such as size, membrane capacitance, and cytoplasm conductivity can affect its "impedance spectrum" providing a probeless approach for cell profiling using microfabricated cytometers [64]. The approach has been successfully applied to cellular analyses [64–69]. The promise is that DEP methods may be more sensitive than current approaches for the detection of changes in cellular viability or obtain informative signatures at earlier stages of commitment to cell death [70]. Cell separation can also be achieved by isodielectric separation (IDS), for sorting cells based on electrically distinguishable phenotypes. IDS involves separating cells and particles in an electrical conductivity gradient using dielectrophoresis and can be applied in a continuous-flow device and is label free [71, 73]. Cell separation by differential mobility cytometry (DMC) provides a unique separation approach employing oscillating flow and differential imaging to analyze cells as they retard and adhere to an affinity surface such as a capture antibody. DMC has the advantages of high cell capture efficiency and the tracking of cell adhesion [73].

2.4 ADVANCES IN PLATFORM FORMATS: MICROFLOW AND NEW LIGHT SOURCES

The prospects for the application of nanotechnology in healthcare and for the development of personalized medicine appear to be bright [74–76] although in the case of cellular analyses there are clear problems in cell handing that require microscale solutions.

2.4.1 Microflow Cytometry

Miniaturization of cytometers for main laboratory-based activities is not of high value if it merely repeats the performance envelope of already compact instruments. However, advances in microfluidics bring new physical principles to bear at the microscale that can have a real impact on the types of analyses. Measuring the DNA content of eukaryotic cells is a fundamental task in biology and medicine. The changes in capacitance generated by passing individual cells across a 1 kHz electric field can be used to detect differences in DNA content of eukaryotic cells, leading to the development of microfluidic "capacitance cytometry" [77].

Microfabrication also lends itself to production advantages and the possibility of assays being configured by the design of a given microfluidic chip. Recent advances in miniaturized microfluidic flow cytometry for clinical use have been described [78] and there are numerous options for photonically coupling chips for fluorescence event detection. Recent developments in microflow cytometry have involved focusing the particles to be analyzed in the microfluidic channel, miniaturization of the fluid handling components, miniaturization of the optics, and integration and applications development. Strategies for focusing particles in a narrow path as they pass through the detection region include the use of focusing fluids, nozzles, and dielectrophoresis [79], as described above, but must deal with the microscale fabrication issues. An example of the microfabrication solution to control fluid flow in such systems is the recent development of groove-generated sheath flow directing sheath fluid from the sides to the top and bottom of the channel, completely surrounding the sample stream [80]. Strategies for optics range from the use of microscope objectives to polymer waveguides or optical fibers embedded on-chip [79]. Chip-based solutions employing standard reagents promise cost-effective solutions for the analysis of drug responses [81], for example, in the application of SYTO probes for the detection of cell death using automated microfluidic chip-based cytometry [82].

2.4.2 Compact and New Light Sources

New, convenient, and inexpensive light sources are having an impact upon microscopy and the exploration of efforts to improve signal-to-noise ratios. Flow cytometry platform design and versatility are also benefiting from the availability of new lasers [83–85] with the incorporation of diode and DPSS lasers emitting a variety of wavelengths, for example, in the analysis of time-gated luminescence in a system employing a UV light emitting diode (LED; 100 mW) to excite fluorescence from a europium chelate immunoconjugate with a long fluorescence lifetime ($\tau > 100\,\mu m$) with an improvement in the distinction of labeled Giardia cysts in an autofluorescent background [86]. The integration of time-gated detection of long-lifetime (1–2000 μs) luminescent labels into flow cytometry systems is highly attractive but has to deal with the need for pulsed operation mode with continuous sample flow. A particular configuration has been studied using LED excitation to detect europium dye-labeled targets with the demonstration of the feasibility of detection of europium calibration beads in a UV LED-excited flow cytometer [87–89].

A recent evaluation of new solid-state laser systems (DPSS 580, 589, and 592 nm sources), integrated into a cuvette-based flow cytometer (BD LSR II) and a stream-in-air cell sorter (FACSVantage DiVa), demonstrated the efficient excitation of yellow, orange, and red excited fluorochromes in comparison to red fluorescent protein HcRed [90]. The availability of supercontinuum white light lasers [91] also provides new excitation source for flow cytometry (e.g., 480 nm to the long red (>700 nm)) and "fine-tuning" of optimal excitation for particular probes without the restriction of specific laser line emission, as described by Telford et al. [91].

Recent semiconductor microtechnology has greatly reduced laser size down to the scale of single cells. The integration of ultrasmall lasers with biological systems makes it possible to create microelectromechanical systems that might revolutionize cytometry [92]. The integration of biology and photonics will be achieved using semiconductor optoelectronic crystals. The versatility of semiconductor optoelectronics with its customizable optical and electronic properties makes it an ideal technology for biophotonics. Unsurprisingly therefore semiconductor laser and photodiodes have been widely used in biosensing and analysis systems. True "biophotonic" integration, that is, embedding the living system within the semiconductor photonics, is relatively unexplored [92–95]. A format using an intracavity, vertical geometry biolaser for cell analysis was used to establish a new technique to rapidly assess the properties of cells flown through a nanolaser semiconductor device with a detection of biophotonic differences in normal and transformed mouse liver cells by using intracellular mitochondria as biomarkers for disease [96].

2.5 REAGENTS AND APPLICATIONS

A wide range of information sources exist for the selection of molecular probes, matched light sources, and optical filter requirements [97]. Exploitation of the functionality of particular fluorochromes, in terms of cellular penetration, compartmentalization, and affinity for cellular components, is a well-established approach in flow cytometry. In high-content image-based screening assays, there is a demand for segmentation with respect to major cellular compartments to allow translocation of specific signals to be assessed through algorithm-driven routines. Here we consider some selected areas in particular recent aspects of nanoparticle exploitation and cell analyses relevant to drug discovery and therapeutics.

2.5.1 Micro- and Nanoparticles

Cell separations predominantly rely upon labeled cell migration within a magnetic field termed magnetic-activated cell sorting (MACS) providing flow-through, immunomagnetic cell separation. Antibodies reacting with specific cell surface markers can be bound to magnetic beads and used to specifically capture cells exhibiting the marker. Theoretical calculations showed that the throughput can be increased to 10^6 cells/s by a scale-up of flow-through systems incorporating cell sorting in a quadrupole field [98–100]. Differences in magnetically induced motion of diamagnetic, paramagnetic, and superparamagnetic microparticles are detected by cell tracking velocimetry [101].

There are a number of different fluorescent probes and labels that are commonly used for detection and tracking of a fluorescence signal from the cells, many of which are organic fluorophores. The advantages of using nanoparticles in place of traditional organic fluorophores have been widely reported [102, 103] with an impact on diagnostics [104]. Advances in detector technology have also enabled single-molecule imaging using quantum dots [105]. First, they are photostable allowing long-term labeling of live cell populations [106]. Second, nanoparticles present major advantages over conventional organic fluorophores as they are chemically stable and are not metabolized by the cell. Quantum dots (eight colors) have been used for multicolor immunophenotyping, where with conventional fluorophores at least 17 emission windows have been identified to reveal the phenotype of multiple antigen-specific T-cell populations, hence setting the gold standard for polychromatic flow cytometry [107].

An optical cell coding approach has been described by Machleidt et al. using quantum dot encoded reporter cell lines to enable multiplexed reporter cell-based assay with a beta-lactamase reporter gene readout as demonstrated for the analysis of tumor necrosis factor alpha (TNF-alpha), interleukin 6 (IL-6), and interferon gamma (IFN-gamma) [108]. Sukhanova et al. described the first application of nanocrystal-encoded microbeads to clinical proteomics [109]. Two-color complexes, consisting of encoded, antigen-covered beads, anti-antigen antibody or clinical serum samples, and dye-tagged detecting antibodies, were analyzed by flow cytometry exhibiting the narrow distributions of fluorescence signals, thus obviating the need of any fluorescence compensation [109]. A combination of fluorescence and magnetic properties for nanobeads provides new applications in simultaneous bioseparation and biosensing. Recently, Di Corato et al. have described spherical nanobeads (30–400 nm) with a core magnetic domain of iron oxide sheathed in fluorescent oligothiophene molecules engrafted to an amphiphilic polymer with properties suitable for cell sorting applications [110].

Overall, there is continuing optimism over the applications enabled by the use of engineered nanocrystals with studies suggesting that it may be possible to exploit a cell's active transport machinery for delivery to specific nuclear and cytoplasmic compartments [111]. Inhibitor studies with flow cytometry have recently revealed that the mechanism of fluorescent "nanodiamond" (FND) uptake is energy-dependent clathrin-mediated endocytosis with further evidence for biocompatibility for cell tracking [112].

2.5.2 Cell Death and Cell Cycle Arrest

Simple applications of flow cytometry aid clarification of cell responses to small molecule screens. For example, the promotion of apoptosis in cancer cells remains an attractive approach for targeted therapies [113] with multiple flow cytometric assays being available. In a novel cell- and caspase-based Anticancer Screening Apoptosis Program (ASAP) HTS assay, 4-aryl-4*H*-chromenes have been identified as potent apoptosis inducers [114]. High-throughput flow cytometry has been applied in a screen for apoptosis-activating proteins in transiently transfected cells. In this case, the assay is based on the detection of activated caspase-3 with a specific antibody in cells overexpressing proteins tagged C- or N-terminally with yellow fluorescent protein [115]. Cell cycle analyses where capture of cells into suspension is problematic, high-content analysis (HCA), can be achieved by automated microscopy-based technologies such as the Cellomics's ArrayScan reader, although in general multiple descriptors of the cell cycle are required to aid reliable and accurate quantification of cell cycle phases [116]. New nuclear/cytoplasmic discriminating probes have been developed (DRAQ5 and CyTRAK Orange [117]) with spectral separation from the commonly used reporter proteins (e.g., eGFP, YFP, mRFP) and fluorescent tags such as AlexaFluor 488, fluorescein, and Cy2 [118]. DRAQ5 has recently found application in tracking the pharmacodynamic responses to cell cycle perturbing agents [119, 120]. Clearly, there is a move toward assays in which there is retention of functionality of cells and cross-platform analysis solutions. Here we outline a new approach (P.J. Smith et al., unpublished data) for tracking cells undergoing long-term cell cycle arrest using CyTRAK Orange in studies on chemoresistance and drug targeting (Figure 2.2).

New solutions are being sought to the problem of inefficient drug delivery, a contributor to tumor chemoresistance and drug failure. For example, enhancing tumor access by depleting tumor-associated stromal tissue through the inhibition of the hedgehog cellular signaling pathway has recently been proposed [121]. Indeed, it is recognized that anticancer drugs are unable to penetrate into poorly oxygenated tumor tissue masses. Furthermore, low oxygen concentration may in itself represent a key

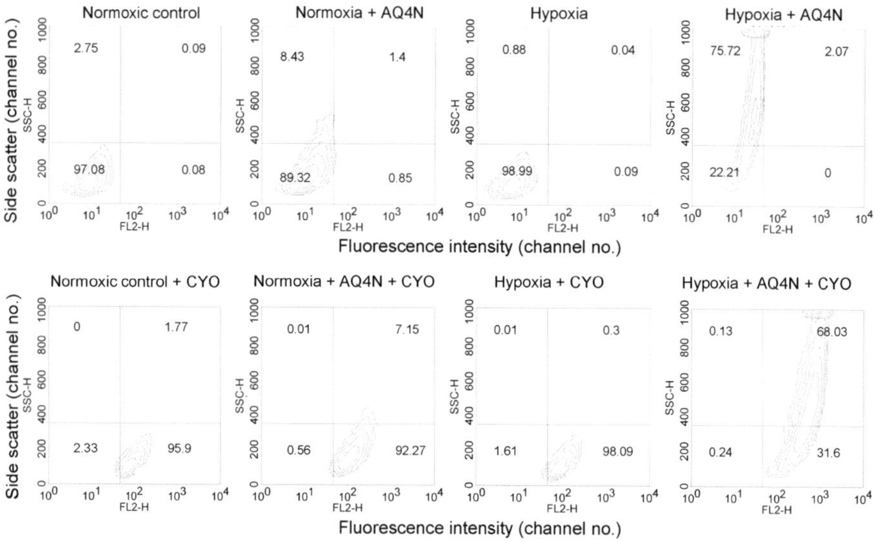

FIGURE 2.2 Analysis of cell cycle arrest in human A549 lung carcinoma cells induced by the candidate anticancer prodrug AQ4N (100 nM exposure) under normoxic or hypoxic conditions (1% O_2, 4 days). The prodrug AQ4N is converted to a cytotoxic form under hypoxic conditions generating cell cycle arrest and a concomitant increase in cellular light scatter and CYO staining (20 μM CyTRAK Orange, 10 min). CYO staining permits the differentiation, for example, by quadrant analysis, of arrested cells (UR) versus nonarrested (LR) cells compared to unstained cells (LL and UL). Quadrants of the flow cytometric contour plots show percent gated cells (P.J. Smith et al., unpublished data).

factor for tumor aggressiveness and metastasis by the activation of signal transduction pathways and gene regulatory mechanisms [122]. Under such conditions, the metabolic environment of the tumor cell is altered and an adaptation response is mounted through HIF-1 (hypoxia-inducible factor-1) induction [122]. Recent attention has been directed toward hypoxia-targeted therapeutics, for example, through a prodrug concept exemplified by the aliphatic amine *N*-oxide AQ4N (Oncotherics Ltd, UK) [123]. The operating principle is that AQ4N, a di-*N*-oxide anticancer prodrug with little intrinsic cytotoxicity but having tissue penetrant properties [124], will undergo bioreduction in hypoxic regions of solid tumors and micrometastatic deposits generating the active parent amine [125]. In early clinical trials, intratumoral concentrations of AQ4 exceeded those required for activity in animal models and support the evaluation of AQ4N as a novel tumor targeting agent [126]. Further, there is evidence that AQ4N possesses antiangiogenic effects in normoxic conditions, which could potentiate overall antitumor activity [127]. Flow cytometry has had an important role in the tracking of cell targeting of such prodrugs [128] and the intended pharmacodynamic effects in terms of long-term cell cycle arrest generated under hypoxia-driven prodrug activation. Figure 2.2 outlines the analysis of long-term arrest that cannot be tracked using cellular DNA content alone. The concept is to take into account the increase in cellular biomass that accompanies unresolved cell cycle interruption and here we describe the application of a novel dye, CyTRAK Orange (CYO, Biostatus Ltd, UK), capable of staining live cells (e.g., leukocytes and epithelial cells). It was developed as a cell-masking agent for the differentiation of cell features in HTS imaging applications. In flow cytometry, CYO has the advantage of not being excited by a 638 nm laser, allowing its use with APC, APC conjugates, and PE-Cy7 and analogues, but importantly provides a convenient integral of cell size and DNA content. Figure 2.2 shows the impact of hypoxia-mediated AQ4N conversion to a cell cycle arresting agent in human A549 lung cancer cells tracked using side scatter and CYO fluorescence. Convenient methods of identifying such events in live cell preparations permit sorting or additional functional assays to be performed with resolution for arrested and nonarrested subpopulations.

2.5.3 Multidrug Resistance and Cancer Stem Cells

It is now well recognized that the ATP binding cassette (ABC)-containing family of proteins has a pharmacological impact on efficacy of many drugs in current use [129]. High-throughput flow cytometry provides a sensitive functional approach for the assessment of selective inhibitors of key transporters such as ABCB1, ABCC1, and ABCG2 [130]. Different computational methods and models have been evolved to predict ABC transporter substrate properties of drug-like compounds with the potential for pharmacological profiling of compound series, in particular their ADME/Tox properties [131]. The P-glycoprotein, encoded by the ATP Binding Cassette B1 (ABCB1) gene, contributes to multidrug resistance (MDR) and continues to be subject to a search for novel inhibitors. Using ABCB1 overexpressed T-ALL cell line, a flow cytometry-based high-throughput screening (HTS) assay has been used to assay ABCB1-mediated doxorubicin efflux, revealing mometasone furoate as a previously unrecognized ABCB1 antagonist [132].

Cancer stem cells (CSCs) may contribute to chemoresistant subfractions in a variety of malignancies [133–135] and underpin the post-therapeutic recovery potential of tumors. Enhanced drug efflux, via ABC transporters, has been linked with CSC biology [135, 136]. The ABCG2 transporter (MXR/BCRP/ABCP1) shows an intriguing link with the CSC phenotype [135] in contributing to the flow cytometric definition of the pluripotential "side population" (SP) [137, 138]. SP cells are capable of efficient exclusion of the DNA minor groove-binding cytotoxic dye Hoechst 33342 as detected by changes in the UV-excited fluorescence emission spectrum of nuclear bound dye [139]. Hoechst 33342 SP analysis is a common method for identifying stem cells in mammalian hematopoietic and nonhematopoietic tissues. Although widely employed for stem cell analysis, a significant drawback of SP analysis has been the expense of UV laser sources, although benchtop analyzer (the Quanta Analyzer, NPE Systems) and platforms using nonlaser UV sources have been described [140]. An alternative is the identification of Hoechst 33342 surrogate molecules with similar pump specificity as Hoechst 33342, but with better violet excitation characteristics (e.g., the DyeCycle Violet reagent (excitation peak 369 nm)), permitting side population analysis on flow cytometers with violet lasers [141].

FACS methods are being applied to a wide range of stem cell research areas. For example, a methodology has been evolved based on the availability of blue, cyan, and yellowish-green fluorescent reporters (based on the jellyfish *Aequorea victoria* green fluorescent protein) and the red fluorescent protein from a *Discosoma* coral. Flow cytometric procedures based on such "rainbow" markers permit simultaneous measurements of transgene expression and phenotypic discrimination of hematopoietic cell subsets is set to aid hematopoietic stem cell transduction protocols for directed gene therapy [142]. Cytometric assays are being used to improve stem cell tests to study embryotoxicity *in vitro* in a drive toward high-throughput screening of compounds [143]. For example, Buesen et al. have recently reported an improvement to the embryonic stem cell test (EST) aimed at revealing teratogenic potency in a developing heart tissue. Quantitative marker protein flow cytometry (FACS-EST) for sarcomeric myosin heavy chain and alpha-actinin expression could be performed at an early point (7 days in mice) in development with a similar sensitivity for embryotoxic potency as that achieved by the morphological end point of beating cell agglomerates assayed at 10 days [144].

2.5.4 Cell Signaling

GPCRs have been proven to be the largest family of druggable targets in the human genome. Receptor function operates through two primary processes, namely, G-protein and beta-arrestin signaling. Recent insight into the operations of GPCRs (seven-transmembrane (7TM) receptors) has led to the appreciation that certain ligands initiate only portions of the signaling mechanism mediated by a given receptor. This understanding is promoting a return to whole system assays and opens new opportunities for drug discovery [145]. Flow cytometry has been used to select clones with desired pharmacological profiles for use with the Tango 7TM receptor assay technology, a high-throughput homogeneous assay method for monitoring beta-arrestin recruitment that uses a live cell fluorescent readout [146]. Interestingly, biosensor-centered label-free cell assay technologies are becoming a very active area of effort for GPCR screening [147].

RGS proteins are important components of signal transduction pathways initiated through GPCRs. The interaction of these regulators of G-protein signaling proteins with the G-protein Galphao has been assessed using a flow cytometry protein interaction assay (FCPIA). This approach permits the measurement of nanomolar binding constants of such protein–protein interactions and Roman et al. have shown the feasibility of targeting RGS/Galpha protein–protein interactions with small molecules to modulate GPCR-mediated signaling processes [148]. The study used a Luminex 96-well plate bead analyzer and a novel flow cytometric protein interaction assay. To increase throughput, a set of 8000 compounds was compressed by combining 4 compounds in a single assay well [149].

The signal transducers and activators of transcription (STATs) family responds to a variety of cytokines, hormones, and growth factors. STATs are activated by tyrosine phosphorylation, which results in their dimerization and translocation into the nucleus where they exert their effect on transcription of regulated target genes. The phosphorylation of STATs is mediated mainly by Janus kinases (JAKs). The JAK/STAT pathway plays a critical role in hematopoietic and immune cell function. The assay formats available to detect, for example, phosphorylated STAT5 (pSTAT5) have relatively low throughput and involve lengthy protocols including those for flow cytometry. Recently, an alternative homogeneous assay has been described. The SureFire (Perkin Elmer, Waltham, MA) pSTAT5 assay utilizes AlphaScreen (Perkin Elmer) technology to detect pSTAT5 in cell lysates, amenable to high-throughput screening applications [150].

2.6 FACING THE CHALLENGE OF HIGH-THROUGHPUT FLOW CYTOMETRY DATA ANALYSIS

Data analysis techniques in the larger flow community remain embryonic and represent the major area in flow cytometry from where new developments will emerge over the forthcoming decade [151]. Typically, current high-throughput methods can generate up to a thousand flow cytometry data files per day and each data file can consist of millions of multiparametric descriptions of individual cells. A simple time series experiment for high-throughput analysis encompasses a screening matrix consisting of typically nine cell types and a treatment regimen requiring 16 treatments with corresponding baseline (controls) and usually sampled at different stimulation time points; one single treatment regimen contains more than 12 million data points. This highlights the principal data challenge facing the field: automated cluster analysis of high-content flow cytometry data.

2.6.1 Manual Analysis of Data Presents Limitations

Manual data analysis (using software packages such as FlowJo (Tree Star Inc.) and FCSExpress (DeNovo Software)) is typically undertaken and appropriate for hypothesis-driven interrogation to extract quantitative distributions and parameter relationships, but is essentially limited by the capacity of the researcher to only visualize and interrogate three-dimensional data. These packages enhance user capacity to organize and group data as well as provide tools to segment (gate) subpopulations in histograms and also cascade appropriate gating using Boolean expressions to different data sets. Manual analysis is, however, not suitable for tackling a drug screen or multicolor clinical investigations; any process that requires manual intervention often results in issues of subjectivity, generation of errors, and low throughput providing a series of hurdles to reproducible data analysis.

2.6.2 Systematic Methodologies for Data Analysis

High-throughput data demand systematic computational approaches to analyze and rationally interpret data screens. Open source packages such as Bioconductor for analysis of flow cytometry data provide an integrated framework for bioinformaticians, statisticians, and computational biologists to develop methods to analyze and interpret flow cytometry data (flowCore) [152]. flowCore consists of tools or packages that provide a workflow for rigorous and robust analysis. Critical steps include data handling import and normalization (flowCore), visualization of suitable data structures that support the application of similar operations to a collection of samples (flowViz) [153], and identification of cell populations while handling the issues of outlier identification and data transformation (flowClust). flowClust offers various tools to summarize and visualize the outcomes of clustering resources [154]. Importantly, the Bioconductor environment provides a developer framework for incorporating tools and models that are also open source; hence, the community is able to both contribute to and benefit from the evolving package and importantly keep a pace with ongoing technological advancement.

The current emphasis for data analysis has been the use of computational models to identify subpopulations and enable unsupervised analysis. Many groups are addressing the development of automated, high-dimensional analytical methods. However, flow cytometry data are noisy, asymmetric, and often contain outliers, and these issues offer the primary hurdles for true automated analysis. Recent studies have explored Gaussian mixture modeling to select an "optimal" number of components in the model and partition data sets [155]. A direct multivariate finite mixture modeling approach has been described, using skew and heavy-tailed distributions, to address the complexities of flow cytometric analysis. The approach permits the handling of high-dimensional cytometric data with the capacity to detect rare populations. The group at the Broad Institute of MIT and Harvard has demonstrated the approach for modeling the presence of outliers and skew, to perform the critical task of matching cell populations across samples to facilitate downstream analysis [156]. These advances in modeling of flow cytometry data operating within a solid data processing pipeline and supported with fast processing speeds provide the means for achieving a true discovery environment for high-throughput flow cytometry data.

2.6.3 Introducing the Need for Data Standards and Formats

The value of data is only as good as its annotation and accessibility: it must be properly curated and archived in a software or machine-readable format (see editorial in *Nature Cell Biology* [157]). The long-standing and enduring goal of the Flow Cytometry Data File Standard (FCS) has been to output data from any flow cytometry instrument and hence facilitate the development of software for reading and writing standardized formatted cytometry data. Listmode files consist of a complete listing of all events corresponding to all the parameters collected, as specified by the acquisition settings. This file follows a format specified by the FCS 3.0 standard [158]. Raw listmode data files can be opened or replayed using any program designed for analysis of flow cytometry data. Data File Standard for Flow Cytometry, Version FCS 3.1 is scheduled for release in 2010 and includes a mechanism for handling time series parameters.

However, in recent years, data handling approaches developed for genomic and proteomic studies are now being applied to flow cytometry supported by data standards and bioinformatics tools that enable robust management and mining of high-throughput data. Raw flow cytometry data (in the form of listmode data files) can be used to cluster data automatically (without any researcher effort or intervention). These methods are very similar to those used to analyze microarray data. Information standards have been used for microarray data [159] and recently for flow cytometry data, the latter providing the Minimum Information about a Flow Cytometry Experiment (MIFlowCyt) [160] developed by a cross-disciplinary international group comprising bioinformaticians to basic research scientists. Establishing the flow cytometry minimal standard was based on understanding the fundamental concepts of data acquisition workflows, including best practice in flow cytometry. The MIFlowCyt standard includes recommendations for describing the specimens and reagents included in the flow experiment, instrument configuration, and the data processing protocols used to interpret the primary output data. MIFlowCyt has been adopted as a standard by the International Society for Advancement of Cytometry (ISAC).

To meet the immediate vision for a global exchange of flow cytometry data and including the need to develop data repositories ensuring open access and data sharing for the healthcare and research community, collaborative efforts are being taken to build suitable data models with common ontologies and vocabulary [161]. The Functional Genomics Experiment (FuGE) formalizes common aspects of comprehensive and high-throughput experiments across different biological technologies [162]. This has been extended to map MIFlowCyt terms into a FuGE object model to accommodate flow cytometry data and metadata. These aspirations are likely to develop alongside other major projects such as The Ontology for Biomedical Investigations (OBI; http://purl.obolibrary.org/obo/obi) project that offers an integrated ontology for the description of biological and clinical investigations. This is a typical endeavor where the ontology supports the consistent annotation of biomedical investigations. It is likely that as approaches such as FuGE roll out to include microscopy and high-throughput imaging technologies, the community meets an ideology for cytometry platform integration.

REFERENCES

1. Imeri F, Herklotz R, Risch L, Arbetsleitner C, Zerlauth M, Risch GM, Huber AR. Stability of hematological analytes depends on the hematology analyser used: a stability study with Bayer Advia 120, Beckman Coulter LH 750 and Sysmex XE 2100. *Clin. Chim. Acta* 2008;397(1–2):68–71.
2. Bouvier S, Machon C, Destenay S, Brun S, Arnaud A, Cochery-Nouvellon E, Lissalde-Lavigne G, Gris JC. A new automated haematology analyser: the Excell 2280. *Ann. Biol. Clin. (Paris)* 2008;66(4):417–426.
3. Sklar LA, Carter MB, Edwards BS. Flow cytometry for drug discovery, receptor pharmacology and high-throughput screening. *Curr. Opin. Pharmacol.* 2007;7 (5):527–534.
4. Chakraborty C, Hsu CH, Wen ZH, Lin CS, Agoramoorthy G. Zebrafish: a complete animal model for *in vivo* drug discovery and development. *Curr. Drug Metab.* 2009;10 (2):116–124.

5. Edwards BS, Kuckuck FW, et al. HTPS flow cytometry: a novel platform for automated high throughput drug discovery and characterization. *J. Biomol. Screen.* 2001;6(2):83–90.
6. Edwards BS, Oprea T, Prossnitz ER, Sklar LA. Flow cytometry for high-throughput, high-content screening. *Curr. Opin. Chem. Biol.* 2004;8(4):392–398.
7. Edwards BS, Bologa C, Young SM, Balakin KV, Prossnitz ER, Savchuck NP, Sklar LA, Oprea TI. Integration of virtual screening with high-throughput flow cytometry to identify novel small molecule formylpeptide receptor antagonists. *Mol. Pharmacol.* 2005;68 (5):1301–1310.
8. Feng Y, Mitchison TJ, Bender A, Young DW, Tallarico JA. Multi-parameter phenotypic profiling: using cellular effects to characterize small-molecule compounds. *Nat. Rev. Drug Discov.* 2009;8(7):567–578.
9. Caldwell JS. Cancer cell-based genomic and small molecule screens. *Adv. Cancer Res.* 2007;96:145–173.
10. Zeisel MB, Barth H, Schuster C, Baumert TF. Hepatitis C virus entry: molecular mechanisms and targets for antiviral therapy. *Front. Biosci.* 2009;14:3274–3285.
11. Ziegler S, Kronenberger B, Albrecht BA, Kaul A, Gamer AL, Klein CD, Hartmann RW. Development and evaluation of a FACS-based medium-throughput assay for HCV entry inhibitors. *J. Biomol. Screen.* 2009;14(6):620–626.
12. Haney SA. Expanding the repertoire of RNA interference screens for developing new anticancer drug targets. *Expert Opin. Ther. Targets* 2007;11(11):1429–1441.
13. Vignali DA. Multiplexed particle-based flow cytometric assays. *J. Immunol. Methods* 2000;243(1–2):243–255.
14. Morgan E, Varro R, Sepulveda H, Ember JA, Apgar J, Wilson J, Lowe L, Chen R, Shivraj L, Agadir A, et al. Cytometric bead array: a multiplexed assay platform with applications in various areas of biology. *Clin. Immunol.* 2004;110(3):252–266.
15. Woods TA, Graves SW, Nolan JP. Microsphere surface protein determination using flow cytometry. *Curr. Protoc. Cytom.* 2005; Chapter 13:Unit 13.2.
16. Watson DA, Gaskill DF, Brown LO, Doorn SK, Nolan JP. Spectral measurements of large particles by flow cytometry. *Cytometry A* 2009;75(5):460–464.
17. Galanzha EI, Shashkov EV, Tuchin VV, Zharov VP. *In vivo* multispectral, multiparameter, photoacoustic lymph flow cytometry with natural cell focusing, label-free detection and multicolor nanoparticle probes. *Cytometry A* 2008;73(10):884–894.
18. Nolan JP, Yang L. The flow of cytometry into systems biology. *Brief. Funct. Genomics Proteomics* 2007;6(2):81–90.
19. Khan IA, Husemann P, Campbell L, White NS, White RJ, Smith PJ, Errington RJ. ProgeniDB: a novel cell lineage database for generation associated phenotypic behavior in cell-based assays. *Cell Cycle* 2007;6(7):868–874.
20. Smith PJ, Khan IA, Errington RJ. Cytomics and drug development. *Cytometry A* 2007;71(6):349–351.
21. Smith PJ, Khan IA, Errington RJ. Cytomics and cellular informatics: coping with asymmetry and heterogeneity in biological systems. *Drug Discov. Today* 2009;14 (5–6):271–277.
22. Shapiro HM. Lasers for flow cytometry. *Curr. Protoc. Cytom.* 2004; Chapter 1:Unit 1.9.
23. Nolan JP, Lauer S, Prossnitz ER, Sklar LA. Flow cytometry: a versatile tool for all phases of drug discovery. *Drug Discov. Today* 1999;4(4):173–180.

24. Lang P, Yeow K, Nichols A, Scheer A. Cellular imaging in drug discovery. *Nat. Rev. Drug Discov.* 2006;5(4):343–356.
25. Shapiro HM. Scanning laser cytometry. *Curr. Protoc. Cytom.* 2004; Chapter 2:Unit 2.10.
26. Basiji DA, Ortyn WE, Liang L, Venkatachalam V, Morrissey P. Cellular image analysis and imaging by flow cytometry. *Clin. Lab. Med.* 2007;27(3):653–670, viii.
27. Kottaridi C, Georgoulakis J, Kassanos D, Pappas A, Spathis A, Margari N, Aninos D, Karakitsos P. Use of flow cytometry as a quality control device for liquid-based cervical cytology specimens. *Cytometry B Clin Cytom* 2010;78:37–40.
28. Galbraith DW. Simultaneous flow cytometric quantification of plant nuclear DNA contents over the full range of described angiosperm 2C values. *Cytometry A* 2009;75 (8):692–698.
29. Ortyn WE, Perry DJ, Venkatachalam V, Liang L, Hall BE, Frost K, Basiji DA. Extended depth of field imaging for high speed cell analysis. *Cytometry A* 2007;71(4):215–231.
30. George TC, Basiji DA, Hall BE, Lynch DH, Ortyn WE, Perry DJ, Seo MJ, Zimmerman CA, Morrissey PJ. Distinguishing modes of cell death using the ImageStream multispectral imaging flow cytometer. *Cytometry A* 2004;59(2):237–245.
31. George TC, Fanning SL, Fitzgerald-Bocarsly P, Medeiros RB, Highfill S, Shimizu Y, Hall BE, Frost K, Basiji D, Ortyn WE, et al. Quantitative measurement of nuclear translocation events using similarity analysis of multispectral cellular images obtained in flow. *J. Immunol. Methods* 2006;311(1–2):117–129.
32. George TC, Morrissey PJ, Cui C, Singh S, Fitzgerald Bocarsly P. Measurement of cytoplasmic to nuclear translocation. *Curr. Protoc. Cytom.* 2009; Chapter 9:Unit 9.28.
33. Henery S, George T, Hall B, Basiji D, Ortyn W, Morrissey P. Quantitative image based apoptotic index measurement using multispectral imaging flow cytometry: a comparison with standard photometric methods. *Apoptosis* 2008;13(8):1054–1063.
34. McGrath KE, Bushnell TP, Palis J. Multispectral imaging of hematopoietic cells: where flow meets morphology. *J. Immunol. Methods* 2008;336(2):91–97.
35. Ortyn WE, Hall BE, George TC, Frost K, Basiji DA, Perry DJ, Zimmerman CA, Coder D, Morrissey PJ. Sensitivity measurement and compensation in spectral imaging. *Cytometry A* 2006;69(8):852–862.
36. Wojakowski W, Tendera M, Kucia M, Zuba-Surma E, Paczkowska E, Ciosek J, Halasa M, Krol M, Kazmierski M, Buszman P, et al. Mobilization of bone marrow-derived Oct-4 + SSEA-4 + very small embryonic-like stem cells in patients with acute myocardial infarction. *J. Am. Coll. Cardiol.* 2009;53(1):1–9.
37. Elliott GS. Moving pictures: imaging flow cytometry for drug development. *Comb. Chem. High Throughput Screen.* 2009;12:849–859.
38. Haynes MK, Strouse JJ, Waller A, Leitao A, Curpan RF, Bologa C, Oprea TI, Prossnitz ER, Edwards BS, Sklar LA, et al. Detection of intracellular granularity induction in prostate cancer cell lines by small molecules using the HyperCyt high-throughput flow cytometry system. *J. Biomol. Screen.* 2009;14(6):596–609.
39. Young SM, Curry MS, Ransom JT, Ballesteros JA, Prossnitz ER, Sklar LA, Edwards BS. High-throughput microfluidic mixing and multiparametric cell sorting for bioactive compound screening. *J. Biomol. Screen.* 2004;9(2):103–111.
40. Young SM, Bologa C, et al. High-throughput screening with HyperCyt flow cytometry to detect small molecule formylpeptide receptor ligands. *J. Biomol. Screen.* 2005;10(4): 374–382.

41. Young SM, Bologa C, Prossnitz ER, Oprea TI, Sklar LA, Edwards BS. High-throughput screening with HyperCyt flow cytometry to detect small molecule formylpeptide receptor ligands. *J. Biomol. Screen.* 2005;10(4):374–382.

42. Naumann U, Wand MP. Automation in high-content flow cytometry screening. *Cytometry A* 2009;75:789–797.

43. Fulwyler MJ. Electronic separation of biological cells by volume. *Science* 1965;150 (698):910–911.

44. Bortner CD, Sifre MI, Cidlowski JA. New approaches for determining apoptotic volume decrease in cells. *Methods Enzymol.* 2007;428:161–181.

45. Krishan A, Cabana R. Flow cytometric analysis of electronic nuclear volume and DNA content in normal mouse tissues. *Cell Cycle* 2004;3(3):380–383.

46. Thomas RA, Krishan A, Brochu M. High resolution flow cytometric analysis of electronic nuclear volume and DNA content in normal and abnormal human tissue. *Methods Cell Sci.* 2002;24(1–3):11–18.

47. Postema M, Gilja OH. Ultrasound-directed drug delivery. *Curr. Pharm. Biotechnol.* 2007;8(6):355–361.

48. Karshafian R, Bevan PD, Williams R, Samac S, Burns PN. Sonoporation by ultrasound-activated microbubble contrast agents: effect of acoustic exposure parameters on cell membrane permeability and cell viability. *Ultrasound Med. Biol.* 2009;35(5):847–860.

49. Dainiak MB, Galaev IY, Kumar A, Plieva FM, Mattiasson B. Chromatography of living cells using supermacroporous hydrogels, cryogels. *Adv. Biochem. Eng. Biotechnol.* 2007;106:101–127.

50. Radisic M, Iyer RK, Murthy SK. Micro- and nanotechnology in cell separation. *Int. J. Nanomed.* 2006;1(1):3–14.

51. Pappas D, Wang K. Cellular separations: a review of new challenges in analytical chemistry. *Anal. Chim. Acta* 2007;601(1):26–35.

52. Dainiak MB, Kumar A, Galaev IY, Mattiasson B. Methods in cell separations. *Adv. Biochem. Eng. Biotechnol.* 2007;106:1–18.

53. Goddard G, Martin JC, Graves SW, Kaduchak G. Ultrasonic particle-concentration for sheathless focusing of particles for analysis in a flow cytometer. *Cytometry A* 2006;69 (2):66–74.

54. Goddard GR, Sanders CK, Martin JC, Kaduchak G, Graves SW. Analytical performance of an ultrasonic particle focusing flow cytometer. *Anal. Chem.* 2007;79(22):8740–8746.

55. Morgan H, Hughes MP, Green NG. Separation of submicron bioparticles by dielectrophoresis. *Biophys. J.* 1999;77(1):516–525.

56. Li Y, Dalton C, Crabtree HJ, Nilsson G, Kaler KV. Continuous dielectrophoretic cell separation microfluidic device. *Lab Chip* 2007;7(2):239–248.

57. Cen EG, Dalton C, Li Y, Adamia S, Pilarski LM, Kaler KV. A combined dielectrophoresis, traveling wave dielectrophoresis and electrorotation microchip for the manipulation and characterization of human malignant cells. *J. Microbiol. Methods* 2004;58(3):387–401.

58. Kua CH, Lam YC, Rodriguez I, Yang C, Youcef-Toumi K. Dynamic cell fractionation and transportation using moving dielectrophoresis. *Anal. Chem.* 2007;79(18):6975–6987.

59. Gambari R, Borgatti M, Altomare L, Manaresi N, Medoro G, Romani A, Tartagni M, Guerrieri R. Applications to cancer research of "lab-on-a-chip" devices based on dielectrophoresis (DEP). *Technol. Cancer Res. Treat.* 2003;2(1):31–40.

60. Coley HM, Labeed FH, Thomas H, Hughes MP. Biophysical characterization of MDR breast cancer cell lines reveals the cytoplasm is critical in determining drug sensitivity. *Biochim. Biophys. Acta* 2007;1770(4):601–608.

61. Errington RJ, Ameer-Beg SM, Vojnovic B, Patterson LH, Zloh M, Smith PJ. Advanced microscopy solutions for monitoring the kinetics and dynamics of drug–DNA targeting in living cells. *Adv. Drug Deliv. Rev.* 2005;57(1):153–167.

62. Huang C, Chen A, Guo M, Yu J. Membrane dielectric responses of bufalin-induced apoptosis in HL-60 cells detected by an electrorotation chip. *Biotechnol. Lett.* 2007;29 (9):1307–1313.

63. Huang C, Chen A, Wang L, Guo M, Yu J. Electrokinetic measurements of dielectric properties of membrane for apoptotic HL-60 cells on chip-based device. *Biomed. Microdevices* 2007;9(3):335–343.

64. Gawad S, Cheung K, Seger U, Bertsch A, Renaud P. Dielectric spectroscopy in a micromachined flow cytometer: theoretical and practical considerations. *Lab Chip* 2004;4(3):241–251.

65. Cheung K, Gawad S, Renaud P. Impedance spectroscopy flow cytometry: on-chip label-free cell differentiation. *Cytometry A* 2005;65(2):124–132.

66. Dopp E, Jonas L, Nebe B, Budde A, Knippel E. Dielectric changes in membrane properties and cell interiors of human mesothelial cells *in vitro* after crocidolite asbestos exposure. *Environ. Health Perspect.* 2000;108(2):153–158.

67. Huang Y, Wang XB, Gascoyne PR, Becker FF. Membrane dielectric responses of human T-lymphocytes following mitogenic stimulation. *Biochim. Biophys. Acta* 1999;1417 (1):51–62.

68. Wang X, Becker FF, Gascoyne PR. Membrane dielectric changes indicate induced apoptosis in HL-60 cells more sensitively than surface phosphatidylserine expression or DNA fragmentation. *Biochim. Biophys. Acta* 2002;1564(2):412–420.

69. Gryczynski I, Gryczynski Z, Lakowicz JR, Yang D, Burke TG. Fluorescence spectral properties of the anticancer drug topotecan by steady-state and frequency domain fluorometry with one-photon and multi-photon excitation. *Photochem. Photobiol.* 1999;69(4):421–428.

70. Ratanachoo K, Gascoyne PR, Ruchirawat M. Detection of cellular responses to toxicants by dielectrophoresis. *Biochim. Biophys. Acta* 2002;1564(2):449–458.

71. Vahey MD, Voldman J. An equilibrium method for continuous-flow cell sorting using dielectrophoresis. *Anal. Chem.* 2008;80(9):3135–3143.

72. Vahey MD, Voldman J. High-throughput cell and particle characterization using isodielectric separation. *Anal. Chem.* 2009;81(7):2446–2455.

73. Wang K, Solis-Wever X, Aguas C, Liu Y, Li P, Pappas D. Differential mobility cytometry. *Anal. Chem.* 2009;81(9):3334–3343.

74. Jain KK. The role of nanobiotechnology in drug discovery. *Drug Discov. Today* 2005;10(21):1435–1442.

75. Jain KK. Personalized clinical laboratory diagnostics. *Adv. Clin. Chem.* 2009;47:95–119.

76. Jain KK. Role of nanobiotechnology in the development of personalized medicine. *Nanomedicine* 2009;4(3):249–252.

77. Sohn LL, Saleh OA, Facer GR, Beavis AJ, Allan RS, Notterman DA. Capacitance cytometry: measuring biological cells one by one. *Proc. Natl. Acad. Sci. USA* 2000;97 (20):10687–10690.

78. Chung TD, Kim HC. Recent advances in miniaturized microfluidic flow cytometry for clinical use. *Electrophoresis* 2007;28(24):4511–4520.

79. Ateya DA, Erickson JS, Howell PB, Jr., Hilliard LR, Golden JP, Ligler FS. The good, the bad, and the tiny: a review of microflow cytometry. *Anal. Bioanal. Chem.* 2008;391 (5):1485–1498.

80. Golden JP, Kim JS, Erickson JS, Hilliard LR, Howell PB, Anderson GP, Nasir M, Ligler FS. Multi-wavelength microflow cytometer using groove-generated sheath flow. *Lab Chip* 2009;9(13):1942–1950.

81. Wlodkowic D, Skommer J, McGuinness D, Faley S, Kolch W, Darzynkiewicz Z, Cooper JM. Chip-based dynamic real-time quantification of drug-induced cytotoxicity in human tumor cells. *Anal. Chem.* 2009;81:6952–6959.

82. Wlodkowic D, Skommer J, Faley S, Darzynkiewicz Z, Cooper JM. Dynamic analysis of apoptosis using cyanine SYTO probes: from classical to microfluidic cytometry. *Exp. Cell Res.* 2009;315(10):1706–1714.

83. Kapoor V, Subach FV, Kozlov VG, Grudinin A, Verkhusha VV, Telford WG. New lasers for flow cytometry: filling the gaps. *Nat. Methods* 2007;4(9):678–679.

84. Telford WG. Small lasers in flow cytometry. *Methods Mol. Biol.* 2004;263:399–418.

85. Shapiro HM, Perlmutter NG. Violet laser diodes as light sources for cytometry. *Cytometry* 2001;44(2):133–136.

86. Connally R, Jin D, Piper J. High intensity solid-state UV source for time-gated luminescence microscopy. *Cytometry A* 2006;69(9):1020–1027.

87. Jin D, Connally R, Piper J. Practical time-gated luminescence flow cytometry. I. Concepts. *Cytometry A* 2007;71(10):783–796.

88. Jin D, Connally R, Piper J. Practical time-gated luminescence flow cytometry. II. Experimental evaluation using UV LED excitation. *Cytometry A* 2007;71(10):797–808.

89. Jin D, Piper JA, Leif RC, Yang S, Ferrari BC, Yuan J, Wang G, Vallarino LM, Williams JW. Time-gated flow cytometry: an ultra-high selectivity method to recover ultra-rare-event mu-targets in high-background biosamples. *J. Biomed. Opt.* 2009;14(2):024023.

90. Kapoor V, Karpov V, Linton C, Subach FV, Verkhusha VV, Telford WG. Solid state yellow and orange lasers for flow cytometry. *Cytometry A* 2008;73(6):570–577.

91. Telford WG, Subach FV, Verkhusha VV. Supercontinuum white light lasers for flow cytometry. *Cytometry A* 2009;75(5):450–459.

92. Gourley PL, Gourley MF. Biocavity lasers for biomedicine. *Trends Biotechnol.* 2000;18(11):443–448.

93. Gourley PL. Semiconductor microlasers: a new approach to cell-structure analysis. *Nat. Med.* 1996;2(8):942–944.

94. Gourley PL. *Nanolasers. Sci. Am.* 1998;278(3):56–61.

95. Gourley PL. Brief overview of BioMicroNano technologies. *Biotechnol. Prog.* 2005;21(1):2–10.

96. Gourley PL, Hendricks JK, McDonald AE, Copeland RG, Barrett KE, Gourley CR, Naviaux RK. Ultrafast nanolaser flow device for detecting cancer in single cells. *Biomed. Microdevices* 2005;7(4):331–339.

97. Shapiro HM. Excitation and emission spectra of common dyes. *Curr. Protoc. Cytom.* 2004; Chapter 1:Unit 1.19.

98. Sun L, Zborowski M, Moore LR, Chalmers JJ. Continuous, flow-through immunomagnetic cell sorting in a quadrupole field. *Cytometry* 1998;33(4):469–475.

99. Lara O, Tong X, Zborowski M, Chalmers JJ. Enrichment of rare cancer cells through depletion of normal cells using density and flow-through, immunomagnetic cell separation. *Exp. Hematol.* 2004;32(10):891–904.

100. Zborowski M, Chalmers JJ. Magnetic cell sorting. *Methods Mol. Biol.* 2005;295:291–300.

101. Jin X, Zhao Y, Richardson A, Moore L, Williams PS, Zborowski M, Chalmers JJ. Differences in magnetically induced motion of diamagnetic, paramagnetic, and superparamagnetic microparticles detected by cell tracking velocimetry. *Analyst* 2008;133(12):1767–1775.

102. Michalet X, Pinaud FF, Bentolila LA, Tsay JM, Doose S, Li JJ, Sundaresan G, Wu AM, Gambhir SS, Weiss S. Quantum dots for live cells, *in vivo* imaging, and diagnostics. *Science* 2005;307(5709):538–544.

103. Ebenstein Y, Gassman N, Kim S, Antelman J, Kim Y, Ho S, Samuel R, Michalet X, Weiss S. Lighting up individual DNA binding proteins with quantum dots. *Nano Lett.* 2009;9 (4):1598–1603.

104. Farias PM, Santos BS, Fontes A. Semiconductor fluorescent quantum dots: efficient biolabels in cancer diagnostics. *Methods Mol. Biol.* 2009;544:407–419.

105. Michalet X, Colyer RA, Antelman J, Siegmund OH, Tremsin A, Vallerga JV, Weiss S. Single-quantum dot imaging with a photon counting camera. *Curr. Pharm. Biotechnol.* 2009;10(5):543–558.

106. Pinaud F, Michalet X, Bentolila LA, Tsay JM, Doose S, Li JJ, Iyer G, Weiss S. Advances in fluorescence imaging with quantum dot bio-probes. *Biomaterials* 2006;27(9):1679–1687.

107. Chattopadhyay PK, Price DA, Harper TF, Betts MR, Yu J, Gostick E, Perfetto SP, Goepfert P, Koup RA, De Rosa SC, et al. Quantum dot semiconductor nanocrystals for immunophenotyping by polychromatic flow cytometry. *Nat. Med.* 2006;12(8):972–977.

108. Machleidt T, Whitney P, Bi K. Multiplexing of pathway-specific beta-lactamase reporter gene assays by optical coding with Qtracker® nanocrystals. *J. Biomol. Screen.* 2009;14(7):845–852.

109. Sukhanova A, Susha AS, Bek A, Mayilo S, Rogach AL, Feldmann J, Oleinikov V, Reveil B, Donvito B, Cohen JH, et al. Nanocrystal-encoded fluorescent microbeads for proteomics: antibody profiling and diagnostics of autoimmune diseases. *Nano Lett.* 2007;7(8):2322–2327.

110. Di Corato R, Piacenza P, Musaro M, Buonsanti R, Cozzoli PD, Zambianchi M, Barbarella G, Cingolani R, Manna L, Pellegrino T. Magnetic-fluorescent colloidal nanobeads: preparation and exploitation in cell separation experiments. *Macromol. Biosci.* 2009;9:952–958.

111. Nabiev I, Mitchell S, Davies A, Williams Y, Kelleher D, Moore R, Gun'ko YK, Byrne S, Rakovich YP, Donegan JF, et al. Nonfunctionalized nanocrystals can exploit a cell's active transport machinery delivering them to specific nuclear and cytoplasmic compartments. *Nano Lett.* 2007;7(11):3452–3461.

112. Vaijayanthimala V, Tzeng YK, Chang HC, Li CL. The biocompatibility of fluorescent nanodiamonds and their mechanism of cellular uptake. *Nanotechnology* 2009;20 (42):425103.

113. MacFarlane M. Cell death pathways: potential therapeutic targets. *Xenobiotica* 2009;39(8):616–624.

114. Cai SX, Drewe J, Kemnitzer W. Discovery of 4-aryl-4*H*-chromenes as potent apoptosis inducers using a cell- and caspase-based Anti-cancer Screening Apoptosis Program (ASAP): SAR studies and the identification of novel vascular disrupting agents. *Anticancer Agents Med. Chem.* 2009;9(4):437–456.

115. Sauermann M, Hahne F, Schmidt C, Majety M, Rosenfelder H, Bechtel S, Huber W, Poustka A, Arlt D, Wiemann S. High-throughput flow cytometry-based assay to identify apoptosis-inducing proteins. *J. Biomol. Screen.* 2007;12(4):510–520.

116. Gasparri F, Cappella P, Galvani A. Multiparametric cell cycle analysis by automated microscopy. *J. Biomol. Screen.* 2006;11(6):586–598.

117. Smith PJ, Blunt N, Wiltshire M, Hoy T, Teesdale-Spittle P, Craven MR, Watson JV, Amos WB, Errington RJ, Patterson LH. Characteristics of a novel deep red/infrared fluorescent cell-permeant DNA probe, DRAQ5, in intact human cells analyzed by flow cytometry, confocal and multiphoton microscopy. *Cytometry* 2000;40(4):280–291.

118. Edward R. Use of DNA-specific anthraquinone dyes to directly reveal cytoplasmic and nuclear boundaries in live and fixed cells. *Mol. Cells* 2009;27(4):391–396.

119. Smith PJ, Marquez N, Wiltshire M, Chappell S, Njoh K, Campbell L, Khan IA, Silvestre O, Errington RJ. Mitotic bypass via an occult cell cycle phase following DNA topoisomerase II inhibition in p53 functional human tumor cells. *Cell Cycle* 2007;6(16):2071–2081.

120. Smith PJ, Chin SF, Njoh K, Khan IA, Chappell MJ, Errington RJ. Cell cycle checkpoint-guarded routes to catenation-induced chromosomal instability. *SEB Exp. Biol. Ser.* 2008;59:219–242.

121. Olive KP, Jacobetz MA, Davidson CJ, Gopinathan A, McIntyre D, Honess D, Madhu B, Goldgraben MA, Caldwell ME, Allard D, et al. Inhibition of Hedgehog signaling enhances delivery of chemotherapy in a mouse model of pancreatic cancer. *Science* 2009;324(5933):1457–1461.

122. Boyle RG, Travers S. Hypoxia: targeting the tumour. *Anticancer Agents Med. Chem.* 2006;6(4):281–286.

123. Tredan O, Grantab R, Dumontet C. Can tumor hypoxia be turned into a chemotherapeutic advantage? *Bull. Cancer* 2008;95(5):528–534.

124. Tredan O, Garbens AB, Lalani AS, Tannock IF. The hypoxia-activated ProDrug AQ4N penetrates deeply in tumor tissues and complements the limited distribution of mitoxantrone. *Cancer Res.* 2009;69(3):940–947.

125. Patterson LH. Bioreductively activated antitumor *N*-oxides: the case of AQ4N, a unique approach to hypoxia-activated cancer chemotherapy. *Drug Metab. Rev.* 2002;34 (3):581–592.

126. Albertella MR, Loadman PM, Jones PH, Phillips RM, Rampling R, Burnet N, Alcock C, Anthoney A, Vjaters E, Dunk CR, et al. Hypoxia-selective targeting by the bioreductive prodrug AQ4N in patients with solid tumors: results of a phase I study. *Clin. Cancer Res.* 2008;14(4):1096–1104.

127. O'Rourke M, Ward C, Worthington J, McKenna J, Valentine A, Robson T, Hirst DG, McKeownSr., Evaluation of the antiangiogenic potential of AQ4N. *Clin. Cancer Res.* 2008;14(5):1502–1509.

128. Smith PJ, Desnoyers R, Blunt N, Giles Y, Patterson LH, Watson JV. Flow cytometric analysis and confocal imaging of anticancer alkylaminoanthraquinones and their

N-oxides in intact human cells using 647-nm krypton laser excitation. *Cytometry* 1997;27 (1):43–53.

129. Schinkel AH, Jonker JW. Mammalian drug efflux transporters of the ATP binding cassette (ABC) family: an overview. *Adv. Drug Deliv. Rev.* 2003;55(1):3–29.
130. Ivnitski-Steele I, Larson RS, Lovato DM, Khawaja HM, Winter SS, Oprea TI, Sklar LA, Edwards BS. High-throughput flow cytometry to detect selective inhibitors of ABCB1, ABCC1, and ABCG2 transporters. *Assay Drug Dev. Technol.* 2008;6(2):263–276.
131. Demel MA, Schwaha R, Kramer O, Ettmayer P, Haaksma EE, Ecker GF. *In silico* prediction of substrate properties for ABC-multidrug transporters. *Expert Opin. Drug Metab. Toxicol.* 2008;4(9):1167–1180.
132. Winter SS, Lovato DM, Khawaja HM, Edwards BS, Steele ID, Young SM, Oprea TI, Sklar LA, Larson RS. High-throughput screening for daunorubicin-mediated drug resistance identifies mometasone furoate as a novel ABCB1-reversal agent. *J. Biomol. Screen.* 2008;13(3):185–193.
133. Reya T, Morrison SJ, Clarke MF, Weissman IL. Stem cells, cancer, and cancer stem cells. *Nature* 2001;414(6859):105–111.
134. Hart LS, El-Deiry WS. Invincible, but not invisible: imaging approaches toward *in vivo* detection of cancer stem cells. *J. Clin. Oncol.* 2008;26(17):2901–2910.
135. Visvader JE, Lindeman GJ. Cancer stem cells in solid tumours: accumulating evidence and unresolved questions. *Nat. Rev. Cancer* 2008;8(10):755–768.
136. Schatton T, Murphy GF, Frank NY, Yamaura K, Waaga-Gasser AM, Gasser M, Zhan Q, Jordan S, Duncan LM, Weishaupt C, et al. Identification of cells initiating human melanomas. *Nature* 2008;451(7176):345–349.
137. Goodell MA, McKinney-Freeman S, Camargo FD. Isolation and characterization of side population cells. *Methods Mol. Biol.* 2005;290:343–352.
138. Hirschmann-Jax C, Foster AE, Wulf GG, Goodell MA, Brenner MK. A distinct "side population" of cells in human tumor cells: implications for tumor biology and therapy. *Cell Cycle* 2005;4(2):203–205.
139. Smith PJ, Nakeff A, Watson JV. Flow-cytometric detection of changes in the fluorescence emission spectrum of a vital DNA-specific dye in human tumour cells. *Exp. Cell Res.* 1985;159(1):37–46.
140. Cabana R, Frolova EG, Kapoor V, Thomas RA, Krishan A, Telford WG. The minimal instrumentation requirements for Hoechst side population analysis: stem cell analysis on low-cost flow cytometry platforms. *Stem Cells* 2006;24(11):2573–2581.
141. Telford WG, Bradford J, Godfrey W, Robey RW, Bates SE. Side population analysis using a violet-excited cell-permeable DNA binding dye. *Stem Cells* 2007;25(4):1029–1036.
142. Hawley TS, Telford WG, Hawley RG. "Rainbow" reporters for multispectral marking and lineage analysis of hematopoietic stem cells. *Stem Cells* 2001;19(2):118–124.
143. Huuskonen H. New models and molecular markers in evaluation of developmental toxicity. *Toxicol. Appl. Pharmacol.* 2005;207(2 Suppl.):495–500.
144. Buesen R, Genschow E, Slawik B, Visan A, Spielmann H, Luch A, Seiler A. Embryonic stem cell test remastered: comparison between the validated EST and the new molecular FACS-EST for assessing developmental toxicity *in vitro*. *Toxicol. Sci.* 2009;108 (2):389–400.
145. Kenakin TP. Cellular assays as portals to seven-transmembrane receptor-based drug discovery. *Nat. Rev. Drug Discov.* 2009;8(8):617–626.

146. Hanson BJ, Wetter J, Bercher MR, Kopp L, Fuerstenau-Sharp M, Vedvik KL, Zielinski T, Doucette C, Whitney PJ, Revankar C. A homogeneous fluorescent live-cell assay for measuring 7-transmembrane receptor activity and agonist functional selectivity through beta-arrestin recruitment. *J. Biomol. Screen.* 2009;14(7):798–810.

147. Fang Y, Frutos AG, Verklereen R. Label-free cell-based assays for GPCR screening. *Comb. Chem. High Throughput Screen.* 2008;11(5):357–369.

148. Roman DL, Talbot JN, Roof RA, Sunahara RK, Traynor JR, Neubig RR. Identification of small-molecule inhibitors of RGS4 using a high-throughput flow cytometry protein interaction assay. *Mol. Pharmacol.* 2007;71(1):169–175.

149. Roman DL, Ota S, Neubig RR. Polyplexed flow cytometry protein interaction assay: a novel high-throughput screening paradigm for RGS protein inhibitors. *J. Biomol. Screen.* 2009;14(6):610–619.

150. Binder C, Lafayette A, Archibeque I, Sun Y, Plewa C, Sinclair A, Emkey R. Optimization and utilization of the SureFire phospho-STAT5 assay for a cell-based screening campaign. *Assay Drug Dev. Technol.* 2008;6(1):27–37.

151. Chattopadhyay PK, Hogerkorp CM, Roederer M. A chromatic explosion: the development and future of multiparameter flow cytometry. *Immunology* 2008;125(4):441–449.

152. Hahne F, LeMeur N, Brinkman RR, Ellis B, Haaland P, Sarkar D, Spidlen J, Strain E, Gentleman R. flowCore: a Bioconductor package for high throughput flow cytometry. *BMC Bioinformatics* 2009;10:106.

153. Le Meur N, Rossini A, Gasparetto M, Smith C, Brinkman RR, Gentleman R. Data quality assessment of ungated flow cytometry data in high throughput experiments. *Cytometry A* 2007;71(6):393–403.

154. Lo K, Hahne F, Brinkman RR, Gottardo R. flowClust: a Bioconductor package for automated gating of flow cytometry data. *BMC Bioinformatics* 2009;10:145.

155. Boedigheimer MJ, Ferbas J. Mixture modeling approach to flow cytometry data. *Cytometry A* 2008;73(5):421–429.

156. Pyne S, Hu X, Wang K, Rossin E, Lin TI, Maier LM, Baecher-Allan C, McLachlan GJ, Tamayo P, Hafler DA, et al. Automated high-dimensional flow cytometric data analysis. *Proc. Natl. Acad. Sci. USA* 2009;106(21):8519–8524.

157. Editorial. Standardizing data. *Nat. Cell Biol.* 2008;10(10):1123–1124.

158. Data file standard for flow cytometry. Data File Standards Committee of the Society for Analytical Cytology. *Cytometry* 1990;11(3):323–332.

159. Brazma A, Hingamp P, Quackenbush J, Sherlock G, Spellman P, Stoeckert C, Aach J, Ansorge W, Ball CA, Causton HC, et al. Minimum information about a microarray experiment (MIAME): toward standards for microarray data. *Nat. Genet.* 2001;29(4):365–371.

160. Lee JA, Spidlen J, Boyce K, Cai J, Crosbie N, Dalphin M, Furlong J, Gasparetto M, Goldberg M, Goralczyk EM, et al. MIFlowCyt: the minimum information about a Flow Cytometry Experiment. *Cytometry A* 2008;73(10):926–930.

161. Qian Y, Tchuvatkina O, Spidlen J, Wilkinson P, Gasparetto M, Jones AR, Manion FJ, Scheuermann RH, Sekaly RP, Brinkman RR. FuGEFlow: data model and markup language for flow cytometry. *BMC Bioinformatics* 2009;10:184.

162. Jones AR, Miller M, Aebersold R, Apweiler R, Ball CA, Brazma A, Degreef J, Hardy N, Hermjakob H, Hubbard SJ, et al. The Functional Genomics Experiment model (FuGE): an extensible framework for standards in functional genomics. *Nat. Biotechnol.* 2007;25(10):1127–1133.

3

INTRODUCTION TO BIOMARKERS

Ole Vesterqvist and Manjula P. Reddy

3.1 INTRODUCTION

The use of biological markers (biomarkers) in diagnosing human diseases is by no means novel. Many biomarkers such as blood pressure, cholesterol, and glucose levels in blood are well established in the scientific community, whereas others have not yet been "accepted" as diagnostic or prognostic biomarkers. The continuous discovery of new biomarkers and the development of new technologies to measure biomarkers have provided us with a plethora of tools to characterize the effects of drugs. We have also seen new biomarkers making their way not only to clinical diagnostics, but also to clinical prognostics and theranostics where biomarkers are used to predict the likelihood of disease progression irrespective of interventional therapy and to predict the response of a patient to a specific therapeutic. Many biomarkers are used in both preclinical and clinical studies facilitating an improved understanding of the difference between mechanism of action of the drug in humans and various animal models used in discovery. This has also allowed us to break down the "wall" between discovery and clinical development, creating a space for translational medicine and biomarkers [1].

Biomarkers have over the past decade become an important part of decision making in both discovery and clinical drug development, especially in the early phases of clinical development as part of a strategy to demonstrate "proof of concept". Health authorities are increasingly expecting drug companies, particularly in oncology, to demonstrate identification of clinical benefits in a subset of patients with particular characteristics as opposed to a one-size-fits-all approach of treatment in light of low response rates of most of the current oncology drugs in the market.

Flow Cytometry in Drug Discovery and Development, Edited by Virginia Litwin and Philip Marder

Although most biomarkers are used to assess the pharmacodynamic effects of drugs to better understand the dose relationship, we are now entering an era where biomarkers will play a role in creating a more personalized approach to drug treatment of patients leading to the development of better and safer drugs for patients. Thus, biomarkers are emerging to be an integral part of drug development and playing an important role in screening, diagnosis, prognosis, therapy selection, and disease monitoring. We will in this chapter discuss the use of biomarkers in different stages of clinical drug development.

3.2 TYPES OF BIOMARKERS

A biomarker is defined as a characteristic that is objectively measured and evaluated as an indicator of normal biological processes, pathogenic processes, or pharmacological responses to a therapeutic intervention [2]. Biomarkers that are typically evaluated during clinical development of a drug fall into four main categories: pharmacodynamic (PD) and mechanism of action (MOA), predictive, prognostic, and surrogate biomarkers [3]. PD biomarkers determine whether a drug interacts with the target and modulates the biological pathway. The PD markers are used to evaluate the MOA of the drug, to perform pharmacokinetic (PK) and PD (PK/PD) correlations, and to determine the dose of the drug and frequency of dosing. PD biomarkers are usually evaluated early in drug development using research-use-only (RUO) tests that need to be validated for clinical implementation. PD markers that are commonly monitored in various therapeutic areas include markers of inflammation, proliferation, and apoptosis.

Predictive biomarkers are those that are used to identify patients most likely to respond or least likely to suffer an adverse event when treated with a drug. Some of the widely used predictive biomarker tests include diagnostic tests: testing for protein expression and gene amplification of human epidermal growth factor receptor 2 (HER2) for Herceptin treatment [4]; test for Philadelphia chromosome (Ph)-positive chronic myeloid leukemia for predicting response to Gleevec [5]; and KRAS mutation test for predicting response to anti-EGFR therapies (panitumumab and cetuximab) in colorectal cancer patients [6]. Greatest impact of predictive biomarkers occurs when they are developed and validated prior to the launch of the drug. However, this will require evaluation of the predictive marker(s) in prospective adaptive trials and in randomized phase II/III studies to determine if the marker(s) is (are) predictive to the specific treatment.

Biomarkers that predict the course of the disease independent of drug treatment fall under prognostic markers. CellSearch (for evaluating circulating tumor cells in prostate [7], breast [8], and colorectal cancers [9]), MammaPrint (for breast cancer metastasis [10]), and Oncotype DX (for breast cancer recurrence [11]) are some of the approved prognostic biomarker tests in use to assess disease prognosis. Biomarkers such as CA 125 or CA 19-9 are widely being used to monitor response to therapy and predict reoccurrence of ovarian, pancreatic, and colon cancer [12]. Surrogate markers are those markers that substitute for clinical approved end points and predict clinical benefit. Some of the common surrogate markers include blood pressure and cholesterol (for predicting therapy benefit in heart disease), CD4 count/viral load (for assessing HIV infection), and prostate-specific antigen (PSA; for prostate cancer screening).

3.3 TECHNOLOGY PLATFORMS

In the past, biomarker assays were mostly quantitative using immunoassays and to a lesser degree quantitative mass spectrometry methods. These methods mostly analyzed a single biomarker such as a circulating protein, peptide, or metabolite. Today, we are seeing more multiplexed assays being developed and a move toward patient stratification biomarkers using genotyping or gene expression. We are also exposed to many new and novel technologies that claim improved accuracy and/or increased sensitivity. Some of these technologies will have significant effects on the measurements of biomarkers. One such example is the Erenna™ platform (Singulex, Inc.) used for measurement of troponin I, which has shown that its limit of detection of 0.2 ng/mL surpasses that of most of the commercial assays [13]. Using the Erenna™ platform for measurement of IL-13 in serum lowered the lower limit of quantitation of IL-13 in serum from 9.8 to 0.07 pg/mL, a 140-fold increase in sensitivity, using identical antibody pairs [14]. A number of recent advances in technology are contributing to the discovery of new biomarkers and to the development of new biomarker tools to measure them in a variety of tissue types. Molecular technology platforms such as arrays for DNA, mRNA, and miRNA, Q-PCR, and sequencing; platforms for measuring proteins such as ELISA and IHC/FISH; multiplexed platforms using Luminex, MesoScale Discovery, mass spectrometry-based technologies, and so on; and cellular platforms such as FACS and CellSearch are being widely used for biomarker research [15–18].

Major challenges with many of the "omics" technologies (proteomics, genomics, metabolomics, etc.) and other multiplexed technologies are the sheer volume of data generated and the lack of clear guidelines from regulatory agencies on the required analytical validation of these methods, especially when using newly developed technologies. Nowhere has the revolution been greater than in DNA sequencing. Some of the most advanced technologies today, such as SOLiD and Illumina GAIIx, are capable of producing 400 million 50 bp and ~200 million 75–100 bp reads, respectively [19]. The cost of performing extensive analytical validation of many of these new technologies is somewhat prohibiting, but nowhere close to the cost of conducting translational medicine clinical studies to qualify a biomarker or a set of biomarkers. FDA has regulatory oversight over all medical devices when it comes to the analytical and clinical validity of a test and considers the most crucial validation to be the clinical performance of a test [20]. Several research papers have proposed a "fit-for-purpose" approach to method development and validation [21, 22] based on the intended use of the biomarker results in discovery and/or clinical development. This approach is gaining acceptance, but seems to be used mostly for internal decision making in the pharmaceutical industry. When using biomarker results for regulatory decision making, both the analytical validation and the qualification of the biomarker will require extensive evaluation and validation in prospective clinical studies.

3.4 BIOMARKERS IN DISCOVERY

Biomarkers are used throughout the discovery phase to assess effects of a drug in animal models or for *in vitro* studies using various tissues. In some cases, experiments are performed to demonstrate *in vitro* effects on a biomarker using human blood or tissue in the presence of a new therapeutic. The use of biomarkers in the discovery phase is a prerequisite for successful transfer of biomarkers into clinical drug development. Ideally, the same biomarker(s) used in discovery should be carried into clinical development to better facilitate comparison of the preclinical results with the clinical results. The key role of translational medicine scientists is to develop biomarker strategies that can be implemented in the discovery phase and carried into clinical development. However, there are many challenges with using biomarkers in discovery.

A major challenge with using biomarkers in discovery is the risk of obtaining misleading results due to the use of biomarkers that are not well characterized and qualified and/or the use of biomarker assays that have undergone only limited analytical validation [23]. The introduction of high-throughput screening of potential new targets and the need to expeditiously move potential new compounds through the discovery pipeline have created a lot of challenges for the use of biomarkers. In addition, many biomarkers in discovery have limited characterization in humans and will challenge the translation from preclinical to clinical development. Despite these challenges, there are many examples of successful transitions of biomarkers from preclinical to clinical development. The example below took place more than 14 years ago at Bristol-Myers Squibb Pharmaceutical Research Institute. This is an example of how collaboration between scientists can bridge discovery and clinical development leading to the successful use of a biomarker in clinical development. In this case, the drug named omapatrilat inhibited two enzymes, neutral endopeptidase (NEP) and angiotensin converting enzyme inhibitor [24].

A key question in planning for potential biomarkers to be measured in the clinical development of omapatrilat was whether to measure atrial natriuretic peptide (ANP) in blood or in urine. A preclinical study in monkeys had been performed in discovery and showed that infusion of ANP for 2 h in vehicle-treated monkeys increased plasma concentration of ANP by about 11-fold, but the urinary excretion of ANP increased by a mere 50% [25]. This indicated that NEP was able to degrade most of the increased circulating ANP, leading to only a 50% increase in the urinary excretion despite the 11-fold increase in the circulating concentrations of ANP. Based on these observations in discovery research and after numerous discussions between the discovery scientists, the biomarker laboratory, and the clinical team, it was decided that the phase I trial would include measurements of the urinary excretion of ANP. This led to a successful characterization of the dose-dependent PD effects of omapatrilat on the urinary excretion of ANP and provided the basis for demonstrating

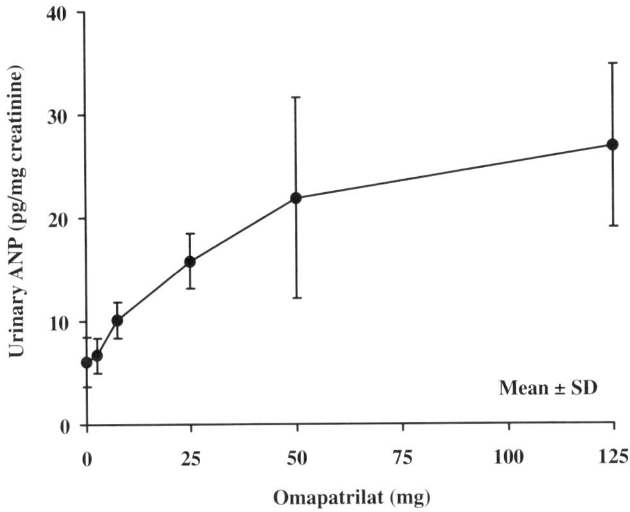

FIGURE 3.1 Mean (±SD) 24 h urinary excretion of ANP after single oral dose of omapatrilat ($n = 6$ per dose group) or placebo ($n = 21$) in healthy volunteers (from Ref. 39, with permission).

the NEP inhibition in a clinical study using a dose of 25 mg of omapatrilat or higher (Figure 3.1).

Discovery of new biomarkers often takes place at academic institutions, but the use of newly discovered biomarkers in the discovery and development of drugs usually takes place at pharmaceutical companies. The concept has been to provide measurements of new and novel biomarkers that could potentially give the drug developer an advantage over other developers. The challenge with using novel biomarkers is the fact that they are "novel" and we therefore have limited understanding of their utility. In addition, many of the assays used to measure novel biomarkers have undergone very limited analytical validation or characterization (see Section 3.5). There is therefore a need for more critical analytical validation of the biomarker assays used in the discovery phase.

Measurements of biomarkers in the late stage of discovery using validated assays would significantly improve the biomarker information available to the clinical teams. In addition, the use of larger animal species, such as nonhuman primates, could provide very valuable information about the PK and PD relationship. Knowledge of the PK/PD relationship from an *in vivo* animal experiment, prior to the first clinical development studies, will considerably improve the planning and designing of the clinical studies when it comes to measurement of biomarkers.

We have numerous times experienced programs where lack of information about the preclinical PD effects in an *in vivo* animal model led to a less than optimal study design in the first clinical study. Often the PD effect was missed because too few samples were collected or the timing of the sample collection was outside the time window of the PD effect.

3.5 ANALYTICAL VALIDATION OF BIOMARKER ASSAYS

The ICH Harmonised Tripartite Guideline of 1996 states that "the main objective of validation of an analytical procedure is to demonstrate that the procedure is suitable for its intended use" [26]. Many papers have been published detailing the parameters (e.g., precision, accuracy, recovery, stability, specificity, etc.) that should be evaluated during assay validation. The majority of these have focused on assay validation in the bioanalytical laboratories that support measurements of drug concentrations [27, 28]. Today most analytical laboratories involved with measurement of biomarkers are quite familiar with these validation parameters and many laboratories document biomarker assay validation in formal validation reports ([29, 30]; Table 3.1). Many pharmaceutical companies have instituted biomarker assay validation committees to oversee the approval process of the analytical validation of proposed biomarker methods. This scrutiny of the validation of the analytical tools has clearly increased the quality of the biomarker methods used in clinical development.

The ultimate outcome of the analytical validation of a biomarker method is to understand the limitations of the analytical method, but this goes beyond the characterization of the analytical method. It is of utmost importance to understand the sources of biological and preanalytical variability, such as the stability of the biomarker after sampling, and the influence of biological variability, such diurnal/seasonal effects, effects of food intake, physical activity, posture, and so on [23, 30]. The biological variability in a biomarker should be assessed in clinical studies where the subjects are well controlled. Some of the most controlled clinical studies are the very first clinical development studies, the so-called single ascending dose (or first-in-human) studies, as well as some of the early multiple-dose studies. All subjects in most first-in-human studies are usually kept under strict surveillance throughout the study; therefore, these studies are ideal for assessing biological variability in the measured biomarker if frequent sampling is acceptable. Figure 3.2 shows the intrasubject variability in a biomarker in 18 healthy subjects treated with placebo over a period of 5 days in a phase I

TABLE 3.1 Validation Parameters for Biomarker Assays

1. Reference material (detailed documentation)
2. Analyte–response relationship
3. Biological matrix-based quality controls
4. Intraassay precision (at least five analytical runs with $\geq$3 different samples)
5. Interassay precision (at least five analytical runs with $\geq$3 different samples)
6. Sample stability (collection, processing, storage, and shipping)
7. Accuracy/specificity
8. Recovery from biological matrix
9. Dilution linearity
10. Reportable range
11. Evaluate potential interference by test drug
12. Reference range for pertinent population(s)
13. Method comparison (when possible)

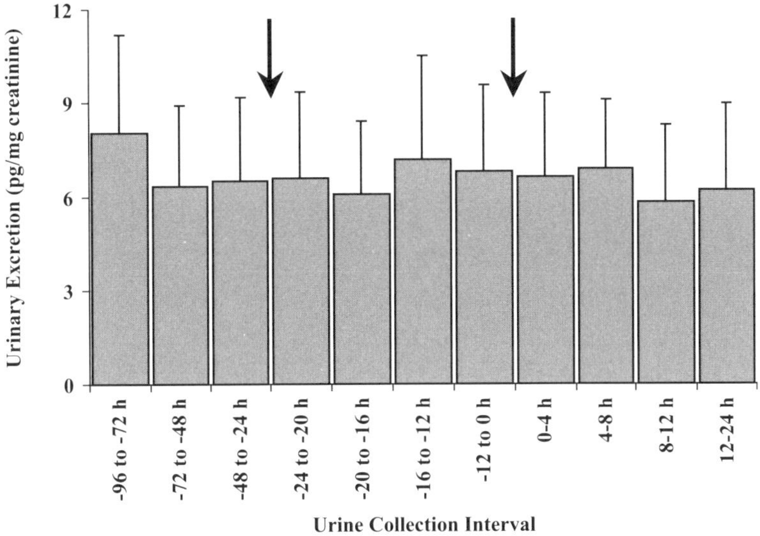

FIGURE 3.2 Urinary excretion of ANP in placebo-treated healthy subjects in a phase I clinical study. At a mean individual urinary excretion of ANP ranging from 3.1 to 13.8 pg/mg creatinine, the overall intrasubject variability in the urinary excretion was 22 ± 9% CV (mean ± SD, $N = 18$) with a range from 10% to 45% CV. *Arrows* indicate placebo administration time (Adapted from Ref. 31, with permission).

clinical study [31]. Although the intrasubject variability of a biomarker in healthy subjects is important to understand, it would be far more important to study the intrasubject variability of a biomarker in the pertinent patient population. Studies of intrasubject variability of biomarkers in patient populations would add valuable information for the design of clinical phase II, III, and IV studies.

Evaluation of preanalytical stability of a biomarker is another key aspect that should be addressed simulating the sample collection and handling conditions. For example, stability of cell surface markers in shipped blood samples and physiological variability of a marker over time are important parameters to examine for studies that include longitudinal monitoring of marker expression. Choice of matrix, anticoagulant, shipment, and storage temperature should all be included in the evaluation of stability of biomarkers that are intended to be used in clinical studies. Understanding the total (biological + preanalytical + analytical) variability in a biomarker is essential for the interpretation of the biomarker results.

3.6 BIOMARKER QUALIFICATION

The qualification of a biomarker is a lengthy process where the biomarker goes through a series of both preclinical and clinical studies with published results made available to the scientific community. This will involve both the analytical validation of the biomarker assay and clinical studies assessing the biological variability. The goal of the qualification is to obtain a consensus in the scientific community on the validity of the biomarker for a particular use. One strategic approach to this is the so-called "fit-for-purpose" biomarkers. Dr. Wagner describes biomarker qualification as being "the evidentiary process of linking a biomarker with biological processes and clinical end points" [22]. The aim of this qualification in clinical development is to ensure that the intended use of the biomarker in a clinical study is supported by evidence from earlier studies.

It will, however, take many years before the scientific community and FDA will accept a newly discovered biomarker as a surrogate end point, and very few biomarkers will ever reach this level of qualification. For this reason, most biomarkers used in clinical development are used to support decision making in early clinical development and will have undergone only limited qualification.

3.7 BIOMARKER IN CLINICAL DEVELOPMENT

Evaluation of biomarkers in clinical studies (early and late phases) is increasingly becoming integral part of drug development, with considerable investment in biomarker sample collection and testing. Development of biomarkers occurs in different phases during drug development. Biomarker hypothesis should ideally be generated in preclinical studies using *in vitro* and *in vivo* models, followed by assay development and validation using healthy donor samples and archived samples from clinical studies (retrospective analysis) [16]. The biomarker assay is further validated in prospective studies and implemented in randomized control trials. The following sections include a discussion of various aspects that play an important role in the development of biomarkers in clinical studies.

3.7.1 Biomarker Strategies

The implementation of a biomarker strategy into a clinical study can often be a complex process. The debate on potential utility of biomarkers that highlight the pros and cons of including biomarker assessment in clinical studies is well known and thought provoking when one has to make investment decisions [32, 33]. Inclusion of biomarker evaluation in clinical studies could potentially affect cost, timelines, and enrollment (if sample collection is mandatory). Therefore, the need and feasibility of biomarker evaluation should be carefully investigated. Investment in biomarker measurement is questionable if the scientific need is not strong. Most important aspect to consider during development of a biomarker strategy is identification of the potential utility of the intended biomarker/s i.e MOA or predictive or prognostic etc. Certain commercial aspects, such as, potential of the biomarker/s to increase the market value of the drug and whether they provide a competitive edge against drugs in the same class are also valid considerations during drug development. The other important factors to consider for implementing a biomarker strategy are the availability of relevant samples for biomarker analysis, stability of the markers in these samples during shipment and storage, and an appropriate assay system to accurately measure these biomarkers. If the requirement of biomarkers for the drug development is recognized and measurement of biomarkers is feasible, investment on the relevant marker(s) is justified.

3.7.2 Implementation of Biomarker Strategy

The operational aspects of clinical studies should be carefully planned and executed for successful implementation of biomarker strategy. Successful biomarker evaluations depend on proper sample collection and preservation of the integrity of samples during sample processing, storage, and shipping. Experienced clinical sites, with proper training and oversight of clinical operations and site staff, play a key role in successful generation of biomarker data from a clinical study. Biomarker group within a pharmaceutical organization must be aligned to provide operational or logistical support for biomarker sample collection and tracking and to identify appropriate assays and vendors for biomarker testing (if in-house testing is not performed). Similar assay qualification or validation requirements apply to tests performed at different vendor laboratories. Further, appropriate sample coding and deidentification procedures (for long-term storage) must be established prior to sample collection. Biomarker sample collection time points must be selected within a time window so that detection of the treatment-associated changes is not missed. Interim testing of biomarker samples should be part of the sample analysis plans so as to provide relevant data that may help development decisions. Relevant bioinformatics support should be available for biomarker data analysis and determining whether changes in biomarkers significantly correlate with clinical outcome measurements. Thus, implementation of biomarker strategy is a collaborative effort between biomarker group, clinical operations, biostatisticians, and clinicians.

3.7.3 Biomarkers in Early Development

The early clinical pharmacology studies play a critical role in the clinical development of a drug. This is where the very first PK and biomarker results are generated and decisions about a new therapeutic have to be made rapidly as soon as the results are available. There is therefore a need to provide real-time biomarker and PK results following each dose cohort and the biomarker laboratories will have to have operating procedures in place to use adequate quality control checks to enable the release of final data sets following each cohort.

The early clinical studies are the first studies in humans that can be designed to generate the very first human safety, PK, and biomarker data. Despite a lot of efforts to predict the PK of a drug in the first-in-human study based on preclinical data, many early clinical studies provide unexpected PK results. There are many examples where the addition of a biomarker or a set of biomarkers in these early clinical development studies has provided valuable information that can explain or support the PK observations. This is especially true in studies where it is difficult to collect a sample from the targeted tissue for direct measurement of PK and/or the biomarker(s) in the tissue. In such cases, the biomarker(s) could provide dose-dependent information.

A critical component of the early clinical development studies is the study design. When study design is done well, it will minimize the effects of preanalytical sources of variability, such as the effects of posture, exercise, diurnal biorhythms, food intake, and so on. If the preanalytical variability has not been characterized in advance of the clinical studies, it is usually necessary to obtain several samples prior to dosing to better understand the intrasubject variability. The early clinical studies are ideal for determining intrasubject variability as some of the subjects are treated with placebo. The intrasubject variability in the measured biomarker will describe the total variability, which consists of the sum of the analytical, preanalytical, and biological variability. Knowledge of the intrasubject variability will ultimately determine the number of subjects needed in a clinical study to statistically observe a certain change in the biomarker and also to generate data that can support decision making.

There is often a delay between the drug exposure and the PD effect, which usually means that samples for biomarker measurements need to be collected well beyond the expected T_{max} (time of maximum drug concentration) for the drug. Sometimes the PD effect will last for days despite a relatively short half-life of the drug and this is especially true for irreversible inhibitors. An example of this is aspirin that has a half-life of about 17 min, but despite this 1.0 g of aspirin (Figure 3.3) will cause an inhibition of platelet thromboxane A_2 for up to 7–10 days [34]. In another example, increases in gene expression of ABCA1 and ABCG1 occurred approximately 2–3 h after the T_{max} of the drug LXR-623, a novel liver X-receptor agonist [35].

Another challenge for study design is a biomarker that shows significant diurnal variation. An example of this is plasma 7-α-hydroxy-4-cholesten-3-one, which actually shows two distinctive peaks (two- to fourfold above baseline) during a

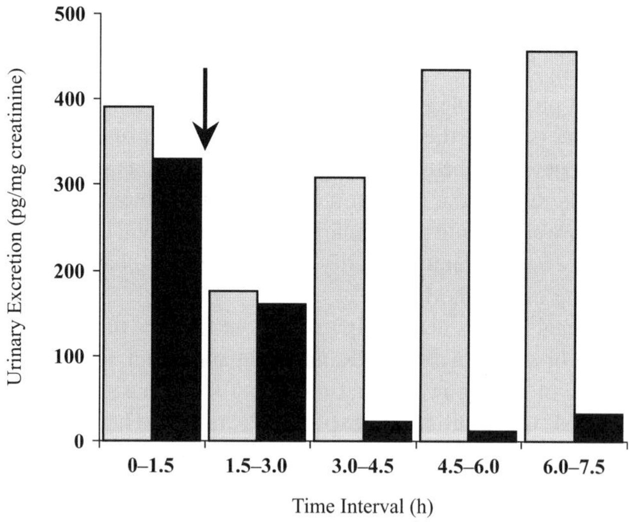

FIGURE 3.3 Urinary excretion of thromboxane (measured as the urinary metabolite 2,3-dinor-TxB_2) following a single dose of 1.0 g of aspirin (Adapted from Ref. 40, with permission).

24 h interval, the first at 1:00 pm and the second at 9:00 pm [36]. In this case, the best design would be to perform measurements of 7-α-hydroxy-4-cholesten-3-one in samples collected before dosing at the same time points as after dosing. Alternatively, the area under the curve can be calculated and the results compared between placebo- and drug-treated subjects.

An interesting concept of clinical studies is the use of adaptive trial design, "adapting by design," where several doses are being explored in a single trial, but data are reviewed in real time to allow for an adoption based on the emerging data [37]. This concept assumes that during the course of a clinical study, with multiple doses and a control arm, the emerging data will guide the clinical team to modify (adapt) the study dose groups in response to the data being collected [38].

In most early clinical development studies, the biomarker is a secondary objective together with the PK measurements. However, in a few studies, the biomarker has been listed as a primary objective of the study. These are usually translational medicine studies where a biomarker hypothesis is tested and study designed based on this hypothesis. For example, based on preclinical evidence if it is likely that subjects with a particular mutation might respond better to a drug that otherwise is expected to have low response rate in that disease setting, primary objective of this phase I or phase 0 study would be to stratify patients into two arms (mutant and wild type) for hypothesis testing and building on this hypothesis based on results of interim analysis. The clinical team should have a clear strategy as to how this biomarker data would be used to support further clinical development. Further, in studies where biomarkers are deemed to be important as primary objectives of the study, collection of biomarker samples (serum, plasma, tissues, etc.) must be made mandatory and should be part of inclusion criteria for enrollment. Clinical sites that demonstrate high success in patient consent for mandatory samples as well as collecting good quality samples should be selected. Relevant biomarker assays must be developed, analytically validated, and carefully scrutinized prior to implementation into the study. The number of samples as well as number of subjects needed for obtaining statistically significant biomarker data must be determined and included in the protocol. Overall, biomarker development in early trials helps to gather proof of concept for drug MOA, provides rationale for drug dose and frequency, may explain clinical outcomes, and may help identify responding patient population.

3.7.4 Biomarker in Late Clinical Development

When drugs enter into late clinical development (phases III and IV and in some cases phase II) and measurements of a biomarker are planned, it is of utmost importance for the clinical team and the biomarker laboratory to discuss the intended use of the biomarker. This discussion should address the importance of the biomarker in the planned study and what kind of laboratory qualification will be necessary to ensure that high-quality biomarker results are delivered. The safety consequences of incorrectly including a patient into the study have to be discussed when the intended use is for inclusion/exclusion. This discussion may also lead to further discussion about the analytical accuracy of the biomarker method. An example of this is genotyping, where assurance of the accuracy of the genotyping method may require confirmation by another technology or by dideoxy sequencing (considered the golden standard) before a patient is enrolled.

As discussed earlier in this chapter, very few biomarkers will ever reach the status of a surrogate end point. In our experience, most biomarkers in late clinical development will be used either for inclusion/exclusion (e.g., genotyping and gene expression) or for demonstrating that the expected effect on the biomarker is detected. Inclusion/exclusion biomarkers are critical for the clinical study enrollment. The information showing the expected PD effect is mostly used to add valuable information to the mechanism of action in an NDA application.

3.8 CONCLUSIONS

Incorporation of biomarkers may add complexity and cost early in drug development; however, biomarkers can help select therapeutic agents with better chances for success. A carefully planned biomarker strategy with proper implementation is the key to increase the success rate of a drug with high probability of clinical benefit.

REFERENCES

1. Feuerstein, GZ, Dormer C, Ruffolo RR, Stiles G, Walsh FS, Rutkowski JL. Translational medicine perspectives of biomarkers in drug discovery and development. Part I. Target selection and validation—biomarker take center stage. *Am. Drug Discov.* 2007;2(5): 36–43.
2. Biomarker Definitions Working, Group. Biomarkers and surrogate endpoints: preferred definitions and conceptual framework. *Clin. Pharmacol. Ther.* 2001;69:89–95.
3. Sarker D, Workman P. Pharmacodynamic biomarkers for molecular cancer therapeutics. *Adv. Cancer Res.* 2007;96:213–268.
4. Wolff AC, Hammond HE, Schwartz NJ, Hagerty LK, Allred CD, Cote JR, Dowsett M, Fitzgibbons LP, Hanna MW, Langer A, McShane ML, Paik S, Pegram DM, Perez EA, Press FM, Rhodes A, Sturgeon C, Taube ES, Tubbs R, Vance HG, van de Vijver M, Wheeler MT, Hayes FD. American Society of Clinical Oncology/College of American Pathologists guideline recommendations for human epidermal growth factor receptor 2 testing in breast cancer. *J. Clin. Oncol.* 2007;25(1):118–145.
5. Yanagi M, Shinjo MK, Takeshita A, Tobita T, Yano K, Kobayashi M, Terasaki H, Naoe T, Ohnishi K, Ohno R. Simple and reliably sensitive diagnosis and monitoring of Philadelphia chromosome-positive cells in chronic myeloid leukemia by interphase fluorescence *in situ* hybridization of peripheral blood cells. *Leukemia* 1999;13(4):542–552.
6. Siena S, Sartore-Bianchi A, Nicolantonio DF, Balfour J, Bardelli A. Biomarkers predicting clinical outcome of epidermal growth factor receptor-targeted therapy in metastatic colorectal cancer. *J. Natl. Cancer Inst.* 2009;101(19):1308–1324.
7. Scher IH, Jia X, de Bono SJ, Fleisher M, Pienta JK, Raghavan D, Heller G. Circulating tumour cells as prognostic markers in progressive, castration-resistant prostate cancer: a reanalysis of IMMC38 trial data. *Lancet Oncol.* 2009;10:233–239.

8. Sledge GW, Jr., Circulating tumor cells in breast cancer: blood will tell. *Clin. Cancer Res.* 2006;12:6321.
9. Di Costanzo F, Pinzani P, Orlando C, Gasperoni S, Vannini L, Mario PA. Circulating tumour cells in colorectal cancer. *Eur. J. Cancer Suppl.* 2008;6(14):52–59.
10. Buyse M, Loi S, van't Veer L, Viale G, Delorenzi M, Glas MA, d'Assignies SM, Bergh J, Lidereau R, Ellis P, Harris A, Bogaerts J, Therasse P, Floore A, Amakrane M, Piette F, Rutgers E, Sotiriou C, Cardoso F, Piccart JM. Validation and clinical utility of a 70-gene prognostic signature for women with node-negative breast cancer. *J. Natl. Cancer Inst.* 2006;98(17):1183–1192.
11. Flanagan BM, Dabbs JD, Brufsky MA, Beriwal S, Bhargava R. Histopathologic variables predict Oncotype DX™ Recurrence Score. *Mod. Pathol.* 2008;21:1255–1261.
12. Ludwig JA, Weinstein JN. Biomarkers in cancer staging, prognosis and treatment selection. *Nat. Rev. Cancer* 2005;5:845–856.
13. Wu AHB, Shea E, Thuc Lu Q, Minyard J, Bui K, Hsu JC, Agee SJ, Todd J. Short- and long-term cardiac troponin I analyte stability in plasma and serum from healthy volunteers by use of an ultrasensitive, single-molecule counting assay. *Clin. Chem.* 2009;55: 2057–2059.
14. Ledger K, Agee SJ, Kasaian MT, Forlow SB, Durn BL, Minyard J, Lu QA, Todd J, Vesterqvist O, Burczynski ME. Analytical validation of a highly sensitive microparticle-based immunoassay for quantitation of IL-13 in human serum using the Erenna immunoassay system. *J. Immunol. Methods* 2009;350:161–170.
15. Burczynski ME, Vesterqvist OE. Quantitative PCR assays in clinical drug development. *Eur. Pharm. Rev.* 2006;5.
16. Kulasingam V, Diamonds PE. Strategies for discovering novel cancer biomarkers through utilization of emerging technologies. *Nat. Clin. Pract. Oncol.* 2008;5 (10):588–598.
17. Isler JA, Vesterqvist OE, Burczynski ME. Analytical validation of genotyping assays in the biomarker laboratory. *Pharmacogenomics* 2007;8:353–368.
18. Hanash MS, Pitteri JS, Faca MV. Mining the plasma proteome for cancer biomarkers. *Nature* 2008;452:571–579.
19. McPherson JD. Next-generation gap. *Nat. Methods Suppl.* 2009;6(11 Suppl.):S2–S5.
20. Philip R, Gutman SI, Chan MM. FDA perspective on validating proteomic biomarkers for *in vitro* diagnostic use. *Am. Drug Discov.* 2008;3(6):30–33.
21. Lee JW, Devanarayan V, Barrett YC, Weiner R, Allinson J, Fountain S, Keller S, Weinryb I, Green M, Duan L, Rogers J, Millham R, O'Brien P, Sailstad J, Khan M, Ray C, Wagner J. Fit-for-purpose method development and validation for successful biomarker measurement. *Pharm. Res.* 2006;23:312–328.
22. Wagner JA. Strategic approach to fit-for-purpose biomarkers in drug development. *Annu. Rev. Pharmacol. Toxicol.* 2008;48:22.1–22.21.
23. Vesterqvist O, Park CH, Rutkowski JL. From biological measures to valid biomarker: predicting success of novel therapeutics. *Am. Drug Discov.* 2008;3(3):56–59.
24. Vesterqvist O, Manning JA, Uderman H, Delaney C, Beierle F, Swanson BN, Liao W. Effects of BMS-186716, a new dual metalloprotease inhibitor, on pharmacodynamic markers of neutral endopeptidase (NEP) and angiotensin converting enzyme (ACE) activity in healthy men. *Clin. Pharmacol. Ther.* 1997;61:230 (abstract).

25. Seymour AA, Benoni EA, Smith PL, Mathers PD, Asaad MM, Rogers WL. Potentiation of natriuretic peptides by neutral endopeptidase inhibitors. *Clin. Exp. Pharmacol. Physiol.* 1995;22:63–69.

26. ICH Harmonised Tripartite Guideline *Validation of Analytical Procedures: Methodology*, November 6, 1996.

27. Shah VP, Midha KK, Findlay JWA, Hill HM, Hulse JD, McGilveray IJ, McKay G, Miller KJ, Patnaik RN, Powell ML, Tonelli A, Viswanathan CT, Yacobi A. Bioanalytical method validation—a revisit with a decade of progress. *Pharm. Res.* 2000;12:1551–1557.

28. DeSilva B, Smith W, Wiener R. Recommendations for the bioanalytical method validation of ligand-binding assays to support pharmacokinetic assessments of macromolecules. *Pharm. Res.* 2003;20:1885–1900.

29. Owens MA, Vall HG, Hurley AA, Wormsley SB. Validation and quality control of immunophenotyping in clinical flow cytometry. *J. Immunol. Methods* 2000;243:33–50.

30. Swanson B. Delivery of high-quality biomarker assays. *Dis. Markers* 2002;18:47–56.

31. Vesterqvist O. Using pharmacodynamic biomarkers to accelerate early clinical drug development. In Ilyin S, editor. *Functional Informatics in Drug Discovery*, CRC Press, Boca Raton, FL, 2008, pp.41–53.

32. Goulart BHL, Clark JW, Pien HH, Roberts TG, Finkelstein SN, Chabner BA. Trends in the use and role of biomarkers in phase I oncology trials. *Clin. Cancer Res.* 2007;13:6719–6726.

33. Ratain JM, Glassman HR. Biomarkers in phase I oncology trials: signal, noise, or expensive distraction? *Clin. Cancer Res.* 2007;13(22):6545–6548.

34. Gréen K, Vesterqvist O. Drug interactions with the *in vivo* synthesis of thromboxane and prostacyclin. *Drug Metab. Drug Interact.* 1989;7:321–350.

35. Katz A, Chandrasekhar U, Ott E, Hickey L, Burczynski ME, Burghart P, Vesterqvist O, Meng X. Safety, pharmacokinetics, and pharmacodynamics of single doses of LXR-623, a novel liver X-receptor agonist, in healthy participants. *J. Clin. Pharmacol.* 2009;49:643–649.

36. Gälman C, Angelin B, Rudling M. Bile acid synthesis in humans has a rapid diurnal variation that is asynchronous with cholesterol synthesis. *Gastroenterology* 2005;129:1445–1453.

37. Krams M, Bruman CF, Dragalin V, Gaydos B, Grieve AP, Pinheiro J, Maurer W, Gallo P. Adaptive designs in clinical drug development: opportunities, challenges, and scope reflections following PhRMA's November 2006 workshop. *J. Biopharm. Stat.* 2007;17:957–964.

38. Brown CH, Ten Have TR, Jo B. Adaptive designs for randomized trials in public health. *Annu. Rev. Public Health* 2009;30:1–25.

39. Vesterqvist O, Reeves R. Effects of omapatrilat on pharmacodynamic biomarkers of neutral endopeptidase and angiotensin converting enzyme activity in humans. *Curr. Hypertens. Rep.* 2001;3 (Suppl. 2):22–27.

40. Vesterqvist O. Measurements of the *in vivo* synthesis of thromboxane and prostacyclin in humans. *Scand. J. Clin. Lab. Invest.* 1988;48:401–407.

PART II

FLOW CYTOMETRY IN THE DRUG DEVELOPMENT PROCESS

VIRGINIA LITWIN AND PHILIP MARDER

The ultimate objective of the drug development process is to match a therapeutic need with a safe and efficacious treatment.

The lifecycle of a new drug entity begins in research laboratories where disease mechanisms are identified. In drug discovery laboratories, molecular targets for disease intervention are established and target-specific screening assays developed. Lead compounds then undergo comprehensive characterization followed by preclinical toxicological assessment. Finally, a small percentage of compounds enters the realm of clinical trials. The purpose of the clinical testing is to evaluate drug safety and effectiveness after the compounds have been administered to healthy volunteers and/or patient populations.

Flow cytometric methods are being employed at all stages of the drug development process:

- In the earliest stages of drug discovery, multiparametric intracellular flow cytometric analysis supports target identification. In addition, high-throughput flow cytometry is used for compound screening so that potential compounds can be selected from large libraries.
- During lead compound characterization, multiparametric analysis enables an extensive investigation of the complex interrelated mechanisms of action.
- *In vitro* and *ex vivo* flow methods employed in toxicology studies assist in identifying and characterizing off-target effects at the single cell level.

Flow Cytometry in Drug Discovery and Development, Edited by Virginia Litwin and Philip Marder

- Flow cytometry is used during clinical testing in order to assess both safety (e.g., antidrug antibody testing) and pharmacodynamic effects (e.g., $CD4^+$ T-cell counts). It is also used in pharmacokinetic assessment to monitor the plasma levels of protein- and peptide-based therapeutics.

Part II of the book is intended to provide detailed descriptions of the various applications of flow cytometry throughout the drug development lifecycle.

4

HTS FLOW CYTOMETRY, SMALL-MOLECULE DISCOVERY, AND THE NIH MOLECULAR LIBRARIES INITIATIVE

LARRY A. SKLAR AND BRUCE S. EDWARDS

4.1 INTRODUCTION AND OVERVIEW

Once commercial flow cytometers became widely established in research and clinical laboratories in the 1980s, sample handling evolved from a single sample tube at a time manual delivery format, first to automated tube handling systems in the 1990s and then to plate-based sampling systems in the 2000 decade. In concert with advances in data acquisition, data analysis was enhanced with automated and batch processing systems. In the 1990s, NIH signaled important changes in federally funded science in the United States. In addition to the Human Genome Project for sequencing the human genome, the NIH initiated a program in Biomedical Engineering Research Partnerships that gave validity to scientific approaches based on novel technology. Early in the next decade, the NIH Roadmap and its Molecular Libraries initiative fully validated discovery research with its emphasis on generating new tools to accelerate biomedical research. With the economic environment culminating in the biotechnology crash of 2000, the recession of 2008, and the slowdown in the pharmaceutical pipeline to produce new drugs from novel chemical entities, opportunities began to emerge for high-throughput screening (HTS) and small molecular discovery in academic laboratories and research institutions.

The purpose of this chapter is to discuss how the changes at NIH, coupled with new economic realities in Pharma and opportunities in flow cytometry, have led to the

Flow Cytometry in Drug Discovery and Development, Edited by Virginia Litwin and Philip Marder

development of high-throughput (HT) flow cytometry, as well as its contribution to small-molecule discovery. We discuss (1) the NIH Molecular Libraries initiative in the context of discovery research and the public Web sites for obtaining information about the initiative; (2) the position of the UNM Center for Molecular Discovery and HT flow cytometry in the Molecular Libraries Probe Production Center Network (MLPCN); (3) flow cytometry platforms for HTS and their performance as well as HT flow cytometry activities in discovery, (4) applications of flow cytometry to cellular and molecular targets and their outcomes, (5) current operational issues in HTS flow cytometry, and (6) a perspective of future developments in HT flow cytometry.

4.2 NIH AS AN ENGINE IN DISCOVERY RESEARCH

The Human Genome Initiative signaled a new era for biomedical research in the United States in several ways. It was a multitiered initiative with phases, milestones, and networking. Rather than following the model of investigator-initiated, hypothesis-driven research, the initiative was driven by a specific goal to be accomplished by a specific date. In the pilot phase of the initiative [1], academic centers were given the opportunity to develop and evaluate multiple strategies for high-throughput sequencing and to share the results within a network of those centers. The production phase [2] was designed to select the most efficient strategies for throughput within a smaller network in which sequencing data would be made promptly publicly available. The initiative required altruism in a collective sense that may be connected to the worldview of one of its leaders, Dr. Francis Collins, recently appointed as NIH Director [3].

In a series of announcements dating back to 1998, NIH also rolled out Bioengineering Research Partnerships [4] defined as a multidisciplinary research team that applies an integrative, systems approach to develop knowledge and/or methods to prevent, detect, diagnose, or treat disease, or to understand health and behavior. The partnership must include appropriate bioengineering or allied quantitative sciences in combination with biomedical and/or clinical components. A BRP may propose design-directed, developmental, discovery-driven, or hypothesis-driven research at universities, national laboratories, medical schools, large or small businesses, or other public and private entities, or combinations of these entities. It is expected that a BRP will have a well-defined goal or deliverable that will be achieved based on objective milestones specified in the initial application. In 2001, the NIH made a related program announcement [5] for Bioengineering Research Grants (BRGs). The BRGs are performed in a single laboratory, by a single investigator, or by a small group of investigators. In 2003, the NIH made another related program announcement [6] for Exploratory/Developmental (R21) Bioengineering Research Grants (EBRGs). The EBRGs are performed in a single laboratory, by a single investigator, or by a small group of investigators and the projects are high risk/high payoff in nature (R21 mechanism) as compared to the R01-type grants supported by the BRP program.

The research objectives of the BRP are summarized in these terms: Many of today's biomedical problems are best addressed using a multidisciplinary approach that extends beyond the traditional biological and clinical sciences. Bioengineering integrates principles from a diversity of technical and biomedical fields and crosses the boundaries of many scientific disciplines represented throughout academia, laboratories, and industry. The creativity of interdisciplinary teams is resulting in new basic understandings, novel products, and innovative technologies for addressing biomedical problems.

Recognizing the importance of bioengineering in public health, the Bioengineering Consortium (BECON) was established in 1997 as a focus for bioengineering activities at the NIH. To facilitate communication between the allied and biomedical disciplines and to provide input from the scientific community on research needs and directions, the BECON has held annual two-day symposia on emerging topics of interest related to bioengineering, including bioengineering (1998), bioimaging (1999), nanotechnology (2000), reparative medicine (2001), biosensors (2002), and team science (2003). Summaries of the proceedings and recommendations of these symposia are available [7].

In concert with the arrival of a new NIH director, Elias Zerhouni in 2002, NIH rolled out the Roadmap, a series of initiatives designed to accelerate biomedical research and generate new tools for discovery, representing approximately 1% of the NIH budget, or \$300 million of the ~\$30 billion in the middle of the decade 2000–2009 [8]. Three groups of initiatives have been promoted (Table 4.1).

The molecular libraries and imaging initiative [9] could be considered a "chemical genomics" follow-up to the Human Genome Project. Although the genome initiative revealed DNA sequences relevant to protein structure and function, chemical genomics was anticipated to reveal connections between the chemical universe and the biological universe, using small molecules to probe the functions of large molecules. In other terms, it can be perceived as the exploration of the intersection of chemical diversity and biological diversity. As in the Human Genome Project, the pilot phase of the program consisted of an evaluation of tools for high-throughput screening. Each of 10 comprehensive centers in the Molecular Libraries Screening Center Network accepted biological targets that had first been approved through competitive review process, screened the target against a library of compounds assembled by NIH, and identified unique chemical entities as probes to regulate

TABLE 4.1 NIH Roadmap Initiatives

New pathways to discovery
Building blocks, biological pathways, and networks
Molecular libraries and imaging
Structural biology
Bioinformatics and computational biology
Nanomedicine
Research teams of the future
High-risk research and NIH Director's Pioneer Award
Interdisciplinary research
Public–Private partnerships
Re-engineering the clinical research enterprise

structures or pathways potentially serving as imaging agents or leads for drug discovery. Data were to be promptly uploaded to the public database [10]. The pilot-phase centers used both commercial and home-grown screening technologies, including high-throughput flow cytometry.

Questions immediately arose regarding the role of NIH in HTS and discovery, previously the domain of the pharmaceutical industry [11]. In brief, these questions included (1) the cost-benefit of the initiative; (2) whether NIH was in competition with the pharmaceutical industry; (3) how the infrastructure could be developed for acquiring and managing the compound library, as well as for data publication, and so on; and (4) the nature of the conflict between public data and commercial value. A summary of the probe output of the pilot phase has appeared [12].

It is noteworthy that the period of 2000–2010 marked significant changes in the pharmaceutical industry itself: (1) a plateau in the pipeline of new chemical entities reaching approval as new drugs; (2) downsizing and mergers contributing to job loss in the discovery sector and medicinal chemistry enterprises; (3) outsourcing these activities to Russia, China, and India; and (4) significant growth in discovery activities in academic as well as freestanding research institutions. There are listings of 66 sites in Society for Biomolecular Sciences Academic Screening Center Directory [13].

Now in the production phase, there are four comprehensive centers in the Molecular Libraries Probe Production Center Network (TSRI, SBIMR, Broad Institute, and National Chemical Genomics Center), three specialty screening centers (UNM, Southern Research Institute, and Johns Hopkins University), and two specialty chemistry centers (University of Kansas and Vanderbilt University). The goal of the MLPCN is to complete screens on 100 targets per year over a 6 year period with a library of $>$350,000 compounds. An output of $\sim$300 probes as well as a comparison of the activities of the library against all the targets in the public database is projected. The University of New Mexico Center for Molecular Discovery (UNMCMD), specializing in HTS flow cytometry, has identified 10 probes so far and uploaded more than 6M screening data points.

As of January 2010, there are several funding mechanisms for discovery projects that are open to investigators throughout the world. The R03 mechanism [14] is for targets that are ready for screening. The funding provides a means for investigators to participate along side the screening center. The R21 mechanism [15] is for the development of targets to be screened, with resources in the first year for reducing the target to practice in a multiwell screening format. The second year of funding is for partnering with MLPCN centers during the screening and follow-up phases of the project. Individual investigators can be funded by R01 mechanisms that have small-molecule discovery and HTS components. The program announcements change frequently and can be readily assessed through web searches. However, the MLPCN also provides a mechanism for investigators developing HTS targets to enter the MLPCN through a Fast Track mechanism [16] as well as progressing the R21 to screening [17]. Funding opportunities are summarized [18].

As a public initiative, the HTS and follow-up studies for the MLPCN are compiled into a searchable database, PubChem, which allows searches of targets and detailed assay protocols, compounds, screening, and follow-up results [10, 19]. One important concept in chemical genomics is the possibility of searching for the activities of hundreds of thousands of molecules across thousands of biological targets. The probe reports arising from the MLPCN [20] along with information on all MLPCN-funded projects, including those listed here, are available [21]. Individual projects can be searched at NIH reporter by investigator, grant number, or keywords [22]. All MLPCN information is accessible though the homepage [23].

The NIH Roadmap is currently supported by a common fund administered under the office of the NIH Director. Support is scheduled to transition in 2012 to individual institutes at NIH, and the National Cancer Institute has already implemented a consolidated drug discovery and development program under the framework of NCI Experimental Therapeutics (NExT) [24] with its own integrated discovery network, the Chemical Biology Consortium (under the division of cancer treatment and diagnosis [25]. Support within the NIH institutes is anticipated to focus target selection into specific disease and physiological areas. With the United States in recession, one of the first actions in 2010 of the new NIH director, Francis Collins, is embodied in a new "RC4" funding announcement [26]:

"This NIH Funding Opportunity Announcement (FOA), supported by funds provided to the NIH under the American Recovery & Reinvestment Act of 2009 ("Recovery Act" or "ARRA"), Public Law 111-5, solicits through this limited competition applications from domestic (United States) institutions/organizations proposing to develop and implement critical research innovations in one or more of the following five thematic areas:

1. Applying Genomics and Other High Throughput Technologies
2. Translating Basic Science Discoveries into New and Better Treatments
3. Using Science to Enable Health Care Reform
4. Focusing on Global Health
5. Reinvigorating the Biomedical Research Community

This program is a trans-NIH effort supported by Recovery Act funds from the Office of the Director. Applicants may propose to address either a specific disease- or technology-related research question relevant to the mission of one or more participating Institutes and Centers, or propose the creation of a unique infrastructure/ resource designed to accelerate scientific progress in the future." Relevant to discovery, for example, are the following areas identified in NIH terms:

4.2.1 Applying Genomics and Other High-Throughput Technologies

In the past, many basic biomedical science projects were limited in scope to some aspect of genetics, cell biology, or physiology. The revolution now sweeping the field is the ability to be comprehensive—for example, to define all the genes of the human, model organisms or the human microbiota, all the human proteins and their structures, or all the major pathways for signal transduction in the cell. Technologies contributing to these advances, many of which became practical at scale only in the last few years, include DNA sequencing, microarray technology, nanotechnology, small-molecule screening capabilities, new imaging modalities, and computational biology. These comprehensive approaches coupled with systems-level integration, analysis, and mining of large datasets now hold the promise of major advances in the understanding of the mechanisms of diseases.

4.2.2 Translating Basic Science Discoveries into New and Better Treatments

Armed with a wealth of basic science discoveries and an understanding of the pathophysiology of various diseases, we are embarking on the next frontier in designing new diagnostic and therapeutic strategies. Molecular and cellular insights into a disease can be developed into screening assays on hundreds of thousands of compounds, and tested in disease models to identify the most promising leads that can sustain the drug development pipeline and attract public–private partnerships for further pursuits. Additional pathways to therapeutics from gene therapy, biologics, and stem cells (including iPS cells) are also showing great promise. The opportunity here is for translational science to develop small molecule-, gene-, protein/peptide-, and cell-based therapies for common as well as rare diseases.

It is worth noting that the NIH Roadmap concept in the United States has taken hold in Europe with the European Strategy Forum on Research Infrastructures and the European Open Screening Platform for Chemical Biology [27]. A membrane protein network directed at drug discovery is also in place [28].

4.3 THE UNMCMD

Flow cytometry in New Mexico has involved an interplay between technology at Los Alamos National Laboratory and the National Flow Cytometry Resource and Biology at UNM, dating back to the time of the immunotoxicology program led by Noel Warner. In 1990, Larry Sklar, PhD, came to New Mexico to direct both UNM Cancer Center's flow cytometry resource and the Los Alamos P41 National Flow Cytometry Resource. The cytometry group at the University of New Mexico as well as colleagues at the NFCR began to consider the implication of new delivery schemes for flow cytometers, as well as opportunities for revealing small-molecule interactions (review in Ref. 29). Bruce Edwards, PhD, joined the UNM flow cytometry shared resource in 1997. Through a series of internal collaborations, industrial partnerships, and NIH grants, the UNM and LANL teams developed both subsecond time resolution and high-throughput flow cytometers, and applied the instruments to cells and particle-based molecular assemblies.

In 1999, in collaboration with Axiom Biosciences, Edwards and Sklar and coworkers [30] at UNM developed hardware and software for the first generation of HT flow cytometry, which we refer to as plug flow cytometry [31]. Plug flow used a reciprocating multiport flow injection valve to deliver 10 end point assays per min, 4 online mixing experiments per min and, in secondary screens, a 15-point concentration gradient of soluble compound in ~2 min. The plug flow technology led to funding of a BRP from NIH to develop the current generation of HT flow cytometry. HyperCyt® uses a peristaltic pump in combination with an autosampler to boost assay throughput. As the sampling probe of the autosampler moves from well to well, a peristaltic pump sequentially aspirates particle suspensions from each well. Between wells, the running pump draws a bubble of air into the sample line resulting in the delivery of a series of bubble-separated samples [32]. Data from microplate wells are collected in a single data file. The time-resolved data, with periodic gaps corresponding to the passage of the sample-separating air bubbles, are analyzed by software. 384-Well plates with multiplexed data sets are routinely sampled in 10 min with sample volumes of 1–2 μLs. We have adapted the particle counting ability of flow cytometry for HT analysis of compound solubility.

These developmental activities ultimately led to NIH Roadmap funding from the Molecular Libraries Initiative. As a comprehensive screening center in the pilot phase, the New Mexico Molecular Libraries Screening Center developed capabilities in target development, HTS, as well as cheminformatics, and synthetic chemistry for optimization of biological activity of leads identified through the screening process. In the production phase, our name changed to UNMCMD to reflect our status as a specialty probe production center focusing on HTS flow cytometry. The screens are intended to identify active molecules, and to chemically optimize those molecules for biological activity. HT flow cytometry tools enable high-throughput sample handling, high content analysis, and real-time measurements of cell response. The HyperCyt platform delivers microliter-sized samples from 384-multiwell plates at typical sampling rates of 40 wells per minute for end point assays with multiplexing of six or more targets simultaneously. Our experience indicates that virtually any molecular assembly or cell response can be displayed in a format compatible with flow cytometry. Moreover, by creating a suspension array of particles, assays and responses can be highly multiplexed or performed on complex cell populations without compromising throughput.

4.4 PLATE-BASED FLOW CYTOMETRY

During 2000 and 2010, plate-based cytometry has become available through most manufacturers of flow cytometers (Accuri, Beckman Coulter, Becton–Dickinson, Guava, and Luminex) as well as flow cytometry device companies (e.g., Cytek). Since flow cytometers are pressurized, the majority of these plate-based systems depend upon two common principles: (1) samples from individual wells are injected into the flow cytometer one well at a time; and (2) the contents of each well are treated as a single FCS file. In most cases, the contents of individual wells have been drawn into a syringe prior to injection; in some cases, the wells have been sealed and pressurized (Cytek) for sample delivery; and recently in one case the contents of a well have been sipped by the flow cytometer (Accuri), where the plate has been repositioned for each well.

HTS has been typically defined as screening rates of 100,000 samples or more per day. Systems that create an FCS file from each well have been reported at rates of

5–10 samples per minute. The HyperCyt platform for HTS flow cytometry, currently available at 30 or more sites around the world through IntelliCyt [33], can acquire samples at rates up to ~40 wells per minute. These sampling rates are currently compatible with throughput from several thousands to many tens of thousands of samples per day, depending upon the number of hours and extent of automation.

Sites actively reporting usage of flow cytometry in a discovery mode for primary screens in addition to UNMCMD include (1) Vivia [34], a successor to Novasite that performs reprofiling of approved drugs; (2) Rigel that screens SiRNA [35]; (3) and the Center for Chemical Genomics at the University of Michigan Life Science Institute. To our knowledge, the Rigel site is unique in having a fully automated and integrated platform that was developed in collaboration with IntelliCyt and Agilent. A partnership between IntelliCyt and Beckman Coulter [36], the international distributor of the HyperCyt platform, links the Beckman sample handling Biomek platform, HyperCyt, and the CyAn flow cytometer that was recently acquired by Beckman Coulter.

Essential opportunities of flow cytometry for small-molecule discovery are summarized in Table 4.2. Flow cytometry is well known as a multicolor technology that can be used in multiplexing or mutiparameter analysis (the analogue to high-content analysis in microscopy), as well as the homogeneous discrimination of ligand binding and macromolecular assemblies.

The versatility of the flow cytometry platform is remarkable and well known. Capabilities range from >10 color immunological assays and extend to most cell-based assays that can be performed fluorimetrically or microscopically, including cell proliferation, differentiation, death, signaling, adhesion, ligand binding, transport, and others. There are published protocols, stains, and markers for surface, cytoplasmic, and nuclear components. Flow cytometry excels at phenotypic assays, such as antibody staining and light scatter (size and shape) discrimination used on complex

TABLE 4.2 Flow Cytometry Features Useful for HTS

Feature	Utility	Application
Homogeneous detection	Real-time discrimination of particle bound from unbound molecules	Ligand binding or molecular assemblies. Conserves precious reagents on beads
Multiparameter	10 or more colors and scatter signals	High content, e.g., phospho-protein analysis
Multiplexing and barcoding	Two or more colors and scatter signals used to distinguish multiple particle targets	Suspension arrays for multiple assemblies or multiple cell targets
Acquisition rate/digital signal processing	50,000 events per second	Thousands of flow cytometric assays accessible to high throughput
Size and shape discrimination	Particle sizes ranging from submicron to tens of micrometers; changes in shape/morphology	Cells, particles, solubility screening of compound libraries

cell populations in hematology, immunology, cancer biology, reproductive biology, marine biology, microbiology, pharmacology, toxicology, and so on. In principle, all these features are compatible with HyperCyt.

In practice, the flow cytometer measures several 100 particles for an accurate determination of mean fluorescence for a particle population. Fluorescence signals are readily detected when there are 10,000 fluorophores per particle. HyperCyt enables 1 μL volumes to be sampled. For liquid handling reasons, we currently require well volumes ~10 μL with particle densities ~1000–10,000 cells or beads per microliter. Fluorescent molecular assemblies with affinities <500 nM can typically be assayed in homogeneous, no-wash format. Expensive epitope-tagged protein reagents are conserved when associated with beads that use <1 pmol per assay. Less precious reagents—fluorescent small molecules, peptides, and oligos—can be supplied at concentrations above their K_d in contrast to other homogeneous binding methods requiring both sets of reagent concentration near K_d. These elements are discussed elsewhere in more detail [30].

4.4.1 The UNMCMD Pipeline

The diversity of targets for flow cytometry is illustrated in part by the pipeline of projects that have entered the UNMCMD through the Molecular Libraries Initiative (Table 4.3). Pilot-phase projects are described briefly below along with a perspective on practical lessons learned from each project. Projects initially entered through an X01 review mechanism without funding for the target provider that was rapidly replaced by the R03 mechanism with funding for the target provider. There have been three review cycles each year, which is often the case for NIH funding initiatives. Screening milestones varied as the MLSMR grew during the course of the pilot phase. For example, in the first year of the initiative, the library had grown to ~65,000 compounds. Our initial screens were carried out against a ~10,000 compound diversity set augmented by a 17,000 compound subset of the library provided by our collaborator ChemDiv. We were allowed to use this subset by NIH to enhance the possibility of chemical optimization through synthesis by ChemDiv. Extensive data for each target are available at PubChem that can be searched by target, center, compound, and so on. Probes are defined by criteria identified by the target provider, but typically molecular activity must be ~100 nM and cellular activity ~1 μM.

4.4.2 Assay for Formyl Peptide Receptor Ligands: 1R03 MH076381-01

4.4.2.1 Suspension Cell Multiplex The assay for formyl peptide receptor (FPR) is a cell-based assay that measures competitive displacement of a FITC fluorescent receptor ligand by test compounds. The FPR family has been implicated in inflammatory responses in leukocytes (FPR) as well as neurological diseases (FPRL1), and more recently in metastasis in glioma and ovarian cancer. The assay was performed as a duplex, simultaneously measuring test compound effects on two family member targets, FPR and FPRL1 in two cell lines, one of which had been stained with a red fluorescent dye [37]. As part of the assay development, a novel fluorescent peptide

TABLE 4.3

Cycle	Title	Award Mechanism	Target Provider/UNM Target Team Leader	#Targets × compounds screened, PubChem summary AID#
1	Primary HTS assay for formylpeptide receptor (fpr) ligands and primary HTS counter-screen assay for formylpeptide-like-1 (FPRL1) ligands	X01MH07638101	B. Edwards, UNM/S. Young	2 × 25 K, #805, #1202
2	Disassembly of the 26S proteasome (ATP hydrolysis dependent)	X01MH07761301	D. Skowyra, SLU/P. Simons	1 × 25 K, #1824
2	MLSCN assay for ligands of GPR30 and classical estrogen receptors	X01MH07762701	E. Prossnitz, UNM/M. Dennis, M. Haynes	3 × 640, #1989
2	MLSCN HTS assay for allosteric ligands for the VLA-4 integrin	X01MH07763801	L. Sklar, UNM/A. Waller	2 × 25 K, #1998
3	MLSCN assay for activators of prostate cell differentiation	X01MH07893701	T. Thompson, UNM/M. Haynes	1 × 25K, #1260
3	Bacterial quorum sensing inhibition assay for MLSCN 10 K and 17 K ST1 compound sets combined	X01MH07895201	P. Gresham, UNM/S. Young	1 × 25 K, #1206
4	HTS for Bcl-2 family multiplex	X01MH07985001	L. Sklar, UNM, J. Reed, BIMR/P. Simons	6 × 225 K, #1693
5	HTS to identify specific small-molecule inhibitors of Ras and Ras-related GTPases	X01MH08123101	A. Wandinger-Ness, UNM/Z. Surviladze	6 × 225 K, #1772
R21 in lieu of cycle 6	Multiplexed flow cytometry screens for RGS inhibitors	1R21NS05701401	R. Neubig, U. Michigan/ Y. Wu, M. Haynes	5 × 225 K, #1504

7	Multiplex screening for ABC transporter inhibitors	1 R03MH07894601	R. Larson, UNM/I Ivnitski-Steele, J. Strouse	2 × 200K, #1818
9	TR-FRET HTS assay for inhibitors of MEKK2-MEK5 PB1 domain interactions	1R03MH08483001	K. Nakamura, UNC/Z/Surviladze	3 × 300K, #1683
10	HTS for developing T Cell immune modulators	1 X01MH08570701	I. Hwang, TSRI/M. Haynes, Y. Wu	1 × 325 K, #2087
11	Chemical screen of TOR pathway GFP fusion proteins in *S. cerevisiae*	1R03MH08645001	M. Werner-Washburne, UNM/J. Chen, S. Young	5 × 325 K, #1908
11	A yeast HTS for caloric restriction mimetics that inhibit age-related superoxide	1 R03MH08743901	W. Burhans, Roswell Park Cancer Institute/C. Prudom, A. Jensen	1 × 325 K, in process
12	Identification of broad spectrum antifungal efflux inhibitors	R03MH087406-01A1	R. Cannon, U. Otago/J. Strouse	3 × 325K, in process
Fast Track	HTS for identification of VLA-4 allosteric modulators	1 R01HL08106201	L. Sklar, A. Chigaev UNM/P. Simons, S. Young	1 × 330 K, in process
13	RNA aptamer-based screen for selective inhibitors of GRK2	1R03MH08937801 (approved)	J. Tesmer, U. Michigan/J. Chen	1 × 330 K, in process
R21	HTS to identify small molecules that disrupt Rab GTPase activator and effector protein interactions	1 R21NS06643501 (approved)	A. Wandinger-Ness, UNM/tbd	tbd × 330 K, projected

was developed that cross-reacts against both receptors with low nanomolar affinity [38]. Secondary assays used the calcium response of the cell lines to distinguish receptor agonists and antagonists. Probe reports have been prepared using this peptide to identify selective antagonists for each receptor that are posted at the MLPCN Web site [39].

Lessons Learned: In cell-based multiplex assays, it is desirable to use the same detection reagent for all the populations in the multiplex.

4.4.3 Cycle 2. ATP Hydrolysis-Dependent Disassembly of the 26S Proteasome: 1X01MH077613-01 (Bead-Based Singleplex)

The proteasome assay was developed by D. Skowyra who had reported a cycle of proteasome degradation earlier [40, 41]. The proteasome assay uses FLAG-tagged proteasome for capture and display on streptavidin beads with biotinylated anti-FLAG antibody. Since GFP is associated with the end caps of the proteasome, a change in the fluorescent signal results when the caps fall off. Under physiological conditions, caps are proposed to fall off and reassemble during each degradative cycle. The assay identifies inhibitors of the proteasome degradation cycle as indicated through inhibition of the release of the caps. Several small molecules with low micromolar activity dose–response characteristics were identified. During follow-up dose–response curves, we eliminated fluorescent molecules that interfered with the assay using uncapped proteasome on beads to measure background fluorescence. We used a novel secondary cell-based flow cytometry assay as a follow-up to the primary bead-based screen for the ATP hydrolysis-dependent disassembly of the proteasome in the presence of copurified endogenous substrates from yeast. The secondary assay monitors proteasomal degradation of the yeast Sic1 protein, which we have previously linked to the disassembly of the proteasome via *in vitro* analysis. Sic1 is an inhibitor of the S-phase *c*yclin-*d*ependent *k*inase (CDK) and the prototype substrate of the Cdc34/ SCF^{Cdc4} ubiquitination pathway. Sic1 accumulates at the end of mitosis as a stable protein and is rapidly degraded in response to its phosphorylation at the G1/S transition. Two SMR molecules defined as active in the bead-based primary assay were active in the secondary assay by delaying degradation.

Lessons Learned: Although the bead assay typically yielded a signal/background of 4/1 or higher, the fluorescence emitted from the complex was often obscured by fluorescent compounds in the test wells. The secondary assays turned out to be qualitative in the sense that delayed degradation rather than inhibition was observed.

4.4.4 Assay for Ligands of GPR30 and Classical Estrogen Receptors: 1X01MH077627-01 (Permeabilized Cell Assay)

We have characterized a novel estrogen receptor (ER) and a novel probe (G1) in *Science* and *Nature* publications [42–44]. Although targets have been accepted by the MLSCN in cycle 2, HTS on a permeabilized whole-cell binding assay has remained problematic. This assay requires daily transient transfection of GPR30 or ER. The reformatting of the assay to 384-well plates created a challenge with respect to cell number due to surface to volume ratio of 384-well plates as compared to the 96-well format in which the assay was submitted. We received approval to perform a virtual screen to cherry pick from the MLSMR. Since the Center at Emory has previously screened another ER assay, we used our GPR30 assay to test for relevant molecules identified in their PubChem uploads as well as other molecules determined by virtual screening to have similarity with G1. As we had previously reported on a GPR30-specific antagonist, the screen along with synthetic chemistry has also generated an antagonist (G15) [44].

Lessons Learned: We recognized early in the initiative that HT flow cytometry was not well suited to targets that depended on adherent cells and cell permeabilization when screening a large library was required. To mediate this difficulty, a virtual screening approach was taken to search for estradiol analogues in the library followed by screening a small library subset.

4.4.5 Assay for Allosteric Ligands for the VLA-4 Integrin: X01MH077638-01 (Suspension Cell Singleplex)

Allosteric regulators have proven to be important tools in the regulation of integrin function. This assay was intended to detect direct binding inhibitors as well as allosteric regulators. The assay was based on the binding of a fluorescent peptide to VLA-4 in its resting state or modulated to high affinity by the divalent cation Mn^{2+}. The signal to background of the assay was 2/1 for the resting state and 4/1 for the high-affinity state. The screen detected only direct inhibitors with scaffolds similar to families of already known inhibitors [45, 46]. A novel secondary assay was developed to help classify the activity of the competitor molecules. This assay uses an antibody HUTS-241 that is sensitive to VLA-4 conformation. The novel ligand-induced binding site (LIBS) mAb assay is a result of the interplay between R01HL081062 that has taken advantage of compounds identified through the X01MH077638 that screened a portion of the MLSMR. A Fast Track proposal for screening the entire MLSMR is using the improved assay developed to search for novel integrin ligands that block the binding of traditional ligands and do not in themselves induce the LIBS mAb binding site.

Lessons Learned: In this assay, we attempted to read each plate twice, once before and once after the addition of divalent cation. This turned out to be difficult and we wound up preparing each plate separately.

4.4.6 Small-Molecule Inhibition of *Staphylococcus aureus* Virulence: 1X01MH078952-01 (Suspension Cell Singleplex)

The bacterial virulence assay contributed by P. Gresham is a cell-based assay that measures the ability of test compounds to inactivate the quorum sensing pathway of *S. aureus*, a pathway responsible for the production of bacterial virulence-associated factors. This assay was identified as a result of our outreach effort and was developed collaboratively. The RNAIII promoter, activated in the late stages of the pathway, is fused to a GFP expression vector. An active test compound is one that prevents GFP expression in the presence of a pathway-activating peptide. Secondary assay evidence suggests that the two most active compounds from different families act at different points in bacterial quorum sensing pathways as described in the probe reports.

Lessons Learned: HTS for suspension bacterial assays is readily compatible with flow cytometry when the cells express adequate levels of fluorescent GFP marker.

4.4.7 Activators of Prostate Cell Differentiation Target: 1X01MH078937-01 (Adherent Cell Phenotypic Assay)

The lymph node cancer of the prostate (LNCaP) human prostate cell differentiation model was used to determine the ability of test compounds to induce a phenotype with increased side scatter caused by intracellular granule accumulation in prostate cells in the assay contributed by T. Thompson. Significant optimization was required for this assay due to the adherent nature of the cells, the passage-dependent alteration of the responsiveness of the androgen-dependent target cell line (LNCaP), and the reformatting from 96 to 384-wells that reduced the cell number in the assay relative to the assay volume. Two new secondary assays, one involving an androgenic activity assay measured by GFP expression in the flow cytometer [47] and the other involving androgen-independent granularity increases in PC3 cells, have been developed in collaboration with the target provider. Hits from the screens are capable of increasing prostate cell granularity and inducing cell death. Probe criteria have been met for a phenotypic assay, yielding micromolar activities in LNCaP and PC3 phenotypic assays [48]. The target provider is following up on identifying the mechanism of action of increased granularity, presumably autophagy.

Lessons Learned: Light scatter responses are readily adapted to HT flow cytometry. After the screen was conducted, further studies have helped to elucidate the granule accumulation mechanism as autophagy. The PC3 cells were less adherent than the LNCaP cells and were readily resuspended for flow cytometry.

4.4.8 Multiplexed High-Throughput Screen for Small-Molecule Regulators of Bcl-2 Family Protein Interactions: NIH 1X01MH079850-01 (Bead-Based Multiplexes)

Apoptosis is regulated in part by the balance of antiapoptotic and proapoptotic Bcl-2 family members. In humans, six genes have been identified that encode antiapoptotic proteins characterized by the presence of conserved motifs designated as three Bcl-2 homology (BH) regions, BH1, BH2, and BH3. These domains form a hydrophobic cleft in tertiary structure. Proapoptotic family members contain BH3 regions that form an amphipathic helix, and these helices bind in the clefts of the antiapoptotic proteins. The binding of fluorochrome-conjugated BH3 peptides (including Bim) to Bcl-2 family members thus provides the basis for construction of fluorescence-based assays amenable to flow cytometry high-throughput screening for small-molecule regulators of these interactions. The high-throughput screen was a multiplexed assay to identify small-molecule regulators of protein interactions between the BH3 peptide of Bim and the following six Bcl-2 family members: Bcl-XL, Bcl-W, Bcl-B, Bfl-1, Mcl-1, and Bcl-2 (the eponymous founding member of the Bcl-2 family). HTS has been completed and reported in PubChem and we continue to characterize the activities of the molecules, some of which may be chemically reactive, in secondary assays.

Lessons Learned: The setup for this screen depended on the unexpected observation that when GST is arrayed at high densities on microspheres, the attachment is stable due to rebinding [49]. Even though the monovalent affinity of GSH to GST is $\sim 10\,\mu M$, individual proteins can be attached to separately color-coded beads.

4.4.9 Low Molecular Weight G-Protein Multiplex Assay: 1R03MH081231-01 (Bead-Based Multiplexes)

The low molecular weight G-protein assay was submitted by A. Wandinger-Ness [50]. A multiplex assay screened the binding of fluorescent GTP to G-protein–GST fusion proteins on GSH beads. A set of six G-proteins (Rac1 wt, Rab7 wt, Rac1 activated, Ras wt, Rab2 wt, and Cdc42 wt) were arrayed. The dose–responses have also included activated mutants of Ras and Cdc. Dose–response curves have been performed on a total of nine proteins. The screens have identified a pan-activator series (probe report, publication in preparation), a pan-inhibitor series (undergoing optimization), a rho family selective series [51], and a Cdc42 selective inhibitor (probe report and publication in preparation).

Lessons Learned: A successful multiplex screening campaign has the potential of revealing molecules active against one or more of the targets.

4.4.10 HTS to Identify Small-Molecule Regulators of RGS Family Protein Interactions: NIH R21NS057014 (Bead-Based Multiplex)

The RGS assay [52, 53] submitted by R. Neubig and coworkers involves the assembly of five biotinylated RGS proteins on streptavidin beads. Regulators of G-protein signaling (RGS) proteins are a diverse set of intracellular proteins that modulate G-protein-coupled receptor (GPCR) signaling. As a subfamily of GTPase Accelerating Proteins (GAPs), RGS proteins bind directly to Gα-GTP, accelerate the rate of GTP hydrolysis, and shorten the lifetime of active G-proteins up to several 1000-fold. Thus, RGS-Gα interactions are central to the downstream regulation of GPCR signaling events. Since GPCRs control numerous physiological processes in diverse tissues, including brain, heart, liver, and lung, modulation of the RGS/G-protein interaction has become an attractive target for drug discovery. The primary screen was a high-throughput multiplexed flow cytometry assay based on RGS binding to fluorescently labeled Gα. AlexaFluor 488-conjugated Gα_0 (AF-Gα_0) is incubated with purified biotinylated RGS proteins that are immobilized onto streptavidin microspheres. Five RGS family members were studied: RGS4, RGS7, RGS8, RGS16, and RGS19. As a small number of inhibitors identified have the potential of covalently modifying the cysteine residues, the current goal of this project is to identify reversible small-molecule inhibitors of RGS.

Lessons Learned: Multiplexing was extended to biotinylated proteins and streptavidin beads.

4.4.11 Multiplex Screening for ABC Transporter Inhibitors: 1R03MH078946-01

Upregulation of ATP binding cassette (ABC) transporter expression results in the selection of resistant cancer cells and cancer multidrug resistance to diverse transporter substrates. The assay submitted by R. Larson was designed to identify inhibitors of three human ABC transporters, ABCB1, ABCC1, and ABCG2 that have been shown to contribute to cancer multidrug resistance. The proposed screen involved two separate duplex assays: one in which ABCB1 and ABCG2 transporters are evaluated in parallel using fluorescent JC-1 as substrate and the other in which ABCB1 and ABCC1 transporters are evaluated in parallel using fluorescent calcein acetoxymethyl ester (CaAM) as substrate [54, 55]. An active test compound is one that prevents test cells from transporting fluorescent substrate out of the cytoplasm and is detected by an increase in cell fluorescence intensity. Cells expressing different transporters are color coded to distinguish which transporter is affected by a compound. Although we hoped that it would be possible to formulate a cell-based triplex for three pumps (ABCB1, ABCC1, and ABCG2), the target assay involved duplexes, involving two dyes, that each recognized two of three transporters. As a screen for ABCC1 inhibitors was previously performed elsewhere, we performed HTS on the ABCB1/ABCG2 duplex assay. For the primary screening protocol, a duplex assay was constructed in which ABCB1 and ABCG2 transporters were evaluated in parallel using the fluorescent J-aggregate-forming lipophilic cation 5,5′,6,6′-tetrachloro-1,1′,3,3′-tetraethylbenzimidazolcarbocyanine iodide (JC-1) as substrate. ABCB1-expressing cells (CCRF-Adr) were color coded with FarRed DDAO CellTrace SE (Invitrogen) to distinguish them from ABCG2-expressing cells (Ig-MXP3) via a red fluorescence emission wavelength of 665 $+/-10$ nm (635 nm excitation). The output of the primary screen included compounds that inhibited pumps and were themselves nontoxic. The compound family is now being evaluated for selectivity in anticipation of a probe report.

Lessons Learned: HTS flow cytometry is well suited to ABC transporter assays involving suspension cells. Although the approach can be multiplexed, the substrate selectivity of the transporters is such that there is no guarantee that a single fluorescent substrate will be applicable to the collection of transporters assembled in a multiplex.

It is also worth noting that because ABC transporters have multiple binding sites, different fluorescent substrates may identify different sets of inhibitors. We have profiled more than 100 fluorescent substrates against this trio of transporters (publication in preparation).

4.4.12 TR-FRET HTS Assay for Inhibitors of MEKK2–MEK5 PB1 Domain Interactions: 1R03MH084830-01

This project developed as a singleplex TR-FRET assay [56] by K. Nakamura aimed to identify small-molecule modulators of PB1 domain interactions between MEK5 and MEKK2. It was reformulated at a triplex bead-based assay, which also included MEKK3, and a MEKK2 mutant with a lysine to alanine mutation in the PB1 domain that results in altered binding to MEK5 and a loss of downstream activation of ERK5. For each target, the components of the multiplex consist of a streptavidin-functionalized polystyrene bead, a biotinylated MEKK-fusion protein target, and a fluorescent peptide probe (AlexaFluor 488-GST-MEK5 WT). The similarly sized bead sets (~5 μm) are distinguished by emission at 750 nm with excitation at 635 nm.

Lessons Learned: Although the conversion of TR-FRET assay to a bead-based assay was straightforward, the screen was plagued by a large number of compounds that appeared to represent high background fluorescence. No unequivocally active molecules have yet been identified.

4.4.13 HTS for Developing T Cell Immune Modulators (1X01MH085707-01) (Pimary Cell-Based Assay)

The LFA-1 integrin plays important roles in T cell immunity. Along with its ligand partner ICAM-1, this integrin is involved in the immune synapse that stabilizes the interaction between naïve T cells and antigen presenting cells (APCs), promoting T cell proliferation and the generation of effector functions. The complex nature of T/APC interaction (due to involvement of multiple receptor/ligand pairs), however, makes it difficult to use APCs to study the role of individual receptor/ligand interactions in T cell activation. The screening campaign currently underway is to find small chemical compounds that modulate the LFA-1/ICAM-1 interaction. The assay developed by I. Hwang [57] and coworkers detects the interaction of exosome-like membrane vesicles (eMVs) that express murine ICAM-1, B7.1, and L^d (transgenically engineered surrogate APC prepared from S2 drosophila cells) with murine $CD8^+$ T lymphocytes isolated from 2C TCR transgenic mice. 2C TCR Tg T cells actively absorb eMVs that carry the cognate QL9 antigen. The binding of eMVs to T cells is detected by dual staining using anti-CD8 and anti-B7.1 monoclonal antibodies. Small molecules of interest could act as inhibitors or potentiators of the extracellular molecular interactions or intracellular pathways.

Lessons Learned: This assay is an example of the potential to use flow cytometry for screening primary cells from transgenic animals. Considerations involve breeding animals, testing transgenic individuals, preparing lymph nodes for primary cell isolation, and allowing cells and vesicles to interact in the presence of compounds and staining agents. Due to the length of the sample preparation, well contents were fixed for analysis on a subsequent day. The data were displayed in quadrants with positive and negative cell populations, each with the potential of binding the vesicles. Data analysis required not only identifying individual wells but also quantifying quadrants for each of the 325,000 compound sets.

4.4.14 Chemical Screen of TOR Pathway GFP Fusion Proteins in *S. cerevisiae* (1R03MH086450-01, Cell-Based Multiplex)

TOR proteins (target of rapamycin) are ser/thr protein kinases phylogenetically conserved from yeast to man. Rapamycin-sensitive TORC1 signaling pathway promotes protein synthesis and other anabolic processes, while inhibiting macro-autophagy and other catabolic and stress–response processes. As a central controller of cell growth, TORC1 has emerged as a target for drug discovery. Recently, many components or readouts of TORC1 signaling pathway have been identified through classic genetic or chemical genetic approaches. Through the screening of the whole GFP-tagged yeast strain collection (4159 strains), M. Werner-Wasburne and C. Allen identified several yeast strains showing GFP expression change upon rapamycin treatment. The rapamycin responsive genes are classified as readouts of four branches of TORC1 signaling pathway mediated by RTG1/RTG3, GLN3, MSN2/MSN4 transcription factors, and YAK kinase. We used these yeast GFP strains to carry out a 5-plex high-throughput assay. The yeast GFP strains are differentially stained with AlexaFluor 405 and AlexaFluor 633 and incubated with the compound for 3 h. The GFP signal is measured by HyperCyt/CyAn high-throughput flow cytometry. The purpose of this HTS screen is to identify small-molecule modulators of protein targets in the TORC1 pathway. The screen will detect structurally distinct, but functionally rapamycin-like compounds as well as the compounds that regulate individual or multiple branches of TORC1 signaling pathway.

Lessons Learned: Individual yeast strains have been color coded using AlexaFluor dyes excited at 405 and 633 nm. GFP is excited at 488 nm. In principle, because yeasts are heterogeneous, two intensities of staining at 405 and 633 each, along with their single- and double-strained combinations lead to detection of nine populations. However, in a 3 h exposure to compounds, the populations blur as unstained daughter cells are produced. We have found that a fiveplex can be followed (405 low, 405 high, 633 low, 633 high, and 405/605 high. Batch staining of cells for screening each day has turned out to be tedious.

4.4.15 A Yeast HTS for Caloric Restriction Mimetics That Inhibit Age-Related Superoxide (1R03MH087439-01, Cell-Based Singleplex)

Reactive oxygen species (ROS) and activation of growth signaling pathways are important components of aging and contribute to a variety of age-related pathologies, including neoplastic, cardiovascular, and neurodegenerative diseases. Recent findings point to DNA replication stress downstream of growth signaling by ROS and other molecules as an equally important factor in aging and the development of these diseases. The screen provided by W. Burhans is designed to identify compounds that mimic caloric restriction by downregulating growth signaling pathways, thus reducing the accumulation of superoxide anions and promoting a tighter growth arrest in G1. The primary readout for this phenotypic screen is the measurement of superoxide using the fluorogenic compound dihydroethidium (DHE).

Lessons Learned: The cell incubation is 1 week long during which time yeast cells grow and settle, and liquid from the wells evaporates as well. The plates are covered and placed in a humidified incubator and then diluted and resuspended for HTS flow cytometry. The data analysis involves the detection of populations of cells with high and low level of staining.

4.4.16 Identification of Broad-Spectrum Antifungal Efflux Inhibitors: R03MH08740601A1 (Phenotypic Cell-Based Multiplex)

Fungal infections, exemplified by oral and invasive candidiasis, cause morbidity and mortality in an ever-increasing human population: the immunocompromised. Treatment of patients with fungal infections is severely hampered by the development of antifungal drug resistance. The goal of the project, submitted by R. Cannon, is to address this health need by discovering novel therapeutic agents that overcome clinical drug resistance by inhibiting a major cause of antifungal resistance, the drug efflux pump molecules in the fungal cell membrane. Lead compounds will be identified by high-throughput screening in which the strains express individual fungal efflux pumps that will be the targets for the proposed HTS. The yeast cell-based multiplex approach developed involves the use of three "sentinel" strains in an HTS that measures efflux of the fluorescent pump substrate Nile red and follow-up microtiter fluorescent assays with the other panel strains to identify pump inhibitors.

Lessons Learned: The strains provided by R. Cannon highly overexpress the pumps of interest interfereing with the use of stains to color code the cells. The profiling of fluorescence substrates at the UNMCMD revealed that since Nile red is pumped by all the strains [58], mixtures of strains allow detection of transporter inhibition reflecting one, two, or three strains in the primary assay. However, dose–response studies on individual strains with active compounds will be required to identify the strains that respond.

4.5 OPERATIONAL ISSUES

When we entered the pilot phase of the MLI, we had experience with a single target and screening ~10,000 compounds in 96-well plates. Once we transitioned to 384-well plate multiplexes, we can create up to 50,000–100,000 screening data points per day. Because each plate generates a single data file, the multiplex data must be resolved. Moreover, there is a considerable variation in bead- and cell-based assays. The cell-based assays vary according to the source and treatment of the cells. With the creation of some 8 million data points over the course of the MLI pilot and production phases, we have recognized the considerable effort required to balance the needs of target development, HTS, and data analysis ("screening informatics") for plate-based flow cytometry. In particular, cell samples in HyperCyt do not always resolve as well as the bead-based samples, so identifying individual wells has required the development of automated binning algorithms in order that the increase in throughput gained by sampling a plate as a file is not lost in data analysis.

Not unexpectedly, there is variability in sample preparation. In bead-based assays, it is often possible to prepare plates in batches, so the plates for an entire day's screen can be prepared consecutively, stored at the desired temperature, all independent of data acquisition. In some cell-based targets, batches of 10 plates may be prepared with screening interleaved. In some cases, cells can be fixed and batch analysis performed on a separate day. In some cases, the cell plate preparation is timed and screening and plate preparation must be coordinated. Finally, some cell-based incubations with compounds are several days long, in which case batch preparation may precede HTS on live cells by periods as long as a week during which the cells must retain their ability to transit the flow cytometer. It is not surprising that strongly adherent cells compromise convenient HTS.

Because UNMCMD functions in a production mode, with expected throughput of ~6 targets per year for a library of 350,000 compounds, we are finding it useful to incorporate the following operation elements:

4.5.1 Target Teams

Projects typically involve a target team, with a leader responsible for the screen and at least one team member collaborating in the screen. UNMCMD target teams "own" projects through the period of target development, HTS, dose–response follow-up, and often through secondary assays and compound optimization.

4.5.2 Scheduling

We can interleave screens so that two screens are in process if necessary. Or we can alternate days of HTS data acquisition and analysis. This may vary according to the nature of the target. Three flow cytometers are available for the primary screen along with dedicated sample preparation automation. A fourth flow cytometer is available along with a secondary sample preparation station for target development and implementation prior to HTS and for dose–response assays, when the primary instrumentation should be in use for another project.

4.5.3 Accessibility

Integration of HyperCyt has been achieved with essentially all commercial flow cytometry systems. However, consideration needs to be given to the geometry of the flow cytometer relative to the placement of HyperCyt. As noted above, acquisition time for individual plates must be coordinated with the throughput of plate preparation.

4.5.4 Data Analysis

The software IDLeQuery developed to address the issue of well identification has been expanded to include automated binning and error checking in the commercial release called HyperView.

4.5.5 Carryover

The tubing for sample delivery results in carryover of particles and fluids. A thin film of sample fluid is left behind, which amounts to about 10% of the total sample volume. The carryover of active compounds can contribute to the behavior of the next sample if the contents of the well are not at a stable end point or a binding reaction is reversible. The carryover of particles in bead-based assays is typically less than 1%, but can vary according to the adherence characteristics and size of cells.

4.5.6 Beads

Bead-based molecular assays are now commonplace in flow cytometry. Our experience suggests that the approach is generalizable to most molecular assemblies, as long as one molecule is stably attached to the bead and the binding of the other molecule results in a fluorescent signal. For screening large libraries of compounds, assay conditions and costs need to be evaluated in conjunction with reagent manufacturers to ensure access to required reagent quantities at appropriate costs. In addition to targets on beads described here, GPCR [59], kinase [60], and metalloprotease targets [61, 62] have been addressed outside the MLPCN.

4.5.7 Protein–Protein Interaction Targets

Protein–protein interaction (PPI) targets have so far met with variable success with the Molecular Libraries Small Molecule Repository. Two targets described here, Bcl-2 family and RGS family multiplexes, have both identified reactive molecules, whereas the MEKK and proteasome targets identified significant numbers of fluorescent compounds.

4.5.8 Chemical Probe Development Plan and Secondary Assays

The MLPCN production phase has installed a rigorous process of Chemical Probe Development Plan (CPDP) in which timelines and responsibilities for primary and secondary assays are assigned. Independent labs undertaking screening campaigns should have detailed plans in advance for verifying that compound activities identified in HTS represent the desired biological effect rather than a physical or chemical artifact.

4.5.9 Use of PubChem

The value of PubChem cannot be underestimated. We have found it remarkably useful to follow the activity of molecules in UNMCMD screens across the MLPCN. This tracking capability identifies bona fide activity in screens on other targets. The library has also been characterized with respect to compounds that aggregate, fluoresce, or are otherwise reactive species.

4.6 FUTURE OPPORTUNITIES

As we see it, flow cytometry in a small-molecule discovery environment will benefit from ongoing development of software and hardware. The instrumentation leader for the UNMCMD, Bruce Edwards now has an NIH-funded R01 for improved sample handling and automation (1R01HG005066-01 Advancing High Throughput Flow Cytometry). The work is coordinated with the overall automation of the HyperCyt platform to improve sample detection, transition to 1536-well sampling, and integrate a direct sampling system for the flow cytometer.

4.6.1 Bubble Technology: Optical Time Stamping System for Robust Indexing of HyperCyt Flow Cytometry Data

Edwards has been developing an air bubble detection system to supplement our current sample indexing protocols. As a sample-separating bubble passes through the point of analysis of the flow cytometer, a flash of light is produced due to laser light scattering by the air–liquid interface. Both flow cytometry data and bubble data are being recorded in the same FCS format, facilitating integration of the two data streams in the analysis software. We are now in process of developing an optical fiber interface that will extend the range of instruments amenable to bubble data collection.

4.6.2 Adaptation of HyperCyt for Processing of 1536-Well Plates

We are developing 1536-well plate processing capabilities so that a single HyperCyt system can operate four sampling probes per 1536-well plate to deliver 384-well segments simultaneously to four Accuri C6 flow cytometers. Accuri Corporation has released an upgraded version of CFlow software called "CFlow for the HyperCyt System." This software provides an interface by which the C6 cytometer can be remotely controlled over a network via TCP/IP. Also, the software allows collection of up to 96 million events per round of data acquisition, a significant advance over the original 1 million event limit. We have developed TCP/IP software to communicate with the CFlow interface so that a laptop PC controls all four flow cytometers in parallel via an Ethernet network in addition to its control of the HyperCyt autosampler. Multiplex samples in 1536-well plates have been delivered in 11 min.

4.6.3 XYZ Minisampler

The HyperCyt system uses a peristaltic pump to transport a segment stream of samples through ~1 m of tubing. This involves a transit time of ~30 s and a thin film of material that coats the tubing from one sample to another. We have developed a prototype version of a plate transport system to deliver samples directly to the Accuri C6 flow cytometer. Since this flow cytometer sips sample without a need for pressurization or external pump, sample wells can be delivered to the sample intake line. The reduction in path length reduces carryover and transit time. The contents of a 384-well plate have been delivered in less than 20 min.

4.6.4 Platform Integration for Complex Sample Handling

To our knowledge, the only fully integrated HT flow cytometry system is the Rigel SiRNA screening system that is dedicated to a highly specialized task involving repetitive procedures of treating cells with SiRNA, incubating for a particular period of time, staining the cell as appropriate, and then performing flow cytometric analysis with HyperCyt. Centrifugation, wash, and resuspension are incorporated as necessary. Because the SiRNA library is considerably smaller than the MLSMR compound library and because the demands of screening vary dramatically from target to target, automation is likely to require modular flexibility. Given that overall integration of sample preparation and flow cytometry has now been demonstrated, the opportunities for complex sample handling and complex cell suspension are unique to the flow cytometry platform. From hematopoietic stem cells to blood cells in leukemia, immunology, inflammation, and infection disease, the possibilities are endless for HTS flow cytometry based on surface as well as intracellular markers.

4.6.5 Complex Target Data Analysis and Color Coding

Customization of the analysis software for multiplexed targets or complex cell suspension, and the additional genetic color coding of cells, along with antibody immunophenotypic staining and conjugated dyes, should compete the overall requirements for HTS flow cytometry and ensure its niche in the discovery arena.

ACKNOWLEDGMENTS

Work at the UNMCMD was supported by NIH Grants U54MH074425 and U54MH084690. The authors would like to acknowledge the contributions of all the target providers, the grant support to their projects, and to all team members of the UNMCMD listed as well as those not listed for their contributions.

REFERENCES

1. Human Genome Project Pilot Phase: http://www.genome.gov/10000511.
2. Human Genome Project Production Phase: http://www.genome.gov/10001792.

3. Collins, FS. The language of god. A Scientist Presents Evidence for Belief, Simon and Schuster, 2006, p. 320.
4. Bioengineering Research Partnerships: http://grants.nih.gov/grants/guide/pa-files/PAR-04-023.html.
5. Bioengineering Research Grants: http://grants.nih.gov/grants/guide/pa-files/PA-02-011.html.
6. Exploratory Bioengineering Research Grants: http://grants.nih.gov/grants/guide/pa-files/PA-03-058.html.
7. Bioengineering Consortium Symposia: http://www.becon.nih.gov/becon_symposia.htm.
8. NIH Common Fund –about the NIH Roadmap: http://nihroadmap.nih.gov/aboutroadmap.asp.
9. Molecular Libraries Program: http://mli.nih.gov/mli/mlpcn/.
10. PubChem. http://pubchem.ncbi.nlm.nih.gov/.
11. Couzin J. NIH dives into drug discovery. *Science* 2003;302(5643):218–222.
12. Oprea TI, Bologa CG, Boyer S, Curpan RF, Glen RC, Hopkins AL, Lipinski CA, Marshall GR, Martin YC, Ostopovici-Halip L, Rishton G, Ursu O, Vaz RJ, Waller CL, Waldmann H, Sklar LA. A crowdsourcing evaluation of the NIH chemical probes. *Nat. Chem. Biol.* 2009;5:441–447.
13. Academic Screening Center Directory: http://www.sbsonline.org/i4a/member_directory/feResultsListing.cfm?directory_id=7&viewAll=1.
14. Solicitation of Assays for High Throughput Screening (HTS) in the Molecular Libraries Probe Production Centers Network (MLPCN) (R03) PAR 09 129.
15. Assay Development for High Throughput Molecular Screening (R21): http://grants.nih.gov/grants/guide/pa-files/par-10-182.html.
16. Fast Track Notice of Opportunity for R01: http://grants.nih.gov/grants/guide/notice-files/NOT-RM-09-011.html.
17. Fast Track Notice of Opportunity for R21: http://grants.nih.gov/grants/guide/notice-files/NOT-RM-09-008.html.
18. Molecular Libraries Fund Opportunities: http://mli.nih.gov/mli/funding-opportunities/.
19. National Center for Biotechnology Information: http://www.ncbi.nlm.nih.gov/.
20. Molecular Libraries probes: http://mli.nih.gov/mli/mlp-probes/.
21. Molecular Libraries funded research: http://nihroadmap.nih.gov/molecularlibraries/fundedresearch.asp.
22. NIH Reporter database: http://projectreporter.nih.gov/reporter.cfm.
23. Molecular Libraries home: http://mli.nih.gov/mli/mlpcn/.
24. NExT Program: http://ncitranslates.nci.nih.gov/docs/presentations/Doroshow 11052009_508.pdf.
25. NCI Chemical Biology Consortium: http://dctd.cancer.gov/CurrentResearch/Chemical BioConsortium.htm.
26. NIH Director's Opportunity for Research in Five Thematic Areas (RFA OD-10-005): http://www.nidcd.nih.gov/funding/ARRARC4.htm..
27. European Strategy Forum on Research Infrastructure: http://cordis.europa.eu/esfri/. and European Infrastructure for Open Screening Platforms for Chemical Biology: http://www.eu-openscreen.de/.

28. European Membrane Protein Network: http://www.mepnet.org/.
29. Sklar LA, Carter MB, Edwards BS. Flow cytometry in drug discovery, receptor pharmacology, and high throughput screening. *Curr. Opin. Pharmacol.* 2007;7:527–534.
30. Edwards BS, Young SM, Saunders MJ, Bologa CG, Oprea TI, Ye RD, Prossnitz ER, Graves SW, Sklar LA. High throughput flow cytometry for drug discovery. *Expert Opin. Drug Discov.* 2007;2:685–696.
31. Edwards BE, Kuckuck F, Sklar LA. Plug–flow flow cytometry: an automated coupling device for rapid sequential flow cytometric sample analysis. *Cytometry* 1999;37:156–159.
32. Kuckuck F, Edwards BS, Sklar LA. High throughput flow cytometry. *Cytometry* 2001;44:83–90.
33. Intellicyt: http://www.intellicyt.com/.
34. Viviabiotecch: http://www.viviabiotech.com/equipo.html.
35. Rigel Automation of HyperCyt: http://www.intellicyt.com/download/RigelSBS Automation Poster.pdf.
36. Intellicyt–Beckman Coulter Agreement: http://www.beckmancoulter.com/HR/Press Room/oc_pressReleases_detail.asp?Key=8962.
37. Edwards BS, Young SM, Oprea TI, Bologa CG, Prossnitz ER, Sklar LA. Biomolecular screening of formylpeptide receptor ligands with a sensitive, quantitative, high-throughput flow cytometry platform. *Nat. Protoc.* 2006;1:59–66.
38. Strouse JJ, Young, SM, Mitchell HD, Ye RD, Prossnitz ER, Sklar LA, Edwards BS. A novel fluorescent cross-reactive formylpeptide. *Cytometry* 2009;75(3):264–270.
39. Young SM, Bologa CM, Fara D, Bryant BK, Strouse JJ, Arterburn JB, Ye RD, Oprea TI, Prossnitz ER, Sklar LA, Edwards BS. Duplex high-throughput flow cytometry screen identifies two novel formylpeptide receptor family probes. *Cytometry A* 2009;75:253–263.
40. Deffenbaugh AE, Scaglione KM, Zhang L, Moore JM, Buranda T, Sklar LA, Skowyra D. Release of ubiquitin-charged Cdc34-S-Ub from the RING domain is essential for ubiquitination of the SCF(Cdc4)-bound substrate Sic1. *Cell* 2003;114:611–622.
41. Babbitt, SE, Kiss, A, Deffenbaugh, AE, Erdjument-Bromage, H, Tempst, P, Bailly, E, Buranda, T, Sklar, LA, Baumler, J, Gogol, E, Skowyra, D. ATP hydrolysis-dependent disassembly of the 26S proteasome is part of the catalytic cycle. *Cell* 2005;121: 553–565.
42. Revankar C, Cimino D, Sklar LA, Arterburn J, Prossnitz, ER. An intracellular 7-transmembrane G-protein coupled estrogen receptor mediates rapid cell signaling. *Science* 2005;307:1625–1630.
43. Bologa CG, Revankar CM, Young SM, Edwards BS, Arterburn JB, Kiselyov AS, Parker MA, Tkachenko SE, Savchuck NP, Sklar LA, Oprea TI, Prossnitz ER. Virtual and biomolecular screening converge on a selective agonist for GPR30. *Nat. Chem. Biol.* 2006;2:207–212.
44. Dennis MK, Burai R, Ramesh C, Petrie WK, Alcon SN, Nayak TK, Bologa CG, Leitao A, Brailoiu E, Deliu E, Dun NJ, Sklar LA, Hathaway HJ, Arterburn JB, Oprea TI, Prossnitz ER. *In vivo* effects of a GPR30 antagonist. *Nat. Chem. Biol.* 2009;5:421–427.
45. Njus BH, Chigaev A, Waller A, Wlodek D, Ostopovici-Halip L, Ursu O, Wang W, Oprea TI, Bologa CG, Sklar LA. Conformational mAb as a tool for integrin ligand discovery. *Assay Drug Dev. Technol.* 2009;5:507–515.

46. Chigaev A, Waller A, Amit O, Halip L, Bologa CG, Sklar LA. Real-time analysis of conformation-sensitive antibody binding provides new insights into integrin conformational regulation. *J. Biol. Chem.* 2009;284(21):14337–14346.
47. Dennis MK, Bowles HJC, MacKenzie DA, Burchiel SW, Edwards BS, Sklar LA, Prossnitz ER, Thompson TA. A multifunctional androgen receptor screening (MARS) assay for the identification of compounds that modulate androgen receptor activity using the high-throughput HyperCyt® flow cytometry system. *Cytometry* 2008;73 (5):390–399.
48. Haynes MK, Strouse JJ, Waller A, Leitao A, Curpan RF, Bologa C. Oprea TI, Prossnitz ER, Edwards BS, Sklar LA, Thompson TA. Detection of intracellular granularity induction in prostate cancer cell lines by small molecules using the HyperCyt® high throughput flow cytometry system. *J. Biomol. Screen.* 2009;14:596–609.
49. Tessema M, Simons PC, Cimino DF, Sanchez L, Waller A, Posner RG, Wandinger-Ness A, Prossnitz ER, Sklar LA. Glutathione-S-transferase-green fluorescent protein fusion protein reveals slow dissociation from high site density beads and measures free GSH. *Cytometry A* 2006;69:326–334.
50. Schwartz SL, Tessema M, Pylypenko O, Rak A, Simons P, Surviladze Z, Sklar LA, Wandinger-Ness A. Flow cytometry for real-time measurement of guanine nucleotide binding and exchange by Ras-like GTPases. *Anal. Biochem.* 2008;381:258–266.
51. Surviladze Z, Waller A, Wu Y, Romero E, Edwards BS, Wandinger-Ness A, Sklar LA. Identification of a small GTPase inhibitor using a high-throughput flow cytometry bead-based multiplex assay. *J. Biomol. Screen.* 2010;15(1):10–20.
52. Roman DL, Ota S, Neubig RR. Polyplexed flow cytometry protein interaction assay: a novel high-throughput screening paradigm for RGS protein inhibitors. *J. Biomol. Screen.* 2009;14(6):610–619.
53. Blazer LL, Roman DL, Muxlow MR, Neubig RR. Use of flow cytometric methods to quantify protein–protein interactions. *Curr. Protoc. Cytom.* 2010;51:13.11.1–13.11.15
54. Winter SS, Lovato DM, Khawaja HM, Edwards BS, Ivnitski-Steele ID, Young SM, Oprea TI, Sklar LA, Larson RS. High throughput screening for daunorubicin-mediated drug resistance identifies mometasone furoate as a novel ABCB1 reversal agent. *J. Biomol. Screen.* 2008;13:185–193.
55. Ivnitski-Steele I, Larson RS, Lovato DM, Khawaja HM, Winter SS, Oprea TI, Sklar LA, Edwards BS. High-throughput flow cytometry to detect selective inhibitors of ABCB1, ABCC1, and ABCG2 transporters. *Assay Drug Dev. Technol.* 2008;6(2):263–276.
56. Nakamura K, Zawistowski JS, Hughes MA, Sexton JZ, Yeh LA, Johnson GL, Scott JE. Homogeneous time-resolved fluorescence resonance energy transfer assay for measurement of Phox/Bem1p (PB1) domain heterodimerization. *J. Biomol. Screen.* 2008;13(5):396–405.
57. Kim K, Wang L, Hwang I. A novel flow cytometric high throughput assay for a systematic study on molecular mechanisms underlying T cell receptor-mediated integrin activation. *PLoS One* 2009;4(6):e6044.
58. Ivnitski-Steele I, Holmes AR, Lamping E, Monk BC, Cannon RD, Sklar LA. Identification of Nile red as a fluorescent substrate of the *Candida albicans* ATP-binding cassette transporters Cdr1p and Cdr2p and the major facilitator superfamily transporter Mdr1p. *Anal. Biochem.* 2009;394(1):87–91.

59. Simons P, Shi M, Foutz T, Lewis J, Buranda T, Lim WK, Neubig R, Garrison J, Prossnitz ER, Sklar LA. Beads coated with epitope tagged G proteins assemble complexes of ligand-receptor–G protein for detection by flow cytometry. *Mol. Pharm.* 2003;64:1227–1238.
60. Simons PC, Young SM, Gibaja V, Lee W-C, Josiah S, Edwards BS, Sklar LA. Duplex, bead-based, competitive assays for inhibitors of protein kinase. *Cytometry* 2007;7:451–459.
61. Saunders MJ, Kim H, Woods TA, Nolan JP, Sklar LA, Edwards BS, Graves SW. Microsphere based protease assays and screening application for lethal factor and factor Xa. *Cytometry A* 2006;69:342–352.
62. Saunders M, Edwards BS, Sklar LA, Graves S.High throughput multiplex flow cytometry screening for botulinum neurotoxin type A light chain protease inhibitors, assays and target development. *Assay Drug Dev. Technol.* 2010:8, 37–46.

5

A MULTIPARAMETER APPROACH TO CELL CYCLE ANALYSIS AS A STANDARD TOOL IN ONCOLOGY DRUG DISCOVERY

Carmen Raventos-Suarez and Byron H. Long

5.1 INTRODUCTION

In the early part of the twentieth century, the drug discovery process focused first on animals to prove a pharmacological effect. However, focus shifted to cells as cell culture became more practical, expedient, and cost-effective and as animal welfare concerns increased [1]. A more recent shift to molecular targets occurred for almost the same reasons, but mainly for cost and high throughput [2]. Another factor impacting this shift was the rapid discovery of new physiological functions afforded by advances in molecular biology. These advances gave the opportunity to screen compounds directly against designated target molecules in a high-throughput mode and as single molecules tested across parallel assays [3]. Cellular and animal models have since then evolved into parallel development as the new animal testing models moved into xenografts and genetically engineered cancer models [4].

Because of these advances, it is almost forgotten that biological macromolecules can have affinities for any small molecule. The role of specific affinities and the possibility of unexpected cross-reactivity for any biological macromolecule require broader focus, especially at the cellular level. In most cases, cross-affinities will be so small that they can afford to be ignored. However, when a compound with a strong affinity for a specific therapeutic target also has an unexpected affinity for another

Flow Cytometry in Drug Discovery and Development, Edited by Virginia Litwin and Philip Marder

macromolecule, this cross-reactivity needs to be identified properly, since it can provide a toxic response or unexpected and unwanted physiological responses.

Living eukaryote cells grown in culture have been studied for years in the context of structure and function as individual entities. Even though these cells are grown *in vitro* and not *in situ*, it is generally assumed that cells maintained in such conditions closely resemble cells in their natural environment. Since any concern about this assumption exceeds the focus of the present chapter, we believe this topic should be left to those better able to address such a controversy. Instead, our focus will be on the use of cells grown in culture as a model to support drug discovery in relation to target validation and how flow cytometry can be a valuable tool in this validation process.

Flow cytometry is a technology that creates a dynamic approach for analyzing cell population distributions in samples with mixed populations of cells or analyzing cells with mixed responses obtained upon sequential evaluations along specific time points of a reaction. The expanded use and validation of these kinds of studies will allow cells in culture to play an increasingly important role, especially in the field of cancer chemotherapy drug discovery.

Recently, Nolan questioned the validity of drug screening that utilized only one parameter to evaluate an effect [5]. His conclusions are based on the common practice of using several assays to study only one parameter at a time with each assay. However, several parameters can be simultaneously measured at the level of single cells when using flow cytometry. The single-parameter experimental design does not allow an understanding of the impact that some specific activities occurring in a subpopulation may produce as a network cascade on overall results. His approach using flow cytometry demonstrates not only the value of the distribution response on several parameters measured simultaneously per cell but also how these different components can serve as a predictable tool for disease progression when primary cells are used.

This chapter will focus on the multiparameter approach and the advantages that knowledge of cell–population distributions can provide when addressing cell cycle profiles in cell–drug interaction studies in relation to proliferation arrest and apoptosis. This multiparameter approach honors the concept of network interaction, whether this network concept exists between signaling pathways within cells or between subpopulations of cells present in the tumor. This chapter will also focus on how understanding the techniques that illustrate these events can help in the evaluation of experimental compounds and their selection to become drugs approved for clinical development. Examples of individual components in a series of interactions will be addressed.

5.2 CELL CYCLE PROFILE COMPONENTS

5.2.1 The Compound Factor

The first step the pharmaceutical industry generally takes to develop a drug to treat a particular disease or medical condition is to identify a specific macromolecular target against which a screen is designed, and then screen their compound deck for hits against this target. The second step is usually the desire to characterize the resulting hits to identify practical lead candidates that may be chemically tractable or otherwise amenable to analogue synthesis. And the third step is an understanding of the effects these chemical hits may have on cells before embarking on an extensive and expensive analogue campaign. It is especially interesting to characterize up front some of the specific interactions that these compounds may have on cells where the final target is present.

Use of analytical cell cycle profiles in the evaluation of compounds were considered until recently only a secondary step in the selection of compounds and not an appealing screening tool due to the low-throughput attainable by flow cytometers. The rate-limiting step was sample acquisition. Lately, the capacities of the instrumentation have been expanded to include plate collection systems that speed the process considerably. Included with these advancements are some additional systems, which allow today up to 386-well plates to be handled for sample acquisition [6]. This advance demonstrates that change is happening to flow cytometry and more "screening-like" approaches are just around the corner. The limitations now seem to be more related to what information a sample can provide, such as what number of cells and how many parameters per cell does one need to get a reasonably interpretable answer to an experimental question.

5.2.2 The Cell Factor

Obviously, the most important factor to be considered when using flow cytometers to conduct a study of the interactions of compounds or macromolecules with cells is the health of the cells. One key element for healthy cells, and hence reliable results, is that cell culture conditions need to be stringently controlled to limit the variability in the response of cells to drug stimulation. For example, overgrown cultures or cells just out of a stress condition (frozen storage) provide unreliable sources for compound evaluation.

Also, when looking at the effects that compounds may have on cells, the investigator should also pay particular attention to stress conditions that may be induced by a compound or even the carrier used to solubilize the compound during the evaluation process. This concern becomes especially relevant when addressing the advancement of anticancer agents because of their potential to alter specific events occurring during cell cycle traverse. Likewise, some organic solvents, such as dimethyl sulfoxide (DMSO), formamide, and others are capable of inducing cell differentiation at subtoxic doses [7].

5.2.2.1 Sample Processing for Cell Cycle Analysis Samples for cell cycle analysis from cell culture or animal tissues are based on their DNA content and can be processed to either obtain as individual nuclei (this is usually important when studying ploidies) or as whole individual cells that can allow other measurements, such as associated cytoplasmic marker quantifications. Also, when collecting whole

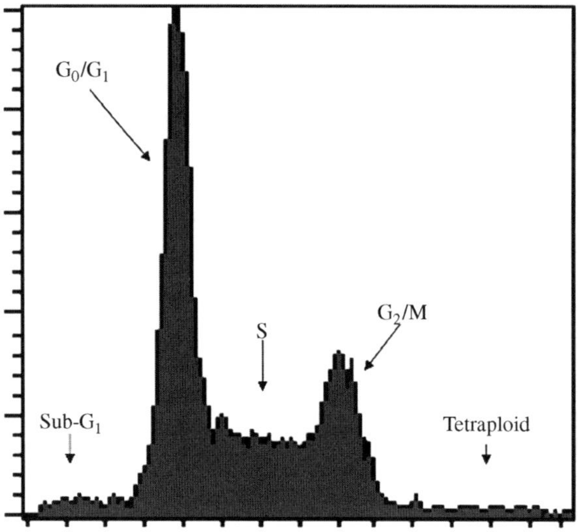

FIGURE 5.1 Standard cell cycle profile histogram.

individual cells, the standard procedures allow preservation of most of cells undergoing apoptosis that otherwise would be lost.

To analyze DNA content in individual cells, as seen in Figure 5.1, we need to start using single-cell suspensions. Unfortunately, most of the procedures for preparing individual cells from tissues or cell culture monolayers have one disadvantage, the presence of a certain number of cells as clumping clusters that need to be discarded in the data processing steps, since they will interfere with the DNA measurements that span only between 2*N* and 4*N* DNA content levels. The advantage of this kind of methodology is that all signals are exclusively associated with individual particles regardless of size and shape and the presence of clusters can be identified by commercially available software, thus giving the investigator the possibility to electronically eliminate clumps of multiple cells from the data pool.

Unfortunately, this crucial simple step is frequently ignored by the experimenter at the time when DNA cell cycle profiles are created during analysis. The cleanup procedure to eliminate these aberrant populations is embedded in the software alternatives as two kinds of gating strategies: the width versus area discrimination and the height versus area discrimination (see Figure 5.2). Despite several opinions in the literature, either approach seems to work with equal efficiency, and it is only a matter to select one approach and stick with it.

*5.2.2.2 **Cell Cycle Analysis Algorithms*** After a single-cell population has been electronically selected, traditional cell cycle analysis algorithms are applied from the included software in analytical packages for flow cytometry. Shapiro mentioned a couple of these programs specially linked to clinical applications [8]. In this chapter

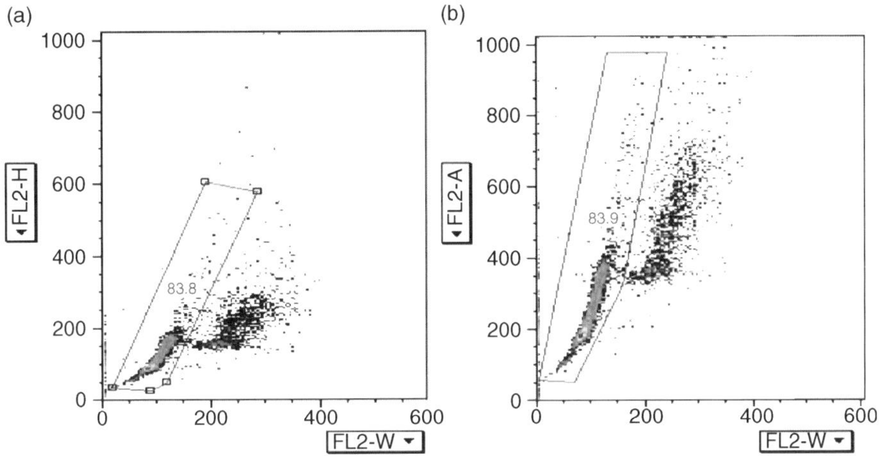

FIGURE 5.2 Gating strategies to remove doublets and other clumping cells. (a) Width versus height and (b) width versus area discrimination in the FACSCalibur (BD).

we will be referring to another one included in Flowjo software. See Section 5.2.3 for details.

5.2.3 The Technologies

When evaluating DNA as a marker for cellular proliferation and cell cycle traverse times (from commitment to divide to the end of DNA synthesis and completion of one division cycle), there are some tools available that provide algorithms to quantify the data in the form of a standard cell cycle profile. Nevertheless, since cells are a dynamic population and their interaction with agents (compounds or biological agents) will result in a variable response inherent to the cell, we need to consider some details. Examples include what information can the standard cell cycle profile provide, what kind of variability can be expected when we go beyond the standard profiles from theory to practical observations (see Figure 5.3), and what can we do to improve our interpretation of the data. For this last idea, the easiest approach is to capture additional data about other biomarkers that can provide more complete information

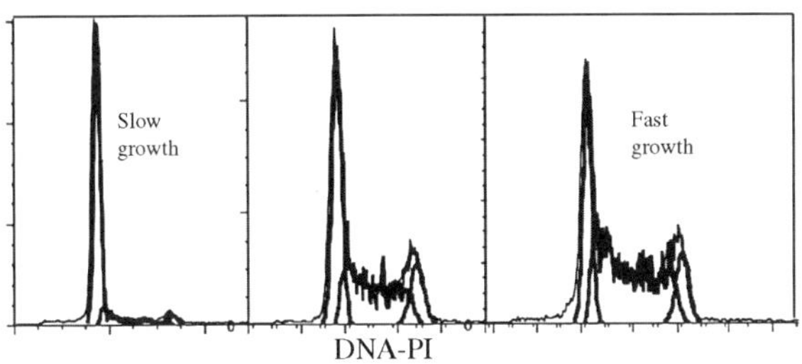

FIGURE 5.3 Variability expected in standard cell cycle profiles of cells growing at different speed rates (doubling times) from left to right with fast growing cycling in less than 24 h.

on other events happening at the sampling time. Some examples include cyclin-dependent proteins and associated phosphoproteins.

Several analytical packages are available to do cell cycle analysis [(Flowjo® (Tree Star) (Tree Star), Modfit® (Verity), Multicycle® (Phoenix Flow Systems)], all of which use more or less modified versions of the Watson or Dean Jett Fox algorithms [9, 10]. Figure 5.4 is a Flowjo analytical example of one algorithm that demonstrates the importance of excluding the clumps defined previously, which otherwise will distort the data.

The results presented in this chapter will be derived using Flowjo package. We are not going to imply that the different algorithms may impact analysis of data except to say that, as with any algorithm, they are based on some assumptions. For example, while the Watson pragmatic is expected to have an overestimation of the S phase, the Dean Jett Fox has an underestimation of the S phase (Figure 5.5). Nevertheless, they both give us tools that only require us to standardize the assays where they need to be applied and chose one over the other, almost as any arbitrary measurement tool would require. Regardless of the choice, it is important to stay with the same algorithm throughout the whole experiment or series of experiments to standardize the system.

5.2.3.1 A Standard Cell Cycle Profile As previously described, a cell cycle profile is a histogram representation of the distribution of DNA content of individual cells in a growing cell population. Figure 5.1 shows that three major subpopulations can be found in a standard DNA profile consisting of two peaks and a valley, as determined by their DNA content. The first peak represents cells at rest (G_0 phase) and/or cells preparing to engage into division (G_1 phase), both of which have the same DNA content ($2N$, where N represents the haploid chromosome number). The second population (the valley, S phase) represents cells that have engaged in DNA replication and their DNA content is higher than the G_0/G_1 peak (higher than $2N$ and lower than $4N$). And the third population includes cells that have doubled the DNA content but not yet engaged in mitosis ($4N$, G_2 phase) and the cells already engaged in the preliminary steps of mitosis ($4N$, M phase).

This overall profile also shows some trailing cells at lower DNA content than G_0/G_1 cells (sub-G_1). The sub-G_1 population contains those cells that had less DNA in the process of apoptosis or are vesicles containing some DNA resulting from a necrotic event. This profile also shows the presence of cells with higher DNA content than the G_2/M peak that identifies a cell population with higher content of DNA per cell than tetraploid cells.

5.2.3.2 Beyond the Standard Profile The standard profile, as defined above, is subject to pronounced variability. This variability can be attributed to the doubling times of specific cell populations where cells that have shorter doubling times (faster division rate) will have a profile different from that of cells with longer doubling times (slower division rates), as shown in Figure 5.3. However, alterations of a specific phase traverse time can also account for this kind of variability.

Another important feature of this distribution is that there is a direct correlation between the frequency distribution of the cell populations at each phase of the cell

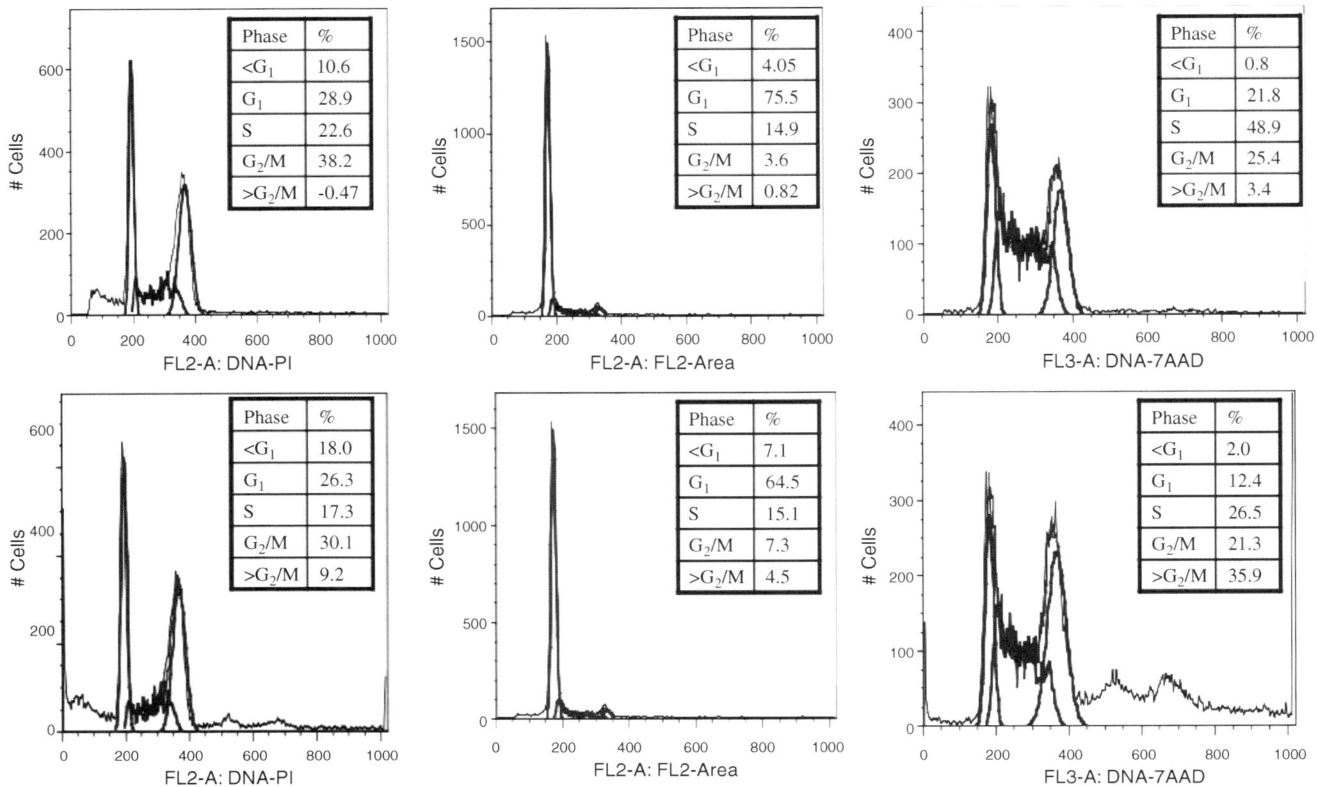

FIGURE 5.4 Cell cycle profiles of cells growing under different conditions and properly evaluated (upper row) compared to alterations induced when doublets and other clumps are not properly eliminated. See table values embedded in the figures.

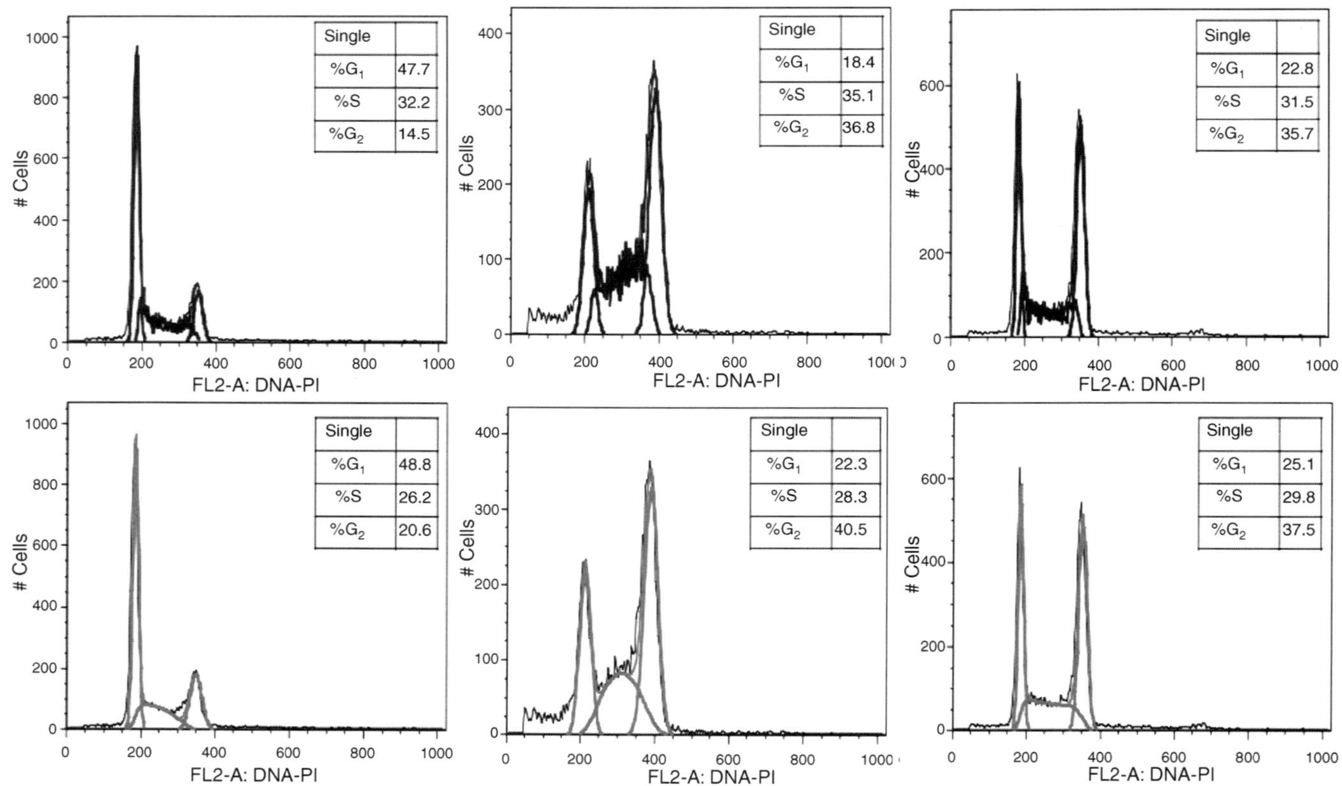

FIGURE 5.5 Watson pragmatic (upper row) and Dean Jett Fox algorithms (lower row) algorithms. See the expected variations between the two algorithms that are due to the mathematical buildup design for each one.

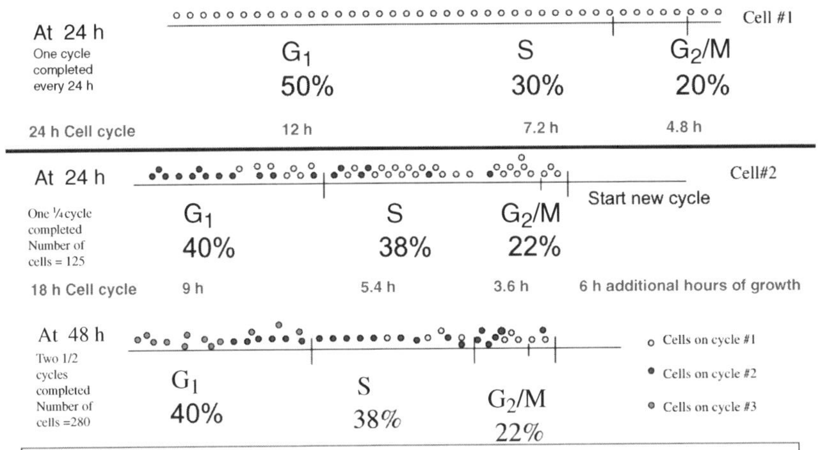

By 72 h
Cells #2 will have completed four cycles, while cell #1 will have completed only three cycles

FIGURE 5.6 Examples of a comparative distribution of cells at the different phases of the cell cycle during a series of doubling times covering a 72 h testing period on two cell lines with different doubling times. (See the color version of the figure in the Color Plates section.)

cycle and the length of time they stay in that specific phase of the cycle. This means that when a cell population shows 60% of cells in the G_0/G_1 phase of the cycle, they also stay 60% of their cycling time at that phase of the cycle.

This link between cycling times and frequency distributions creates another variability that goes beyond the shape of the profile itself and identifies cells that even though are present in one phase of the cycle, in reality are further along in the number of their duplication cycles.

Figure 5.6 is provided to help illustrate this point and shows that on a scheduled experimental sampling sequence, where samples are taken at 24, 48, and 72 h for two cell lines with different doubling times, the cells in the upper row of the figure will have passed through one, two, or three doubling cycles, respectively, for the different samplings. Thus, by 72 h, the cells illustrated in the lower row will have completed four cycles of growth, while the cells in the upper row will had gone only three cycles. This is because the more rapidly dividing cell population will have moved further along the different traverse phases of each cycle since their doubling time is only 18 h. The only way to identify these kinds of movements of a cell population is to add another parameter to these measurements, such as bromodeoxyuridine (BrdU) or EdU that are specifically incorporated into DNA during its duplication time (see below).

5.2.3.3 ***Adding Other Parameters to Identify Movements of Cells Along the Different Phases of Their Cell Cycle Profiles*** There are several different approaches that can be taken using other biomarkers to evaluate cell cycle progression of a growing cell population. We are going to only mention here two examples of biomarkers as main examples. The first example is the addition of analogues of thymidine, such as the derivatives BrdU or EdU [11, 12]. Incorporation of these nucleotides into DNA of cells can be readily detected by FACS using other fluorophore-associated probes. The second example is the evaluation of cyclins, because these proteins participate in the movements of cells along the cycle either in their native forms or as cascades of phosphorylation-dependent events in a series or interactions through regulatory proteins [13, 14].

Identification of Phase Traverse Times: The BrdU/EdU Approach When analyzing cell cycle phase traverse, specific time markers that become incorporated into the DNA during duplication are required. The sole tool used for many years to do this evaluation by flow cytometry has been BrdU, a modified analogue of thymidine (bromodeoxyuridine). BrdU is incorporated in cellular DNA and only cells that have entered S phase or passed through S phase will be susceptible to being stained with an anti-BrdU-specific fluorescent antibody (see Figure 5.7a–c). Only recently has EdU become another analytical tool where the uridine has a different modification [15]. This modification allows the staining of cells that incorporated this marker using a milder technique that avoids the harsh steps required by BrdU. Briefly, we have two methodologies that work in a comparable way to identify DNA synthesis. Both use analogues of Uridine that are incorporated into the DNA with the same efficiency but, the BrdU method requires DNA denaturation in order to expose the BrdU for detection by anti-BrdU antibodies, whereas, with the EdU method, direct staining is possible using a diffusible azide-fluorophore conjugate which binds EdU in the DNA without denaturation. Several algorithms can be applied to the cells containing these analogues to determine the number of cells that have passed through S phase and calculate the traverse times from the beginning of DNA synthesis to the completion of a duplication cycle [16].

Identification of Phase Driver Molecules in the Control of Cell Cycle Traverse Times (Cyclins and Phosphoproteins) Several proteins have been identified that control the movements of cells across the cell cycle, and these are mainly the cyclins and cyclin-dependent kinases [17]. For the purpose of this chapter, we are going to name only a few as characteristic identifiers under special stress conditions when analyzing compounds interfering with the progression of cell proliferation. Figure 5.8 illustrates this point. Cells, as control or after camptothecin treatment, were analyzed for cyclin expression, as previously reported by Traganos et al. [18].

5.3 SPECIFIC ACTIVITIES EVALUATED

With the approaches defined in Section 5.2.3, we now have a set of tools with which to address specific activities that can be followed when compounds interact at the level of a proliferating cell. Thus, these activities can be used to monitor the impact of those compounds on proliferation distributions across the cell cycle. The main advantages

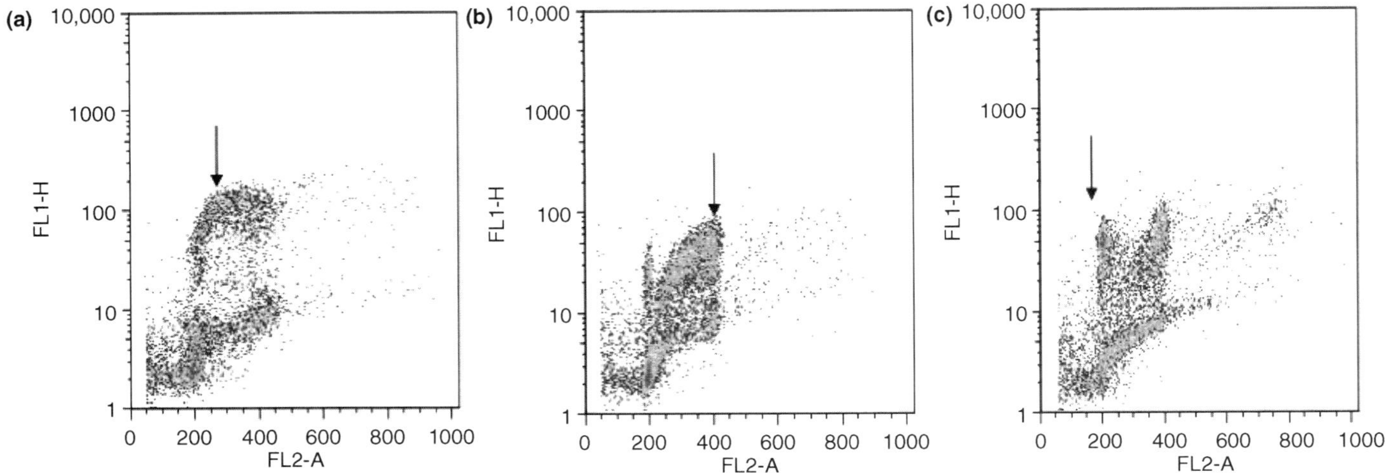

FIGURE 5.7 Cell cycle traverse measured by BrdU uptake. (a) 30 min pulse, all cells in S phase. (b) 5 h after pulse, most cells are getting into G_2/M phase. (c) 8 h after pulse, many cells have already come back to G_0/G_1 phase. See arrows.

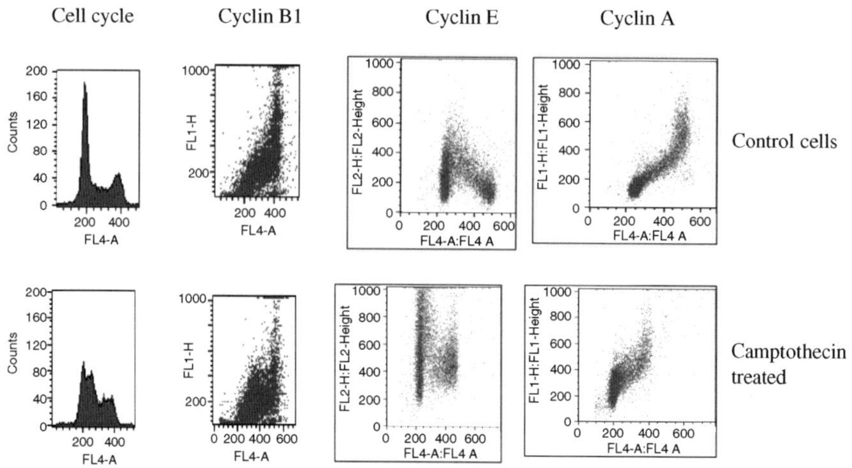

FIGURE 5.8 Cell Cycle and cyclin profiles on cells after Camptothecin treatment.

that these activities provide can be at the level of arrest and/or movement into apoptosis cascades. It is possible to more accurately monitor these activities by including additional parameters capable of being evaluated by flow cytometry.

5.3.1 Arrest

Cell cycle arrest is an activity defined as a pronounced slowdown in traverse times or a stop condition for cells in their movement between one phase and the next during cycle proliferation steps. Arrest can be the end point of a differentiation mechanism, as happens to lymphocytes, which will stay in the G_0 phase of the cycle until specific activation signals can move them into cycling (proliferating) again. Arrest can also be induced by stress or damaging agents, where a slowdown in proliferation will allow cells the time to cope with such a stress condition or repair any damage by activating repair mechanisms that address the insult. Arrest at a specific phase of the cycle can be in place for extensive periods of time, as it usually happens with lymphocytes in G_0 phase of the cycle, or for shorter periods of time, as it happens in most situations of drug-induced arrest at G_2/M phase.

Escape from phase-specific arrest can be the result of either recovery or breakthrough of the block that continues down the path toward death by activation of apoptotic cascades. Arrest is usually evaluated with flow cytometry by looking at the population distribution changes obtained from cell cycle profiles, but these observations provide only a limited interpretation of compounds undergoing evaluation. Simultaneous measurements of other proteins actively controlling the cell cycle traverse times (such as cyclins and certain phosphoproteins) can provide further interpretation of specific mechanisms of action involved in the process when these components are measured simultaneously in cells trapped at specific phase of the cell cycle.

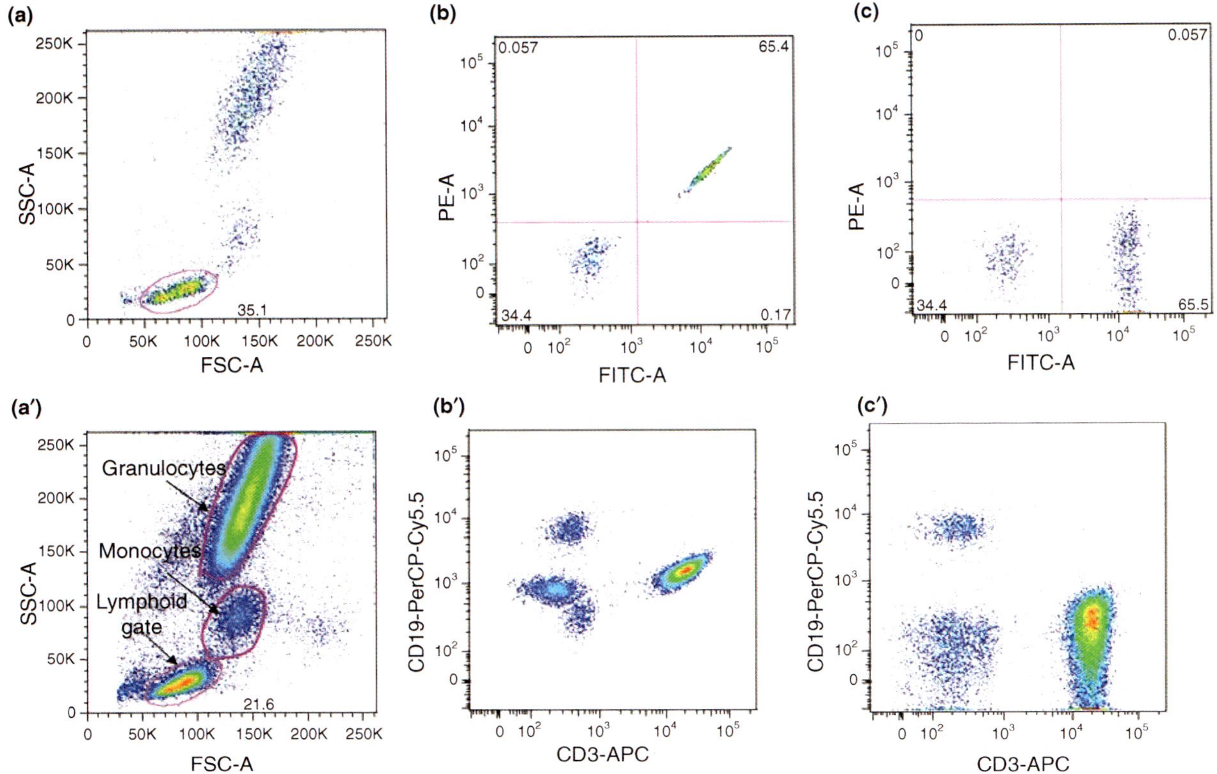

FIGURE 1.5 (See text for details.)

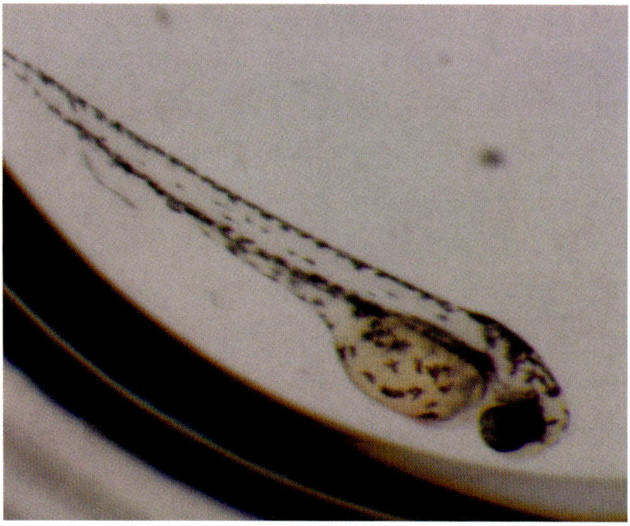

FIGURE 2.1 (See text for details.)

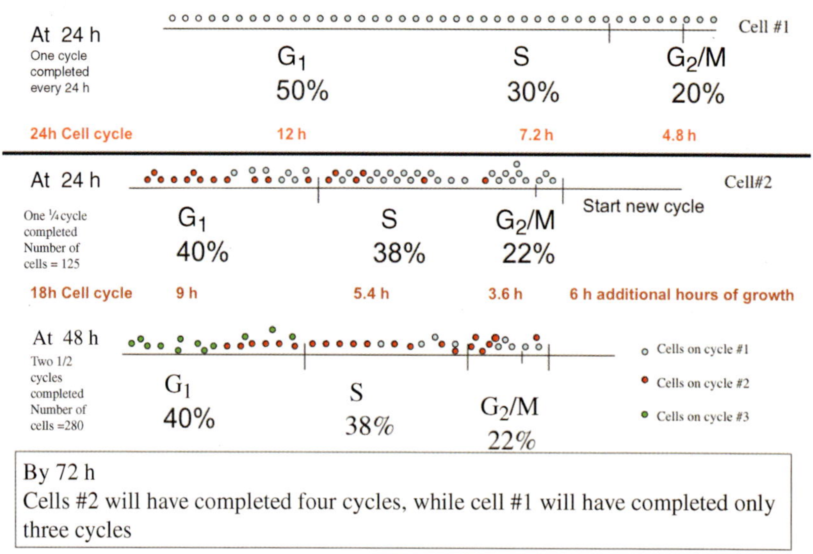

FIGURE 5.6 (See text for details.)

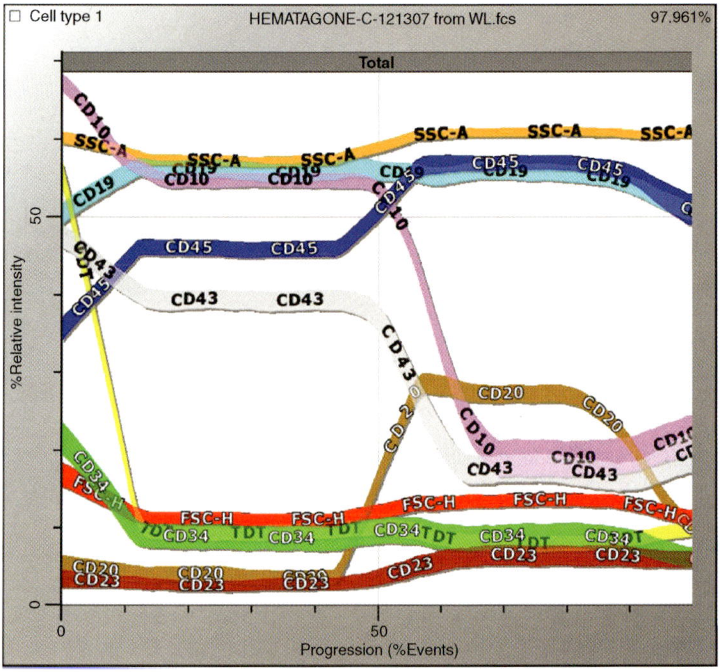

FIGURE 14.11 (See text for details.)

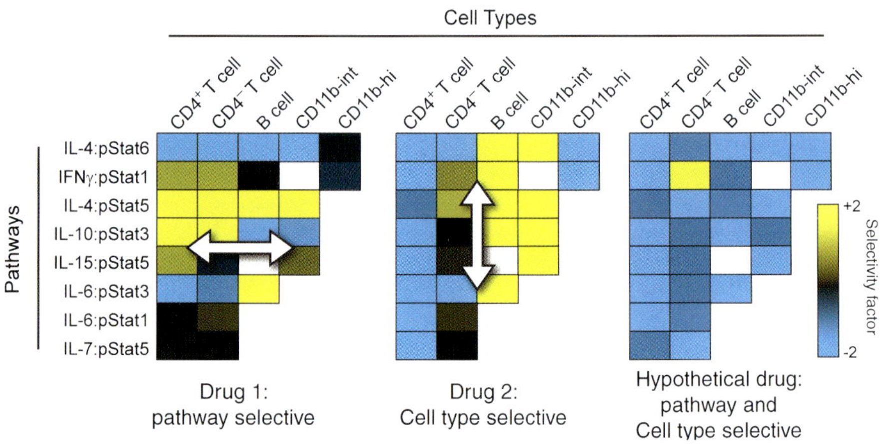

FIGURE 15.5 (See text for details.)

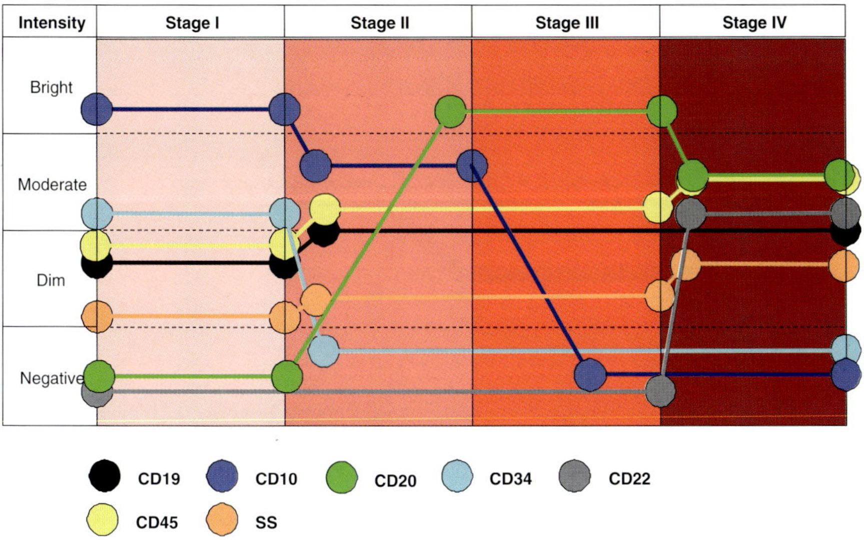

FIGURE 14.2 (See text for details.)

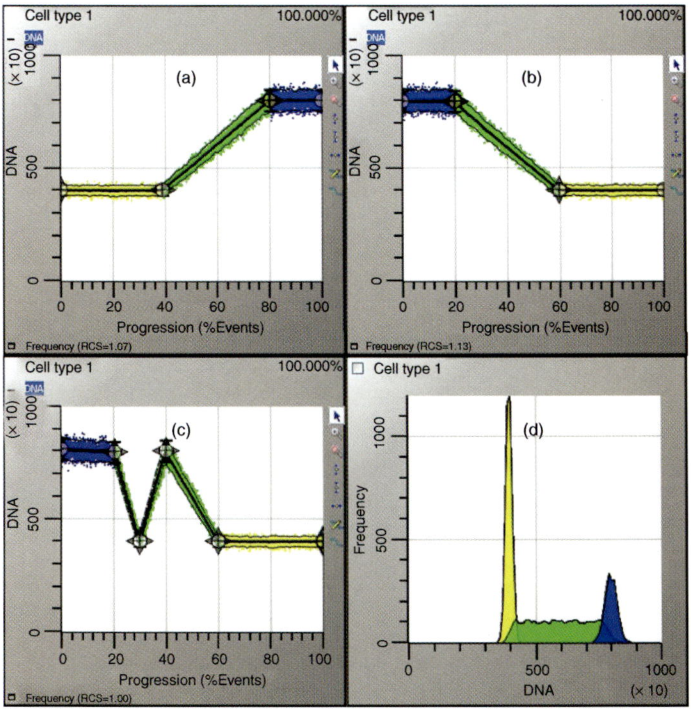

FIGURE 14.8 (See text for details.)

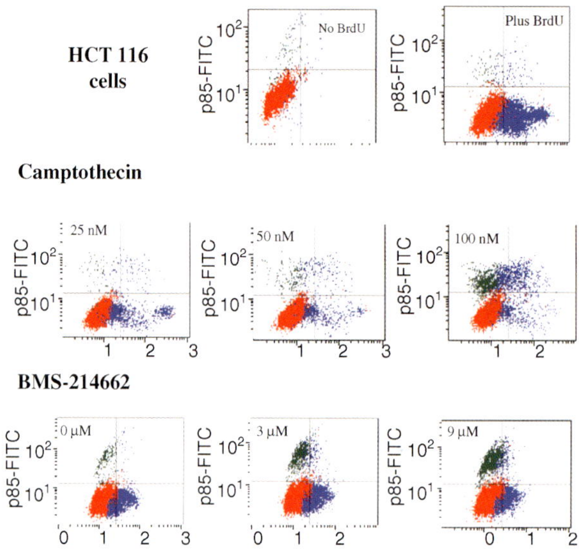

FIGURE 5.13 (See text for details.)

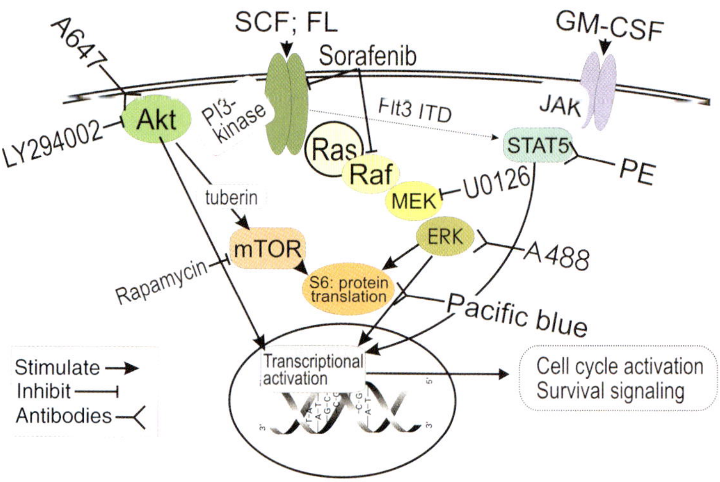

FIGURE 7.3 (See text for details.)

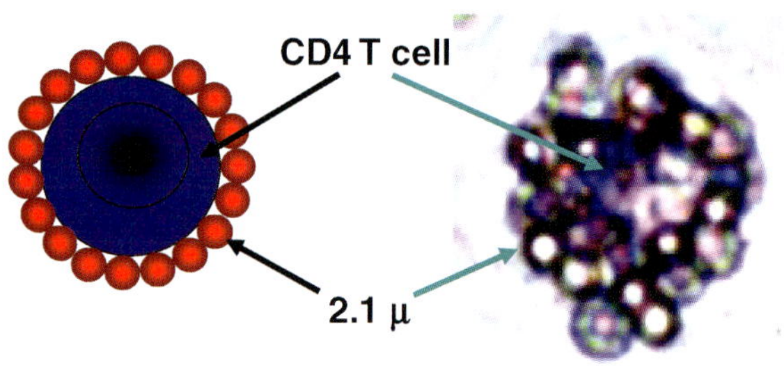

FIGURE 8.3 (See text for details.)

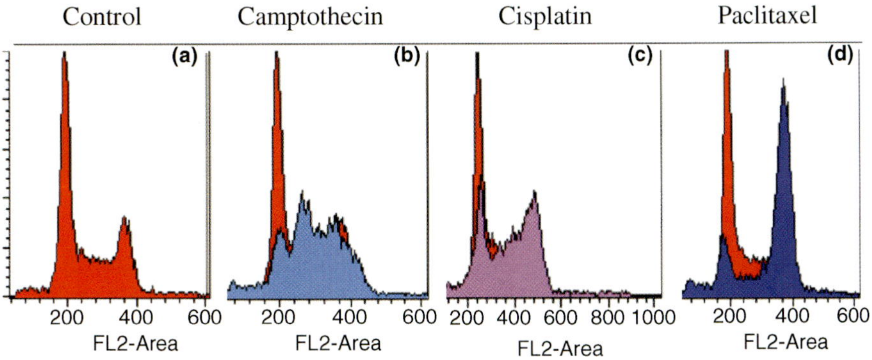

FIGURE 5.9 (See text for details.)

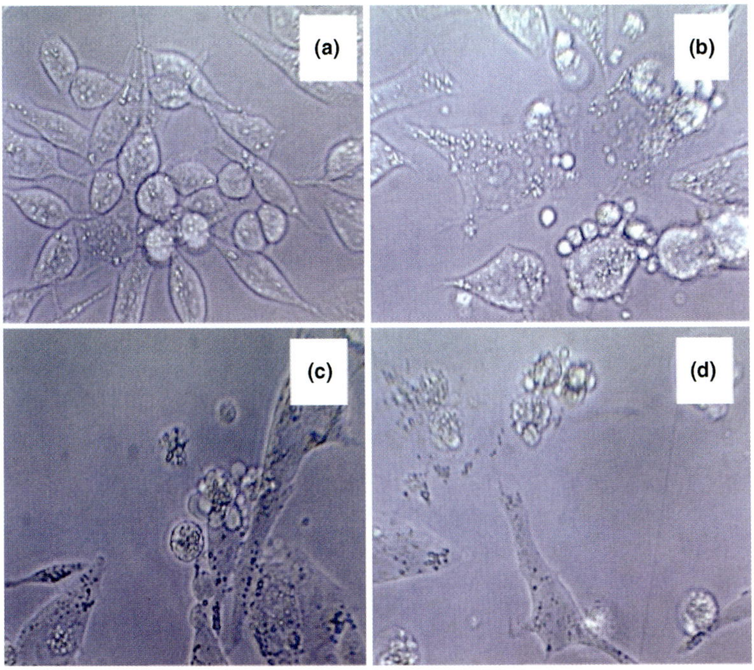

FIGURE 5.10 (See text for details.)

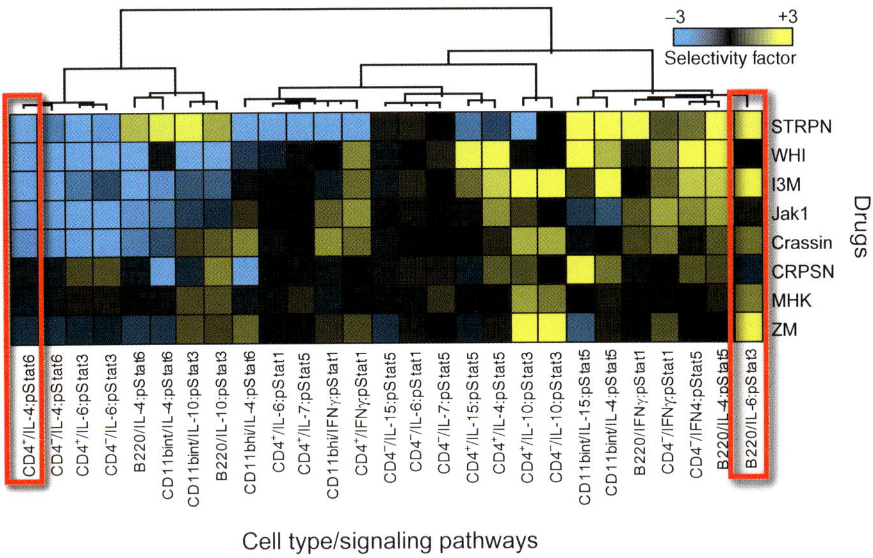

FIGURE 15.6 (See text for details.)

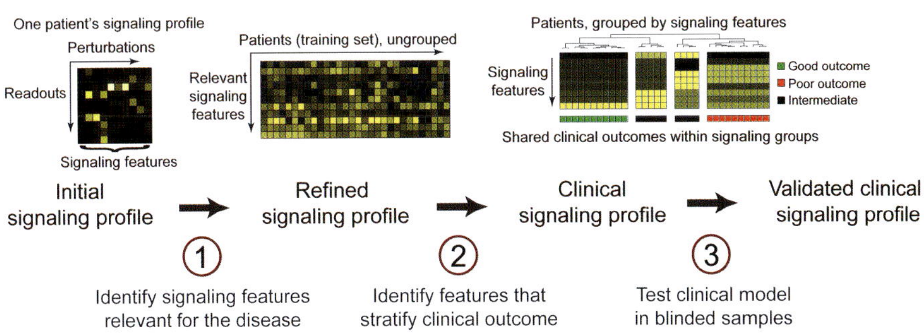

FIGURE 15.7 (See text for details.)

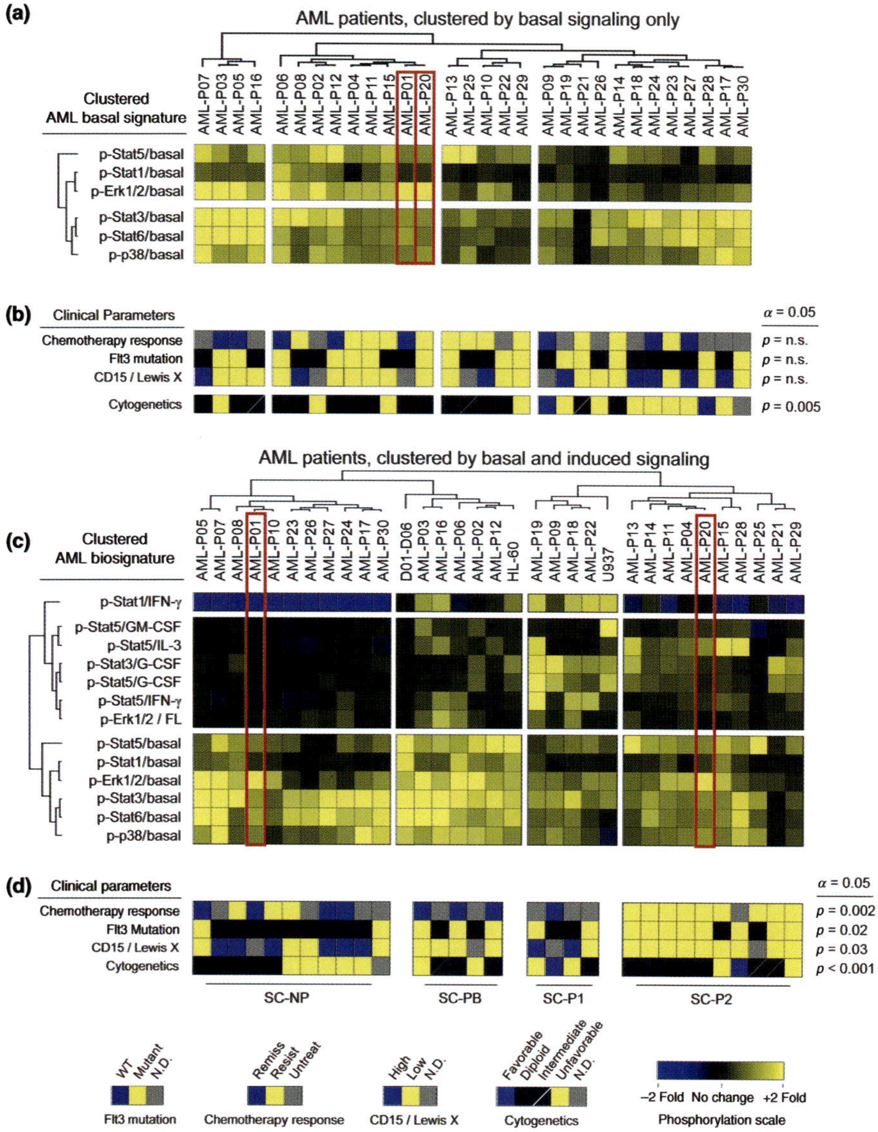

FIGURE 15.8 (See text for details.)

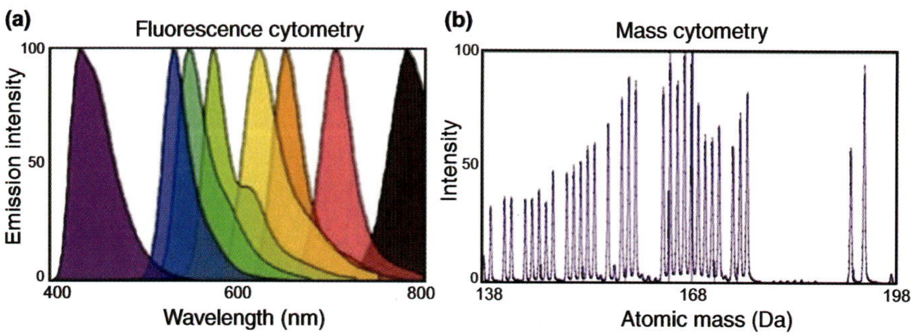

FIGURE 15.9 (See text for details.)

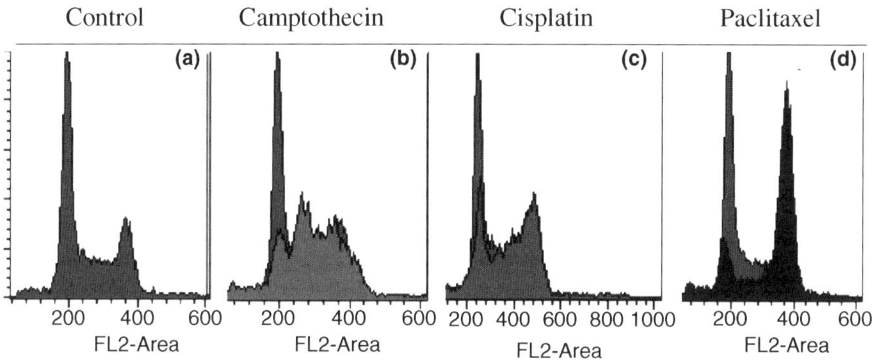

FIGURE 5.9 (a) Control DNA profile on untreated cells compared to overlays of (b) camptothecin; (c) cisplatin and (d) paclitaxel at their optimal arrest concentrations. (See the color version of the figure in the Color Plates section.)

Drugs capable of inducing a specific phase arrest as a response to a known mechanism of action are generally used as models for cell cycle arrest and these cell cycle responses are compared to unknown compounds under evaluation. As shown in Figure 5.9, camptothecin will arrest cells in early S phase, whereas cisplatin will show an effect at late S phase and paclitaxel will specifically impact G_2/M, as previously demonstrated by extensive characterizations done by Traganos et al. [18]. These effects are not as straightforward as seen in the figure. Different movements of the overall distribution of the arrest are dose dependent as well as time dependent, but have a limited range of movement. These dose and time dependencies require that the analysis is done in a titration fashion where the effect of the compound being evaluated can reach enough concentration, not be blocked by repair feedback mechanisms from the cell, or be excluded by other mechanisms involved in the cell detoxification cascades (data not shown).

Following the ideas from the last statement, other considerations need to be incorporated into the experimental design such as the capabilities of the cell to act as a reliable sensor for the specific activities being questioned. Despite the extensive characterization of genome maps to help us understand some of the alterations on functional cascades, the use of established cell lines are still a random approach and preliminary experimentation is required to follow a trend of a response as a proof of principle for each cell line used for a given study (see specific examples in Section 5.4).

5.3.2 Apoptosis

Apoptosis is an active metabolic process that evolved either to eliminate the stressed cells unable to cope with an insult or to eliminate the unneeded cells in the development of multicellular organisms. It may be that apoptosis evolved to serve both functions, and which preceded may be a chicken and the egg argument. Regardless, it may not be appropriate to call apoptosis "a live act of disappearance,"

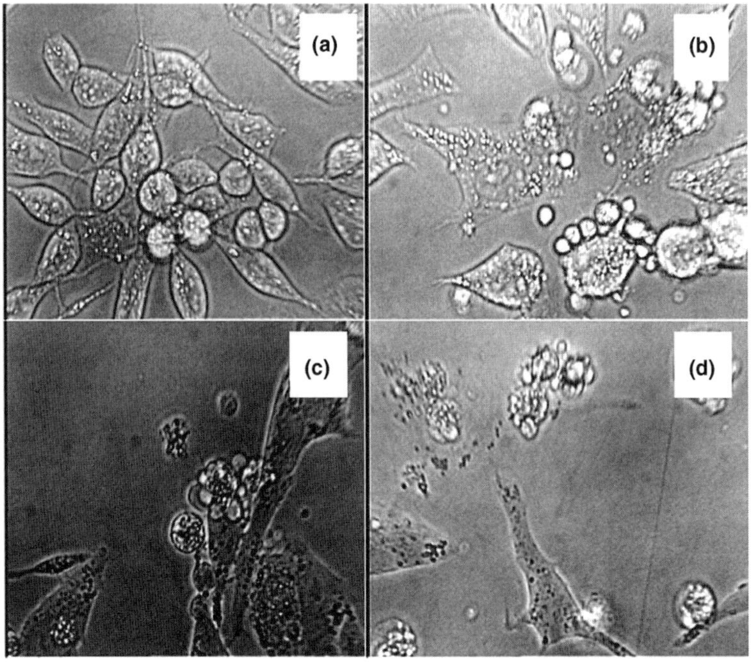

FIGURE 5.10 Control cells (a) and progressive stages of apoptosis (b) to (d). (See the color version of the figure in the Color Plates section.)

but, in reality the cells disappear after a sequence of events that need a metabolically active cell to produce the sequel of events. These events generally include DNA cleavage and formation of small vesicles that will bleb from the surface of the cells until there is no more cell to produce them. These blebs or vesicles will eventually disappear by either secondary necrosis (*in vitro*) (see Figure 5.10) or by being engulfed by phagocytic cells (*in vivo*) [19].

There are several technical approaches available to evaluate the different steps occurring in apoptosis. The initiation step exposes in cells the phosphatidyl serine that can be identified by annexin V signaling and progression can be evaluated by intracellular cytochrome *c* and caspase 3 release, anti-PARP p85 fragment pAb production, and DNA cleavage [20–23]. All are reported to be effective approaches for evaluating apoptosis. Gorczyca et al. also published a comparison of these assays and their efficacy as an evaluation tool [24]. Each technical approach has advantages and limitations due to the very active movements of the population distribution of cells engaging in apoptosis across experimental times. Since cells in culture (or, by the way, also *in vivo*) are never growing in synchronized waves, apoptosis seldom occurs simultaneously in all cells at any given time and while some might have already advanced to bleb into small vesicles or ultimately disappeared under secondary necrosis, other cells in the same culture might have just started to engage in apoptosis. These considerations need to be an integral part of the experimental design for testing compounds to understand as a whole the mechanisms of action taking place in the dismissal of cells by a compound under study.

Here again, the cell line chosen to model a response needs to be considered carefully to provide the best interpretation of screening or further mechanistic data obtained from analytical results. This, together with the previous explanation for the sequence of events described for the apoptosis, can introduce variables that need to be considered carefully. These variables can introduce difficulties and pitfalls in analysis of apoptosis, as pointed out by Darzynkiewicz et al. [25].

5.4 EXPERIMENTAL EXAMPLES

To address the cell cycle profiles as a multiparameter tool in oncology drug discovery, specific models need to be established according to the mechanisms of action in question. Here we present examples for two different scenarios:

1. Compounds inducing a high cell cycle phase-specific arrest (e.g., topoisomerase targeting agents and tubulin targeting agents)
2. Compounds inducing a mild arrest better defined as a slow traverse time across the different phases (e.g., farnesyltransferase inhibitors).

Examples of each of these approaches are presented below.

5.4.1 Topoisomerase Targeting Agents

Topoisomerases are validated targets for cancer chemotherapy and agents capable of interacting with either topoisomerase I or topoisomerase II have demonstrated the potential for increased efficacy in the clinic. Camptothecin analogues that act upon topoisomerase I and the two related analogues etoposide and teniposide that act upon topoisomerase II are clinically active agents with their respective and apparently mutually exclusive mechanisms of action [25–29]. Our earlier efforts to bring a noncamptothecin compounds into the clinic as an anticancer agent lead to the demonstration that topoisomerase I was the target responsible for the cytotoxicity of BMS-250749 [30–33]. We also found that Topo II was the specific target for TAS-103, a proposed topoisomerase I and topoisomerase II dual interactive compound [34, 35].

Both of these compounds were assayed by an experimental design that include the "cell cycle components" and the "specific activities" previously described in this chapter. In addition, each compound was tested on cell line pairs with specific topoisomerase deficiencies. P388/Topo I is a mouse leukemia cell line that is >1000-fold more resistant to camptothecin than parental P388, because it lacks any detectable levels of topoisomerase I [36]. HCT116/Topo II is a human colon carcinoma that is >10-fold more resistant to etoposide than parental HCT116 because it has half-normal levels of topoisomerase II [37]. We also addressed the question of the involvement of p53 in the apoptosis cascade activated either compound by using a human ovarian carcinoma cell line pair containing wild-type and mutant p53 (A2780/wt; A2780/mutp53).

Five drug dilutions were selected for the reference drugs camptothecin and etoposide and the test compound BMS-250749 (0.1×, 1×, 5×, 10×, and 20× the IC_{50} concentrations), which were chosen from their respective IC_{50} values in cytotoxic assays employing MTT as the dye to quantify viable cell number [38]. In addition, two time points at 24 and 48 h were used to cover roughly one cycle or slightly more that two cell cycles for each cell line tested. Figure 5.11a–c demonstrates the efficiency that a multiple compound comparison can achieve across cell lines, dose-dependent treatments, and a couple of time points. Furthermore, Figure 5.11a–c provides a clearer picture of the different dynamics of the response according to the cell background and specificity of the compound. Cell cycle profiles were tested simultaneously with apoptotic cascades evaluation (p85PARP) ([22], data not shown). As a summary, we can say that this experimental design provides the flexibility and speed to see the results at a glance on a screening mode.

The novel quinoline derivative, TAS-103, was reported to interact with topoisomerase I and topoisomerase II in biomolecular assays, but the mechanism of action in mammalian tumor cells had not been fully elucidated. The same kind of experimental approach was taken for TAS-103 as for BMS-250749 above. Specific overall results are shown in Figure 5.12. TAS-103 induced significant apoptosis in P388/Topo I cells at levels consistent with the effects on P388 cells implying that TAS-103 was working by a topoisomerase-I-independent pathway. TAS-103 and VP-16 (etoposide) induced the same cell cycle arrest pattern in HCT116 cells after 24 h. HCT116/Topo II cells, which have lower levels of topoisomerase II, needed 10 times higher concentration of both drugs to show the same cell cycle arrest pattern. Thus, there was no evidence that TAS-103 was circumventing this topoisomerase II resistance effect by acting on topoisomerase I, suggesting only a pure topoisomerase II mechanism of action. TAS-103 induced G_1 arrest and apoptosis in p53 wt cells (A2780wtp53), but only arrested in G_2/M in p53 mutant (A2780mutp53). These three lines of evidence indicate that TAS-103 activity in cells is p53 dependent and mediated by its action on topoisomerase II and not on topoisomerase I [39].

TAS-103 induced hyperchromicity to propidium iodide after 48 h exposure; this effect was absent in VP-16 treated cells, although the arrest pattern was maintained. This hyperchromic effect always occurs at G_2/M arrest, but, unfortunately, there was no explanation for this observation. This was the only difference between TAS-103 and VP-16 and suggests the existence of an additional unknown target that may add or synergize with the response of TAS-103/topoisomerase-II-mediated killing or may contribute to a toxic side reaction. Either possibility will require further studies to identify them. Overall, these studies provide helpful landmarks for the understanding of the activities of these compounds and lower the interest in pursuing TAS-103 for development.

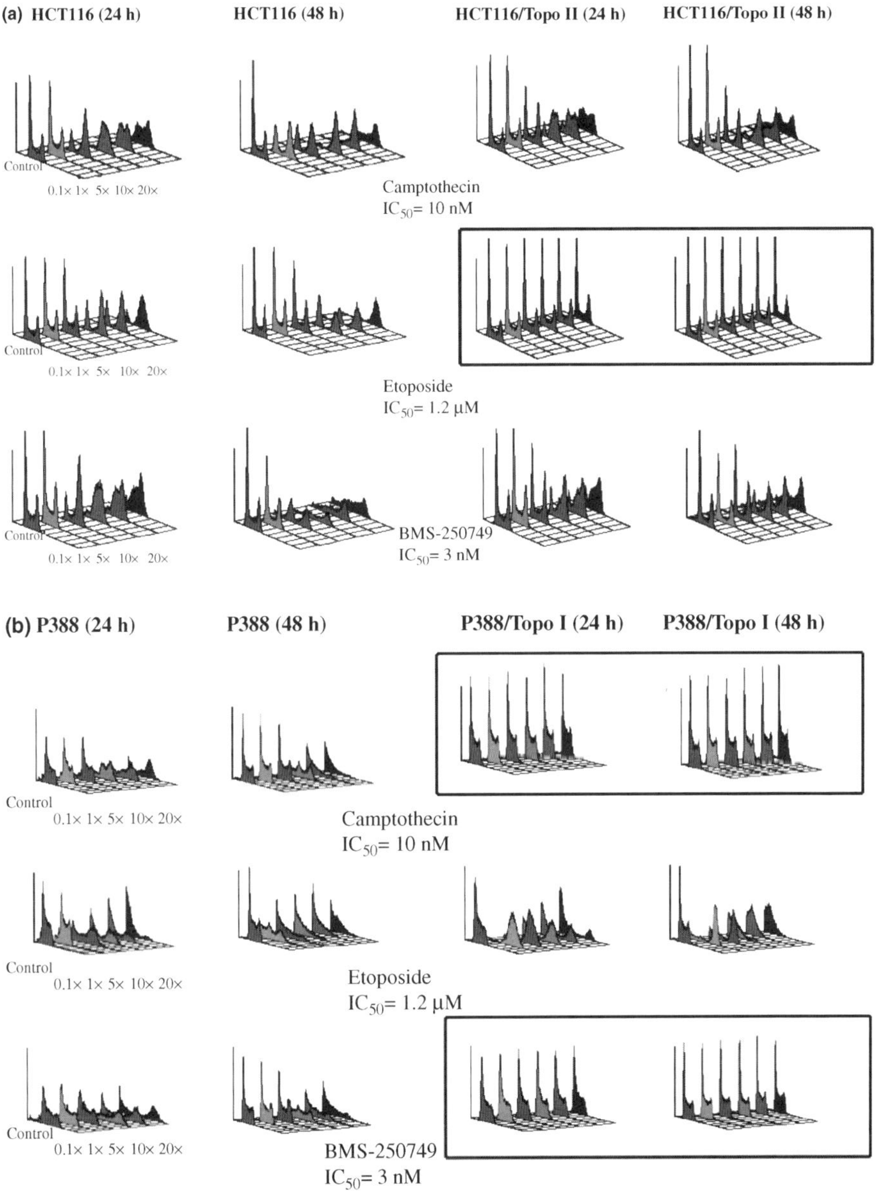

FIGURE 5.11 (a) Cell cycle profiles of paired cell lines deficient of Topo II and the effects of topoisomerase interacting drugs over two time point tests. See the resistance to etoposide of the cell line in the box. (b) Cell cycle profiles of paired cell lines deficient of Topo I and the effects of topoisomerase interacting drugs over two time point tests. See the resistance to camptothecin and BMS-250749 on the drawn boxes. (c) Cell cycle profiles of paired cell lines deficient of p53 remained sensitive to topoisomerase interacting agents.

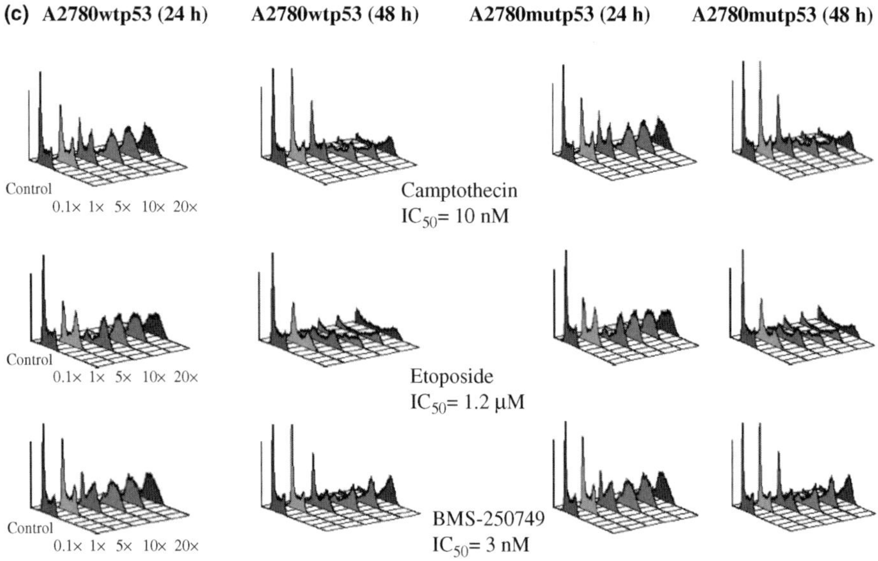

FIGURE 5.11 (*Continued*)

5.4.2 Tubulin Targeting Agents

Tubulin targeting compounds have been very effective in the treatments of cancer (e.g., paclitaxel and docetaxel), but their antitumor activity becomes diminished later in the treatment due to their capacity to induce the expression of multidrug resistance pumps. These pumps actively transport xenobiotic agents out of the cell and impair

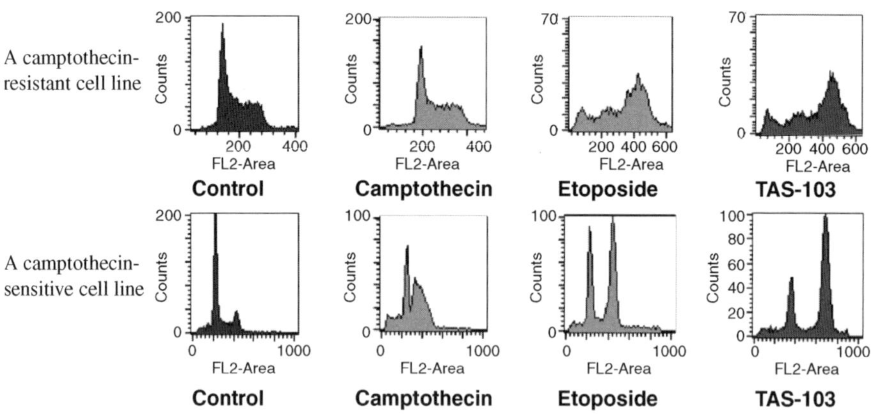

FIGURE 5.12 Comparative display of cells arrested by different topoisomerase interactive drugs. Upper row shows the effects in a Topo-I-deficient cell line and the lower row show the effects in Topo-I-active cell line. TAS-103 has the same profile as the Topo II interacting compound etoposide.

their action [40]. The evaluation of alternative tubulin targeting compounds needs to have an extra focus on their sensitivity to be transported by these pumps; multidrug resistance pumps include a long list of different transmembrane proteins, many of which belong to the ABC cassette transporters (MDR, MRP, BCRP, and also a side population described for stem cells). In addition to the transmembrane pump proteins, tubulin mutations can also cause inactivity of these compounds [41].

Epothilones are compounds that generally lack the ability to interact with transporter pumping proteins. Work in the field of epothilones lead to the discovery of ixabepilone (BMS-247550) as the next generation of microtubule stabilizing agents [42]. BMS-247550 was selected for its superior performance as compared to paclitaxel [43]. In the flow cytometry experiments designed to evaluate this compound against a panel of cell lines presenting multidrug resistance mechanisms, it was demonstrated that this compound was active across all these resistant cell lines. This cell line panel was assembled based upon cell cycle profiles and arrest effects obtained for cell lines expressing MDR activities. These cell lines were assessed for expressing the MDR phenotype by measuring expression of 4E3, as a MDR surface marker (structural detection), as well as quantifying Rh123 transport activity, as another characterization, to corroborate the flow cytometry results evaluating the functional activity on the expressed MDR pump [43]. After the selection of cell lines was done, they were also evaluated in pairs (as shown for the topoisomerase example), where the wild-type strain and a separate strain expressing an alteration or mutation were tested side by side. Some of the pairs used included the HCT116 and HCT116(VM)46 pair previously shown to overexpress MDR [37]. Another cell line pair included A2780 paired in this case with A2780Tax (a cell line with a tubulin mutation). Included in this series was also the Pat-7 (a mammary cell line established from a clinical sample overexpressing the class III β-tubulin) and the H441 lung tumor for evaluation of the ABCG2 pump [44]. Overall, these sets of cell lines demonstrated no disruption of the activity of BMS-247550 in the presence of any of these alterations.

5.4.3 Farnesyltransferase Inhibitors

5.4.3.1 Approaches Utilized for Compound Evaluation Sometimes the evaluation of compounds requires more focus on networking cascades, and this is the case when compounds interact with cells to produce a sequence of events marked by reduction in proliferation but with a very modest cell cycle arrest. This kind of profile is usually accompanied by a high level of apoptosis. In this situation, the cell cycle profile needs to be linked to a time-lapse series for sampling and a number of parameters need to be addressed in the experimental design. We will describe here the evaluation of BMS-214662, which was used as a model where three activities (proliferation, arrest, and apoptosis) were linked in a way that allow to identify the specific sequence of triggering events.

BMS-214662 is a farnesyltransferase inhibitor with high apoptosis induction capabilities [45, 46]. In general, farnesyltransferase inhibitors are not expected to produce any apoptosis in short periods of time, but this compound does. Because of this property, it was necessary to devise an assay that could simultaneously evaluate

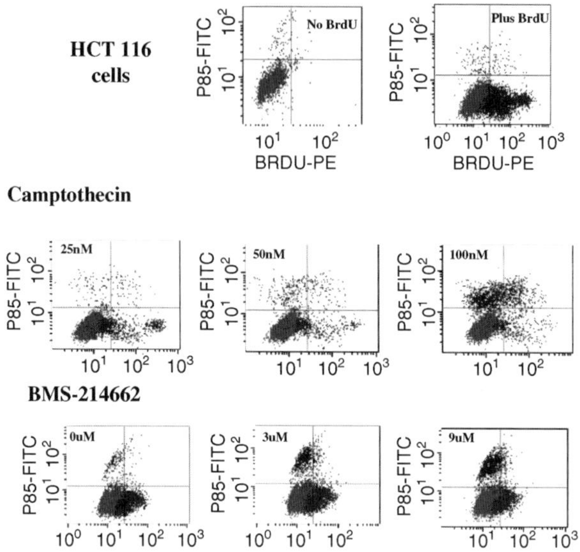

FIGURE 5.13 Apoptotic (green) activity evaluation mapped to proliferating (blue) cells after cisplatin treatments and to quiescent (red) cells after BMS-214662 treatment. (See the color version of the figure in the Color Plates section.)

several different activities on cells. This was done by linking cell cycle profiles (obtained from DNA staining) to proliferating activity measured by incorporation of BrdU and to PARP cleavage. All these activities were measured simultaneously in the same cell along different time points and culture conditions. This experimental design enabled us to demonstrate that those cells dying by the effect of BMS-214662 were at the G_1 phase of the cell cycle (see Figure 5.13) [47]. Furthermore, this observation pointed at the possibility that this drug may target quiescent cells, such as those described as cancer stem cells.

5.5 CONCLUSIONS

The use of a multiparameter approach tailored to be a selective screen, as described in this chapter, provides a broader understanding of the various mechanisms of action exhibited by specific compounds and permits the visualization of multiple effects at a glance in a drug screening approach that can improve the selection of compounds with a greater opportunity for entry into pharmaceutical development.

As this publication goes into print, the status of the compounds used as examples in this chapter is as follows:

BMS-250749 is on hold from further developments due to other priorities.

BMS-214662 is being reevaluated for activities toward cancer stem cells [48].

BMS-247550 is successfully being used in the clinic and is marketed under the name of IXEMPRA.

ACKNOWLEDGMENT

The authors would like to thank Richard Smykla for his invaluable assistance in figure preparation and data transfers.

REFERENCES

1. Alley MC, Scudiero DA, Monks A, Hursey ML, Czerwinski MJ, Fine DL, Abbott BJ, Mayo JG, Shoemaker RH, Boyd MR. Feasibility of drug screening with panels of human tumor cell lines using a microculture tetrazolium assay. *Cancer Res.* 1988;48(3):589–601.
2. Michael S, Auld D, Klumpp C, Jadhav A, Zheng W, Thorne N, Austin CP, Inglese J, Simeonov A. A robotic platform for quantitative high throughput screening. *Assay Drug Dev. Technol.* 2008;6(5):637–657.
3. Pereira DA, Williams JA. Origin and evolution of high throughput screening. *Br. J. Pharmacol.* 2007;152(1):53–61.
4. Suggitt M, Bibby MC. 50 years of preclinical anticancer drug screening: empirical to target-driven. *Clin. Cancer Res.* 2005;11(3):971–981.
5. Nolan GP. What's wrong with drug screening today. *Nat. Chem. Biol.* 2007;3(4):187–191.
6. Ramirez S, Aiken CT, Andrzejewski B, Sklar LA, Edwards BS. High-throughput flow cytometry: validation in microvolume bioassays. *Cytometry A* 2003;53(1):55–65.
7. Bolduc L, Labrecque B, Cordeau M, Blanchette M, Chabot B. Dimethyl sulfoxide affects the selection of splice sites. *J. Biol. Chem.* 2001;276(20):17597–17602.
8. Shapiro H. Using flow cytometers: applications, extensions and alternatives. Practical Flow Cytometry, 4th ed., John Wiley & Sons, Inc., New York, 2003, p. 450.
9. Watson JV, Chambers SH, Smith PJ. A pragmatic approach to the analysis of DNA histograms with a definable G_1 peak. *Cytometry* 1987;8(1):1–8.
10. Fox MH. A model for the computer analysis of synchronous DNA distributions obtained by flow cytometry. *Cytometry* 1980;1(1):71–77.
11. Nüsse M, Jülch M, Geido E, Bruno S, Di Vinci A, Giaretti W, Ruoss K. Flow cytometric detection of mitotic cells using the bromodeoxyuridine/DNA technique in combination with 90 degrees and forward scatter measurements. *Cytometry* 1989;10(3):312–319.
12. Diermeier-Daucher S, Clarke ST, Hill D, Vollmann-Zwerenz A, Bradford JA. Cell type specific applicability of 5-ethynyl-2′-deoxyuridine (EdU) for dynamic proliferation assessment in flow cytometry. *Cytometry A* 2009;75(6):535–546.
13. Juan G, Darzynkiewicz Z. Bivariate analysis of DNA content and expression of cyclin proteins. *Curr. Protoc. Cytom.* 2001; Unit 7.9.
14. Darzynkiewicz Z, Juan G, Traganos F. Cytometry of cell cycle regulatory proteins. *Prog. Cell Cycle Res.* 2003;5:533–542.
15. Buck SB, Bradford J, Gee KR, Agnew BJ, Clarke ST, Salic A. Detection of S-phase cell cycle progression using 5-ethynyl-2′-deoxyuridine incorporation with click chemistry, an alternative to using 5-bromo-2′-deoxyuridine antibodies. *Biotechniques* 2008;44 (7):927–929.
16. Darzynkiewicz Z, Juan G. Analysis of DNA content and BrdU incorporation. *Curr. Protoc. Cytom.* 2001; Unit 7.7.
17. Darzynkiewicz Z, Juan G, Traganos F. Cytometry of cell cycle regulatory proteins. *Prog. Cell Cycle Res.* 2003;5:533–542.

18. Traganos F, Juan G, Darzynkiewicz Z. Cell-cycle analysis of drug-treated cells. *Methods Mol. Biol.* 2001;95:229–240.
19. Wang ZQ, Dudhane A, Orlikowsky T, Clarke K, Li X, Darzynkiewicz Z, Hoffmann MK. CD4 engagement induces Fas antigen-dependent apoptosis of T cells *in vivo*. *Eur. J. Immunol.* 1994;24(7):1549–1552.
20. Vermes I, Haanen C, Steffens-Nakken H, Reutelingsperger C. A novel assay for apoptosis. Flow cytometric detection of phosphatidylserine expression on early apoptotic cells using fluorescein labeled annexin V. *J. Immunol. Methods* 1995;184(1):39–51.
21. Mohr A, Liu J, Meyer LH, Karawajew L, Debatin KM. Identification of deficient mitochondrial signaling in apoptosis resistant leukemia cells by flow cytometric analysis of intracellular cytochrome *c*, caspase-3 and apoptosis. *Apoptosis* 2004;9 (4):457–465.
22. Promega Corporation Anti-PARP p85 fragment pAb Technical bulletin #TB273. Available at www.promega.com.
23. Darzynkiewicz Z, Galkowski D, Zhao H. Analysis of apoptosis by cytometry using TUNEL assay. *Methods* 2008;44(3):250–254.
24. Gorczyca W, Melamed MR, Darzynkiewicz Z. Analysis of apoptosis by flow cytometry. *Methods Mol. Biol.* 1998;91:217–238.
25. Darzynkiewicz Z, Bedner E, Traganos F. Difficulties and pitfalls in analysis of apoptosis. *Methods Cell Biol.* 2001;63:527–546.
26. O'Leary J, Muggia FM. Camptothecins: a review of their development and schedules of administration. *Eur. J. Cancer* 1998;34:1500–1508.
27. Van Maanen JMS, Retèl J, De Vries J, Pinedo HM. Mechanism of action of antitumor drug etoposide: a review. *J. Natl. Cancer Inst.* 1988;80:1526–1533.
28. O' Dwyer PJ, Alonso MT, Leyland-Jones B, Marsoni S. Teniposide: a review of 12 years of experience. *Cancer Treat. Rep.* 1984;68(12):1455–1466.
29. Long BH. Mechanisms of action of teniposide (VM-26) and comparison with etoposide (VP-16). *Semin. Oncol.* 1992; *19* (2 Suppl. 6): 3–19.
30. Long BH, Rose WC, Vyas DM, Matson JA, Forenza S. Discovery of antitumor indolocarbazoles: rebeccamycin, NSC 655649, and fluoroindolocarbazoles. *Curr. Med. Chem. Anticancer Agents* 2002;2:1–10.
31. Saulnier MG, Long BH, Frennesson DB, Ruediger E, Zimmermann K, Eummer JT, St. Laurent DR, Stoffan KM, B. Naidu N, Mahler M, Beaulieu F, Bachand C, Lee FY, Fairchild CR, Linda K. Stadnick LK, William C, Rose WC, Solomon C, Wong H, Martel A, Vyas DM, Kramer R, Langley DR, Balasubramanian BN. Discovery of clinical candidate BMS-250749 with broad spectrum antitumor activity in preclinical tumor models superior to the marketed oncology drug, CPT-11. *J. Med. Chem.* 2005;48(7):2258–2261.
32. Raventos-Suarez C, Class K, Wild R, Menendez A, Long B. *In vitro* specificity profiling of cellular topoisomerase activities using FACS analyses. 2004 ISAC XXII International Congress on Analytical Cytology, Montpellier, France, May 22–27; 2004, Abstract 96207, p. 115.
33. Raventos-Suarez C, Class K, Buczek JL, Balasubramanian N, Long B, Menendez AT. Topoisomerase I is the target responsible for cytotoxicity of BMS-250749: confirmation by flow cytometry. *92nd Annual Meeting, Proceedings of the American Association for Cancer Research*, New Orleans, Mar 24-28, 2001(Abstract 562); *Proc. Am. Assoc. Cancer Res.* 2001;42:103.

34. Hoshi A. Castaner J. TAS-103: antineoplastic topoisomerase I and II inhibitor. *Drugs Future* 1998;23:513–515.

35. Utsugi T, Aoyagi K, Asao T, Okazaki S, Aoyagi Y, Sano M, Wierzba K, Yamada Y. Antitumor activity of a novel quinoline derivative, TAS-103, with inhibitory effects on topoisomerases I and II. *Jpn. J. Cancer Res.* 1997;88:992–1002.

36. Woessner RD, Eng WK, Hofmann GA, Rieman DJ, McCabe FL, Hertzberg RP, Mattern MR, Tan KB, Johnson RK. Camptothecin hyper-resistant P388 cells: drug-dependent reduction in topoisomerase I content. *Oncol. Res.* 1992;4(11–12):481–488.

37. Long BH, Wang L, Lorico A, Wang RC, Brattain MG, Casazza AM. Mechanisms of resistance to etoposide and teniposide in acquired resistant human colon and lung carcinoma cell lines. *Cancer Res.* 1991;51(19):5275–5283.

38. Cory AH, Owen TC, Barltrop JA, Cory, JG. Use of an aqueous soluble tetrazolium/formazan assay for cell growth assays in culture. *Cancer Commun.* 1991;3:207–212.

39. Long BH, Bifano M, Stadnick LK, Fairchild CR, Cornell LC, Raventos-Suarez C, Menendez AT. Cytotoxic mechanism of TAS-103 is related to topoisomerase II mediated DNA cleavage. *91st Annual Meeting, Proceedings of the American Association for Cancer Research,* Mar 24–28, 2000; *Proc. Am. Assoc. Cancer Res.* 2000;41:819.

40. Fojo AT, Menefee M. Microtubule targeting agents: basic mechanisms of multidrug resistance (MDR). *Semin. Oncol.* 2005;32 (6 Suppl. 7): S3–S8.

41. Mozzetti S, Ferlini C, Concolino P, Filippetti F, Raspaglio G, Prislei S, Gallo D, Martinelli E, Ranelletti FO, Ferrandina G, Scambia G. Class III beta-tubulin overexpression is a prominent mechanism of paclitaxel resistance in ovarian cancer patients. *Clin. Cancer Res.* 2005;11(1):298–305.

42. Hunt JT. Discovery of ixabepilone. *Mol. Cancer Ther.* 2009;8(2):275–281.

43. Lee, FYF, Borzilleri R, Fairchild CR, Kim SH, Long BH, Reventos-Suarez C, Vite GD, Rose WC, Kramer RA. BMS-247550: a novel epothilone analog with a mode of action similar to paclitaxel but possessing superior antitumor efficacy. *Clin. Cancer Res.* 2001;7:1429–1437. (*Note*: This paper was published with a typo in the spelling of one name: Reventos instead of Raventos.)

44. Lee FYF, Wen ML, Shen H, She Y, Lenhart R, Covello K, McGlinchey K, Menard K, Wiebeiek A, Raventos-Suarez C, Gan J, Platero S, Kramer R. Ixabepilone overcomes multiple mechanisms of drug resistance including overexpression of class III B tubulin and breast cancer resistance protein. *6th European Breast Cancer Conference (EBCC),* Berlin, Germany, April 2008.

45. Hunt JT, Ding CZ, Batorsky R, Bednarz M, Bhide R, Cho Y, Chong S, Chao S, Gullo-Brown J, Guo P, Kim SH, Lee FY, Leftheris K, Miller A, Mitt T, Patel M, Penhallow BA, Ricca C, Rose WC, Schmidt R, Slusarchyk WA, Vite G, Manne V. Discovery of (R)-7-cyano-2,3,4, 5-tetrahydro-1-(1*H*-imidazol-4-ylmethyl)-3- (phenylmethyl)-4-(2-thienylsulfonyl)-1*H*-1,4-benzodiazepine (BMS-214662), a farnesyltransferase inhibitor with potent preclinical antitumor activity. *J. Med. Chem.* 2000;43(20):3587–3595.

46. Rose WC, Lee FY, Fairchild CR, Lynch M, Monticello T, Kramer RA, Manne V. Preclinical antitumor activity of BMS-214662, a highly apoptotic and novel farnesyltransferase inhibitor. *Cancer Res.* 2001;61(20):7507–7517.

47. Raventos-Suarez C, Class K, Lee F. The pro-apoptotic FT-inhibitor BMS-214662 selectively targets non-proliferating tumor cells. Demonstration by a new flow

cytometric method. *2002 ISAC XXI International Congress on Analytical Cytology,* San Diego, CA, May 2002; *Cytometry* (Suppl.11):72.

48. Copland M, Pellicano F, Richmond L, Allan EK, Hamilton A, Lee FY, Weinmann R, Holyoake TL. BMS-214662 potently induces apoptosis of chronic myeloid leukemia stem and progenitor cells and synergizes with tyrosine kinase inhibitors. *Blood* 2008;111 (5):2843–2853.

6

FLOW CYTOMETRY IN PRECLINICAL TOXICOLOGY/SAFETY ASSESSMENT

David McFarland and Kristi R. Harkins

6.1 INTRODUCTION

6.1.1 Preclinical Toxicology

The primary departmental goal of preclinical toxicology (a.k.a. safety assessment) in drug development is to identify and characterize drug-induced toxicity in test animals by using appropriate animal models. This is accomplished through a large number of individual studies, sometimes referred to as a "tox package," designed to provide specific data on the pharmacology/toxicology of the drug candidate including determination of acute, subacute, and chronic toxicity, carcinogenicity, mutagenicity, teratogenicity, and effects on the reproductive system. Various study designs are employed. Dose ranging and/or dose escalating studies are typically completed early in the safety assessment phase of drug development. This study design is meant to provide maximum tolerated dose (MTD) and no observable (adverse) effect level (NOEL or NOAEL), two important pieces of information that aid in selection of doses for repeat dosing and/or chronic studies. Acute, single-dose experimental designs are also used early in the safety phase to identify organs/tissues that are targets of toxicity. After toxicity is identified, it is important to determine if the effect is reversible given that the animal is allowed to recover for a period of time after administration of the last dose. Reversibility studies may be done separately or a recovery group may be included in a particular study design. It is possible to use varying routes of administration across studies, but it is critical to sufficiently investigate the route intended

Flow Cytometry in Drug Discovery and Development, Edited by Virginia Litwin and Philip Marder

for use in the clinic. In addition to the *in vivo* studies, certain toxicity assessments (e.g., mutagenicity) are carried out *in vitro* using cell culture systems.

The objectives of individual studies are rarely as simple as compiling data that inherently supports/rejects development of the drug candidate. Although some studies can be crafted with such a "go/no go" outcome defined, the goal is more often to better characterize and understand observed toxicities in order to minimize their impact, thus providing a way forward. For example, if a drug designed to treat anemia in cancer patients receiving chemotherapy is oncogenic, that drug series will probably be terminated. However, if an antihistamine causes stomach ulcers, there may be a chance to move forward if this liability can be abrogated in following studies by changing the dosing regimen, formulation, etc.

6.2 FLOW CYTOMETRY AND BIOMARKERS

Toxicology studies include clinical observation, clinical pathology/chemistry end point measurements, and macroscopic/microscopic pathologic examination at and postnecropsy (for terminal studies). Tissues are typically harvested at necropsy and either examined immediately (e.g., blood) or preserved (i.e., chemically fixed and/or frozen) for future examination. Various tissues collected provide many opportunities to apply flow cytometry in preclinical toxicology/safety assessment. Obviously, body fluids such as peripheral blood and urine, routinely collected for clinical pathologic analyses, are amenable to flow cytometric analysis. Perhaps just as obvious, isolation of hematopoietic cells from bone marrow, spleen, and lymph nodes for flow cytometry is simply achieved. In general, as long as a single-cell suspension can be achieved by homogenization and/or digestion protocols, flow cytometric analysis can be performed on any solid tissue as well. Some examples of solid tissues that can be analyzed by flow cytometry include kidney, liver, eye, testis, and pancreas. Even fetal brain and whole embryos can be homogenized and single-cell suspensions prepared with appropriate protocols.

But capability does not equate with applicability; nor does it define benefit. For instance, flow cytometric analysis is superfluous in an MTD study where cell-specific information is not necessary to define dose-limiting toxicity (e.g., gastric ulceration). On the other hand, flow cytometric end points could prove to be extremely useful as correlative data in a dose range-finding study or provide the primary data in cases where a defined toxicity affects hematopoiesis or is immunomodulatory. It is also a useful technique to apply to mechanistic studies in order to identify affected cell types in a heterogeneous tissue or to measure specific effects (e.g., protein expression changes, ion flux, etc.) on select target cells. Given an appropriate analytical end point, one could argue the benefits of including flow cytometric analysis in many study designs. Preclinical toxicology may also be tasked with finding biomarkers of toxicity. It may be possible to move the drug candidate forward if a robust biomarker is available that predicts and/or correlates with pathology, regardless of the type of toxicity observed. For terminal studies in animals, this could be provided by analysis of virtually any tissue that can be obtained at necropsy. But it is highly desirable to develop a biomarker that requires minimal invasive tissue collection in order to provide the opportunity for development as a clinical biomarker as well. Flow cytometric analysis of blood often holds the potential to provide such a biomarker.

6.3 CELL SORTING AND HIGH CONTENT ANALYSIS

Investigative toxicology may also fall under the purview of preclinical toxicology departments as a separate unit or as a portion of responsibility. Investigative studies utilizing flow cytometry can be done *ex vivo* or *in vitro* to answer relatively simple questions of relative toxicity or mechanism. For example, a multiparameter analysis of oxidative stress could be used as a screening tool for multiple compounds in a drug chemical series to aid in candidate selection for *in vivo* experiments. Another example is using a calcium flux analysis on cardiomyocytes to determine if the mechanism of cardiotoxicity is due to perturbation of calcium channels. There are many, many ways to apply flow cytometry to investigative toxicology.

All applications described thus far deal only with simple flow cytometric analysis. But there is further potential for application of cell sorting. If, as described, a single-cell suspension can be achieved for analysis by flow cytometry, then cells of interest can be isolated by cell sorting for further experimentation or analysis. Examples of postsort applications include functional assays, cell culture, gene expression analysis, and imaging of enriched cell populations of interest. Cell sorting with subsequent gene expression analysis can be very powerful and provide data unobtainable in any other manner. An example is gene expression analysis of a rare cell population whose transcriptional profile is buried in the "noise" created by more prevalent cell types. Similarly, enrichment of rare target cells for subsequent culture or functional assays provides an opportunity that is nonexistent when these cells are in a heterogeneous mixture. Even if the cells are not rare, isolation of cells of interest eliminates an inherent flaw in gene expression (i.e., mRNA) analysis of mixed populations. This is due to the fact that an increase in the amount of transcript measured, normalized to some housekeeping gene, is universally interpreted as an increase in transcription, often stated as "upregulation," of a gene of interest. But in an uncharacterized heterogeneous population, this could just as likely be due to an increase in percentage of cells expressing the said gene. Assuming all cells in the starting heterogeneous mixture contain roughly equal amounts of mRNA, enrichment of the cell population of interest to just 90% purity is sufficient for providing a valid transcription signature [1]. A case where this assumption is invalid is isolation of reticulocytes or platelets from whole blood where just a few leukocytes per million target cells can greatly skew results since leukocytes contain a two to three orders of magnitude greater amount of mRNA.

Application of flow cytometric analysis in preclinical toxicity studies provides multiple benefits. Multiparameter immunophenotyping of blood or bone marrow, for example, provides a very rich data set compared to simple hematologic analysis. Instead of simply gaining information about the major hematopoietic populations, data on subpopulations, activation states, proliferation, cell cycle, state of maturation, viability, and so on can be obtained simultaneously. High content analysis such as this can provide increased efficiency and cost reduction since a study may require fewer reagents, animals, and time to completion. Presently available instrumentation and fluorescent reporters make multiparameter (10 + colors) immunophenotyping routine. It is often underappreciated that a single 10-color experiment provides directly correlative data that simply cannot be obtained with two separate five-color panels. In addition, the two-panel approach is inherently inefficient since some antigens are undoubtedly repeated to identify the main populations of interest, hence the savings in resources. There is also the real potential for increased analytical precision. In the case of manual bone marrow analysis, the current "gold standard," it is easy to see how flow cytometry provides increased analytical precision. Basic statistics dictates that the absolute error is larger when counting fewer objects. Manual analysis allows only hundreds of cells to be counted whereas flow cytometry allows from thousands to millions. Flow cytometric cell lineage classification using monoclonal antibodies is also less subjective than manual cytologic methods. A comparison of flow cytometry and manual bone marrow assessment in mice is included in this chapter.

6.4 CHALLENGES

Along with the many opportunities come some challenges. Three major challenges are readily identified. The first is antibody availability. Antibodies raised against murine and human antigens, especially hematopoietic cells, are readily available directly conjugated to a plethora of fluorescent reporters. It is more difficult to find directly conjugated antibodies raised against certain rat, dog, and nonhuman primate antigens. In some cases, antihuman antibodies will cross-react; but this is not always true, not even for nonhuman primate (e.g., some anti-CD3 clones). Antibodies for certain antigens/cell types are not available at all. This makes it difficult, if not impossible, to develop identical panels across species. This can be overcome if you have the time and budget to have antibodies developed, else you must wait until the antibodies become commercially available. The second challenge is related specifically to multiparameter, high-content flow cytometry. Assuming antibodies are available for all antigens of interest, it is fairly easy to develop a three- or four-color panel as most antibody vendors provide three or four conjugation choices. FITC, PE, PECy5, PECy7, APC, and APC-Cy7 as well as AlexaFluor variants with similar fluorescence emissions are common choices. And this could be accomplished with one (488 nm) or two laser (488 nm and ~640 nm) excitation. But building a 10-color panel can prove a little more difficult. It can be more challenging to identify suitable antibody–fluor combinations and may also require additional lasers for excitation. Furthermore, considerable thought should be given to pairing fluors and antigens. "Brighter" flours should be used to label less prevalent antigens. Consideration should also be given to how spectral overlap (a.k.a. spillover) will affect results. The third major challenge is data analysis. A 10-color analysis will likely provide many unique populations, some of them very rare. The wealth of data provided is extremely valuable but does pose a challenge. Manual analysis using multiple two-parameter dot plots is very cumbersome and requires the person performing the data analysis to identify all populations of interest. Although there have been some interesting attempts at automating portions of data analysis [2–5], there is no approach available that is an accepted replacement for a trained cytometrist.

There are also a few lesser challenges that are worth mention. One such challenge is simply getting colleagues in preclinical toxicology to accept flow cytometry-based approaches to analyses that were done using different techniques in the past. Returning to the bone marrow analysis mentioned earlier, where manual evaluation of smears is the commonly accepted practice, it may be difficult to get clinical pathologists and hematologists to endorse a flow cytometric approach. Another challenge could simply be a lack of technical expertise in flow cytometry within the department. An experienced flow cytometrist, especially one with cell-sorter operation experience, is an invaluable asset if flow cytometry is to be a core technique within your organization.

The remainder of this chapter will be devoted to three examples of research performed within the department of safety assessment at a major pharmaceutical company. The first details a multiparameter flow cytometric analysis of oxidative stress for use as an *in vitro* screening tool. The second involves the search for a biomarker for drug-induced vascular injury and will describe how flow cytometry and cell sorting contributed to this effort. The last example describes the development of a flow cytometric method for bone marrow assessment in mice.

6.5 *IN VITRO* OXIDATIVE STRESS ASSAY

Oxidative stress, characterized by disturbances in cellular redox status, is a frequent mechanism by which drugs and chemicals induce toxicity [6]. Therefore, measurement of drug-induced oxidative stress *in vitro* can be useful in assessing toxicity of potential drug candidates. Flow cytometry is the perfect choice for this application since it provides simultaneous data acquisition from multiple parameters allowing analysis of diverse functional changes on an individual cell basis [7]. In addition, a wide variety of fluorescent probes are available for studying pathophysiological changes in cells associated with drug toxicity. The objective of this study was to develop a multiparameter flow cytometric method to assess oxidative stress induction in cultured cells by simultaneously measuring relative intracellular glutathione level, mitochondrial membrane potential, lipid peroxidation, apoptosis, and plasma membrane integrity using fluorescent reporters for these end points.

6.5.1 Methods

6.5.1.1 Cell Lines and Maintenance The promyelocytic leukemic cell line HL-60 was employed as it is commonly used for *in vitro* toxicity assays within the department. HL-60 cells obtained from American Type Culture Collection (ATCC, Gaithersburg, MD), were maintained in RPMI-1640 medium supplemented with 10% heat-inactivated fetal bovine serum, 100 I.U. penicillin,

TABLE 6.1 Multiparameter Oxidative Stress Tool Compounds

Toxicant	Physiological End Point	Concentrations (μM)	Vehicle	Remedy (conc.)
Menadione	Glutathione	0, 10, 30, 100	0.1% DMSO	NAC (1 mM)
FCCP	Mitochondrial membrane potential	0, 0.1, 0.3, 1	0.1% DMSO	Cyclosporin A (20 μM)
Paclitaxel	Apoptosis	0, 10, 30, 100	0.1% DMSO	Z-VAD-FMK (25 μM)
Hydrogen peroxide	Lipid peroxidation	0, 10, 30, 100	Water	Catalase (100 units)

and 100 μg/ml streptomycin at a density of 1×10^6 cells/mL in T150 culture flasks at 37°C in a 5% CO_2 humidified atmosphere. Cells were harvested the day before a planned experiment and replated at a density of 5×10^5 cells/mL.

6.5.1.2 Test Articles and Reporters Table 6.1 defines the model toxicants and remedies evaluated (end point, vehicle, concentrations, and known remedy conditions). Table 6.2 provides information on the fluorescent reporters used (end point, interpretation, concentration, and excitation, and filter collection wavelength information).

6.5.1.3 Experimental Design Figure 6.1 outlines in a flowchart format the following procedure. Approximately, 3×10^7 HL-60 cells were harvested from

TABLE 6.2 Multiparameter Oxidative Stress Fluorescent Reporters

Fluorescent Reporter	Physiological End Point	Interpretation	Concentration	Laser Excitation (nm)	Fluorescence Band-Pass Filter (nm)
5-DAAF	Lipid peroxidation	↓MFI = ↑lipid peroxidation	100 nM	488	530/30
TMRM	Mitochondrial membrane potential	↓MFI = ↓ potential	500 nM	488	575/26
mBCl	Intracellular glutathione	↓MFI = ↓glutathione	40 μM	350	485/22
ASBMS	Loss of plasma membrane integrity	% Positive = % cell death	10 μM	350	530/30
AnnexinV-APC	Phosphatidyl serine exposure	% Positive = % apoptotic	Per vendor's instructions	633	660/20

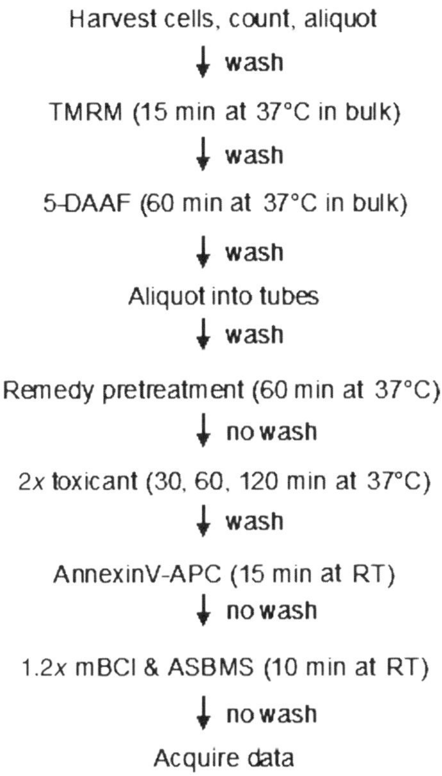

FIGURE 6.1 Multiparameter oxidative stress sample preparation flowchart

culture and an aliquot counted on a hemacytometer to determine cell number and assess viability using trypan blue. An appropriate volume of cell suspension was placed in a 50 mL conical tube and centrifuged at 200 g for 5 min at room temperature (RT). Cells were suspended in a like volume of HBSS and centrifuged again. Cells for single-color flow cytometry compensation controls were removed, treated, and stained separately. TMRM (30 mL of 500 nM in HBSS) was added to cell pellet and vortexed followed by incubation for approximately 15 min at 37°C. Cells were centrifuged as before, resuspended in 30 mL of 5-DAAF (100 nM in HBSS), vortexed, and then incubated for approximately 60 min at 37°C. Sample was removed from incubator, aliquoted among appropriate number of 12 × 75 mm polystyrene tubes, and then centrifuged as before. In some cases, cells were then preincubated with compound appropriate to alleviate the effect of the toxicant to be tested, which will be denoted as remedy compounds. Four toxicant/remedy compound pairs were tested: menadione/NAC, hydrogen peroxide/catalase, FCCP/cyclosporin A, and paclitaxel/Z-VAD-FMK. Cell pellets were suspended in 500 μL of the appropriate remedy compound or HBSS and incubated for approximately 60 min at 37°C. Treatments were added at a 2× concentration and vortexed before proceeding with no centrifugation. Cells were incubated for approximately 30, 60, or 120 min at 37°C and then centrifuged as before. Supernatants were removed and cells were resuspended in 100 μL Annexin V staining solution containing AnV-APC. Samples were incubated at RT in the dark for 15 min, then 500 μL of HBSS containing mBCl and ASBMS was added to each tube and vortexed. After 10 min of incubation at RT, cells were immediately analyzed by flow cytometry (Figure 6.1). Triplicate samples were run for each experiment and an experiment for each compound was run on three separate days.

6.5.1.4 Flow Cytometry The five fluorescent signals (TMRM, 5-DAAF, mBCl, ASBMS, and Annexin V-APC) were collected as outlined in Table 6.2. A minimum of 10,000 events was collected per sample. Median fluorescence intensity (MFI) was recorded for 5-DAAF, TMRM, and mBCl as a measure of lipid peroxidation, mitochondrial membrane potential, and intracellular glutathione content, respectively. Apoptotic cells were enumerated by determining the percentage of events that displayed phosphatidyl serine exposure, yet did not have compromised plasma membranes as determined by positive staining for AnV-APC and dim/negative staining for ASBMS ($AnV^+/ASBMS^-$). Cells with compromised plasma membranes were considered dead/necrotic and were enumerated by determining the percentage of events that stained brightly for ASBMS ($ASBMS^+$).

6.5.2 Results and Discussion

6.5.2.1 Vehicle Effects Vehicle (0.1% DMSO) had a protective effect on relative intracellular glutathione content that increased in magnitude over time. Measured mBCl fluorescence increased to 56.41 ± 9.62 of untreated control at 120 min. No appreciable effect was seen on lipid peroxidation (5-DAAF) or mitochondrial membrane potential (TMRM). Absolute values of both incidence of apoptosis and cell death in vehicle-treated samples and trends of increased incidence over time were similar to those observed in untreated samples.

6.5.2.2 Time Effects All 12 experiments contained untreated control samples. In order to determine effects of time alone, the average change from the 30 min time point was calculated for the 60 and 120 min time points for relative glutathione content, lipid peroxidation, and mitochondrial membrane potential parameters. Relative glutathione content measurement was shown to be very sensitive to time with significant variance from 30 min values at 60 and 120 min. Mean percentage change in mBCl fluorescence from 30 min time point was -40.0 ± 10.1 and -73.8 ± 5.1 at 60 min and 120 min, respectively. Lipid peroxidation and mitochondrial membrane potential measurements were less sensitive over time, but still varied appreciably at 120 min and at 60 and 120 min, respectively. Incidence of apoptosis and cell death both increased slightly over time. Incidence of apoptosis and cell death of less than 5% was considered inconsequential; therefore, only the incidence of cell death at 120 min (9.4 ± 2.3 %) was considered a biologically relevant time-induced stress result. These effects were similar to those seen in cells treated with 0.1% DMSO and therefore were interpreted as time-induced stress effects with little vehicle effect observed.

6.5.2.3 Data Analysis Due to the nature of mechanism of action of the fluorescent reporters 5-DAAF, mBCL, and TMRM, a large interexperimental variation was observed. Therefore, all MFI (median fluorescence) data were normalized as a percentage change from vehicle control and/or percentage change from untreated for comparison of trends among like experiments run on separate days. No normalization of apoptosis and necrosis data was needed since absolute values of percent apoptotic and necrotic can be compared directly. Percentage change from vehicle control and untreated may still be computed easily but is of limited utility. While percentage change analysis may allow a better visualization of trends due to the fact that small absolute changes may translate to fairly large percentage changes, this may also overemphasize a change that is not biologically relevant. For example, an apoptotic index of 2% in treated versus 1% in control is a 100 percent change from control (increase), but reporting this as such misleadingly implies great biological significance. Therefore, it is suggested that only absolute percentages are analyzed and presented for apoptosis and cell death parameters.

Test Pair 1: Menadione and NAC Menadione consistently produced a concentration- and time-dependent decrease in relative intracellular glutathione level, the primary target parameter for this tool compound. Mean percentage change in mBCl fluorescence from control was -84.1 ± 3.5, -89.3 ± 2.9, and -97.7 ± 5.0 at 30, 60 and 120 min, respectively, in cells treated with 100 μM menadione. Lipid peroxidation was evident at 120 min with all concentrations, but inconsistent across experiments. Decreased mitochondrial membrane potential was observed at 30 min with 100 μM; at 60 min with 30 and 100 μM; and at 120 min with 10, 30, and 100 μM. The magnitude of the decrease at 120 min was similar for 10 and 30 μM. A slight increase in mitochondrial membrane potential over untreated control was seen at 60 min with 10 μM similar in magnitude to that seen in the vehicle control. There was no effect on apoptosis. An appreciable increase in cell death was observed with 100 μM at 120 min. NAC greatly abrogated the effects of 100 μM menadione on relative glutathione content. NAC-treated cells lost only 12.7% of mBCl fluorescence at 120 min compared to 97.7 % in cells that did not receive NAC. NAC also greatly reduced the amount of menadione-induced cell death and completely abrogated effects on lipid peroxidation and mitochondrial membrane potential. In fact, measured TMRM and 5-DAAF fluorescences were at levels above untreated.

Test Pair 2: FCCP and Cyclosporin A FCCP consistently produced a concentration-dependent decrease in mitochondrial membrane potential, the primary target for this tool compound. This effect was observed at all time points, but the best separation in effect was at 30 min due to the fact that mitochondrial membrane potential decreased over time in the absence of compound treatment. FCCP effects on relative intracellular glutathione content varied depending on concentration and time, but in all cases values were above untreated controls at the corresponding time point. The greatest difference between treated and control was seen at 60 min. FCCP at all concentrations slightly increased 5-DAAF fluorescence, interpreted as decreased lipid peroxidation, at 60 and 120 min with a peak seen at 60 min. A very slight increase in cell death was observed at all time points with all concentrations. However, even vehicle controls demonstrated relatively high levels of cell death in comparison to vehicle control groups in other compounds.

Cyclosporin A did little to prevent mitochondrial depolarization by FCCP in the current study. Oligomycin may be superior in this regard and is recommended as a possible substitute [8]. Cyclosporin A pretreatment resulted in further elevation of relative intracellular glutathione above untreated control. Lipid peroxidation was also further reduced, as determined by 5-DAAF fluorescence values above those of untreated control on average, but with a very large variation. There was no effect on cell death.

Test Pair 3: Hydrogen Peroxide and Catalase A concentration-dependent increase in lipid peroxidation, the primary target for this tool compound, as determined by a decrease in 5-DAAF fluorescence intensity, was observed at all time points with the best separation at 120 min. Hydrogen peroxide caused an increase in relative intracellular glutathione that increased in magnitude over time. A slight increase in mitochondrial membrane potential was observed with 100 μM at 30 min and 10 μM at 120 min. Slight and dramatic decreases in mitochondrial membrane potential were observed at 120 min at concentrations of 30 and 100 μM, respectively. Cell death increased moderately over time in samples treated with 100 μM.

Catalase pretreatment only slightly abrogated hydrogen peroxide-induced lipid peroxidation. Catalase had no additional effect on relative intracellular glutathione content. Reduction of mitochondrial membrane potential was partially negated. Incidence of cell death was decreased slightly.

Test Pair 4: Paclitaxel and Z-VAD-FMK A concentration-dependent increase in apoptosis, the primary end point for this tool compound, was observed at all time points although no concentration yielded an increase in percent apoptotic over time. Paclitaxel increased mean relative intracellular glutathione at 60 min in a concentration-dependent manner, but with a great variation in magnitude of response at 100 μM. Lipid peroxidation was slightly lower (higher 5-DAAF MFI) at 120 min at concentrations of 10 and 30 μM. Mitochondrial membrane potential was reduced at all time points by 100 μM paclitaxel. A concentration-dependent decrease was observed at 120 min. Cell death was slightly elevated at all time points with a concentration-dependent effect observed at 60 min, but of biologic significance only at 120 min.

Z-VAD-FMK did not appreciably reduce the incidence of apoptosis. Addition of Z-VAD-FMK increased mean relative intracellular glutathione content more than paclitaxel alone. Z-VAD-FMK had no effect on lipid peroxidation, mitochondrial membrane potential, or cell death parameters.

Z-VAD-FMK worked poorly as an inhibitor of paclitaxel-induced apoptosis when used as described in this study. Recent literature reports that inhibitors of caspase-3, -6, or -8 partially reduce paclitaxel-induced apoptosis but caspase-10 inhibition totally abrogates the process [9]. Z-VAD-FMK was chosen for this study since it is purported to be a pan-caspase inhibitor, but in hindsight it appears that a caspase-10-specific inhibitor, such as Z-AEVD-FMK, may better attenuate the effect of paclitaxel [10]. Z-AEVD-FMK or a combination of Z-AEVD-FMK and Z-VAD-FMK as well as a higher concentration of inhibitor (100 μM) is suggested.

6.5.2.4 Assay Design Observations The simultaneous use of ASBMS and mBCl requires significant fluorescence compensation due to a large amount of spectral overlap. Proper compensation using appropriate single-color controls is paramount to achieve accurate fluorescence measurements for mBCl. Multiple titration experiments were performed to determine working concentrations of TMRM, 5-DAAF, and all tool compounds, as well as time points of interest that were appropriate for HL-60 cells cultured as described. Changes in culture conditions of HL-60 cells, the use of a different cell line, and/or the use of primary cells would likely require similar titrations to optimize experimental conditions to achieve desired effects from tool compounds. If using to screen compounds of unknown mechanism of toxicity, a wider range of concentrations may be needed to appreciate toxic effects. This assay as described for use with HL-60 cells is suitable only for short-term experiments due to the great stress effects observed over time, even in untreated cells. In addition, it is conceivable that at least some loss of signal of mBCl, 5-DAAF, and TMRM at later time points is due to leaching of respective fluorescent reporters from cells and not due to an actual decrease in intracellular glutathione, increase in lipid peroxidation, and decrease in mitochondrial membrane potential. Therefore, time points of 15, 30, 45, and 60 min are recommended for future studies with HL-60 cells. It is recommended that tool compounds, such as those described herein, be included as positive controls in all experiments applying this assay. Also, due to variation observed with tool compound treatments, multiple samples should be run simultaneously in each experiment and then results from multiple experiments compared to confirm results. This is deemed especially important if investigating toxicants with unknown modes of action and/or toxicants that may not produce the magnitude of change seen with positive control tool compounds.

Concentration-dependent changes in relative intracellular glutathione content (decrease), lipid peroxidation (increase), mitochondrial membrane potential (decrease), and apoptosis (increase) were induced with the tool compounds menadione, hydrogen peroxide, FCCP, and paclitaxel, respectively, in HL-60 cells. In addition, secondary effects on other stress parameters, both negative and positive, were also observed, suggesting cross talk among parameters and/or propagation of the stress effect to other compartments. This strongly supports the use of multiparameter analysis during assessment of cytotoxicity, especially if mechanism is a goal of the experimental design. Effects of time and vehicle (0.1% DMSO) were also observed and should be considered when performing experiments of this type. Multiparameter flow cytometry is deemed valid for reporting the effects of known toxicants on oxidative stress, apoptosis, and cell death parameters simultaneously in HL-60 cells. This method could be applied as a screening tool to investigate the multistress potential and mechanism of toxicity of uncharacterized toxicants on HL-60 cells.

6.6 DEVELOPING A BIOMARKER-BASED ASSAY FOR DRUG-INDUCED VASCULAR DAMAGE

A robust blood-based preclinical toxicity assay requiring only phlebotomy for sample collection that predicts or corresponds to histopathological findings of vascular injury is highly desirable. In 2005, the Expert Working Group on Drug-Induced Vascular Injury, composed of toxicologists and pathologists from the pharmaceutical industry, academia, and the Food and Drug Administration (FDA), stated that "aside from histological methods, the detection, noninvasively, of acute drug-induced vascular injury in animals or humans is not possible due to the lack of specific and sensitive biomarkers of endothelial and/or vascular smooth muscle injury" [11] and this remains the case as of this writing. Historically, vascular injury was monitored by looking for changes in blood pressure (BP) and/or heart rate (HR) as vascular damage was not observed without concomitant changes in these parameters. It was discovered that some newer drug classes could cause vascular injury in animals without measurable changes in BP or HR. Some examples are endothelin-receptor antagonists, dopamine (DA1) agonists, adenosine agonists, and later generation phosphodiesterase 4 inhibitors. This problem is compounded by the relatively low therapeutic index for some of these drugs. There is also a trend toward increasing frequency of histological vascular lesions in preclinical toxicity studies and increasing pressure from regulatory agencies (FDA and others) to identify reliable biomarkers. Given this, a large multiyear effort consisting of many individual studies and multiple species was undertaken with the task of developing a specific, sensitive, noninvasive method for detection of drug-induced vascular injury that predicts or correlates with histopathology in multiple animal species used in preclinical toxicity studies with a potential for adaptation to a clinical setting.

6.6.1 Endothelial Model

A large portion of this effort was given to enumeration of circulating endothelial cells (CECs) and/or circulating endothelial progenitors (CEPs) from rat blood (control and tool compound treated), sorting of cell populations of interest, and subsequent gene expression analysis (i.e., genechip, PCR, etc.). A rat model was chosen since it is commonly used in preclinical drug development toxicity testing. Familiarity within the department with existing drugs that induce mesenteric vasculopathy in rats was also useful in providing tool compounds for method development and proof-of-concept experiments.

CECs were first described decades ago as a possible indicator of drug-induced vascular damage [12, 13], but few publications correlating CECs and drug-induced injury exist. However, there have since been numerous publications describing correlations between CEC numbers and various human disease states, and CEC enumeration is gaining acceptance as a clinical biomarker of vascular disease [14]. Unfortunately, there is no consensus method for enumeration or immunophenotypic definition of human CEC/CEP [15]. In addition, very few publications exist on flow cytometry and CEC/CEP in animals. Lack of consensus methods made the undertaking particularly challenging since so many factors can impact results including euthanasia (e.g., CO_2, isofluorane, etc.), blood collection method (e.g., blood vessel, needle size, and anticoagulant), blood fraction processed (e.g., lysed whole blood, mononuclear fraction, and platelet-rich plasma), enumeration method (e.g., microscopy after centrifugation, microscopy after magnetic enrichment, and flow cytometry), and primary phenotypic identifier (e.g., cellular morphology alone, DiI-Ac-LDL uptake, lectin binding, and various cell surface proteins). Lack of consensus methods also made it difficult to establish a "normal" range for basal CEC levels since basal levels in humans greatly vary among methods employed [15]. Rat CEC gene expression was also unprecedented although microarray analysis had been done with magnetically isolated human CEC [16].

6.6.2 Identifying CEC/CEP Markers

Initial experiments in untreated animals used 1,1′-dioctadecyl-3,3,3′,3′-tetramethylindocarbocyanine perchlorate-acetylated low-density lipoprotein (DiI-Ac-LDL) that has been widely used as endothelial identifier. DiI-Ac-LDL has been used by other investigators in sorting experiments to isolate endothelial cells from explant cultures [17, 18]. This reagent was useful for early method development work but was logistically impractical for biomarker development and cell sorting with subsequent gene expression analysis experiments since it requires a 4 h incubation period to allow uptake. Since DiI-Ac-LDL is ingested through binding to scavenger receptors, it was postulated that a similar cell population could be identified using an antibody against said scavenger receptor. CD36 is a type B scavenger receptor and the rat ortholog, fatty acid translocase (FAT), has been shown to be present on endothelium [19]. Rat blood samples costained with DiI-Ac-LDL and anti-CD36 exhibited ~80% overlap between labels, therefore anti-CD36 antibody replaced DiI-Ac-LDL in all subsequent experiments. However, CD36/scavenger receptor is also present on the cell surface of other cell types in circulation including macrophage/monocytes, erythrocytes, and platelets. Initially, only monocytic lineage cells were excluded using anti-CD11b/c since ammonium chloride lysis of RBCs was performed and platelets are easily excluded based on size. In hindsight, given the sheer number of erythrocytes in blood, the fact that lysis is not 100% effective, reticulocyte resistance to lysis and very low number of CECs, it should also have been obvious that erythrocyte lineage would comprise a large percentage of remaining $CD11b/c^-/CD36^+$ cells. Negatively selecting for erythrocyte lineage was problematic because the only antirat erythroid antibody available at the time was an IgM isotype. This reagent led to agglutination and severe clump formation. To circumvent this problem, the cell-permeant DNA binding dye Hoechst 33342 (HO) was used to identify nucleated cells allowing negative selection of both erythroid lineage and (free) platelets. Since it was confirmed that platelets often associated with other cell types, it became necessary to add anti-CD42d to exclude platelet-associated cells from analysis [20]. Anti-CD3 and anti-CD45R were included to negatively select for T and B lymphocytes, respectively. A specific reagent to negatively select for NK cells was not necessary as they are $CD11b/c^+$ in the rat strain (Sprague–Dawley) used. Although some investigators choose to use a pan-leukocyte marker (i.e., CD45) rather than a panel for leukocyte subsets, this is not recommended since there is disagreement in the literature regarding CD45 status on progenitors (hematopoietic and endothelial). Our department was also privy to unpublished observations from another laboratory that observed CD45$^+$ immunohistologic staining of endothelium of select vascular beds. Therefore, the minimum immunophenotypic definition used to identify putative CEC/CEP became CD36$^+$/HO$^+$/CD11b/c$^-$/CD3$^-$/CD45R$^-$/CD42d$^-$/ASBMS$^-$. CD3, CD11b/c, CD45R, and CD42d were combined into a single-color hematopoietic lineage "dump" channel to allow room for other endothelial "specific" antibodies. CD36$^+$/Lin$^-$ will be used for brevity throughout the rest of this chapter to refer to the minimum immunophenotype. Representative flow cytometry data and sorting strategy is described in Figures 6.2 and 6.3.

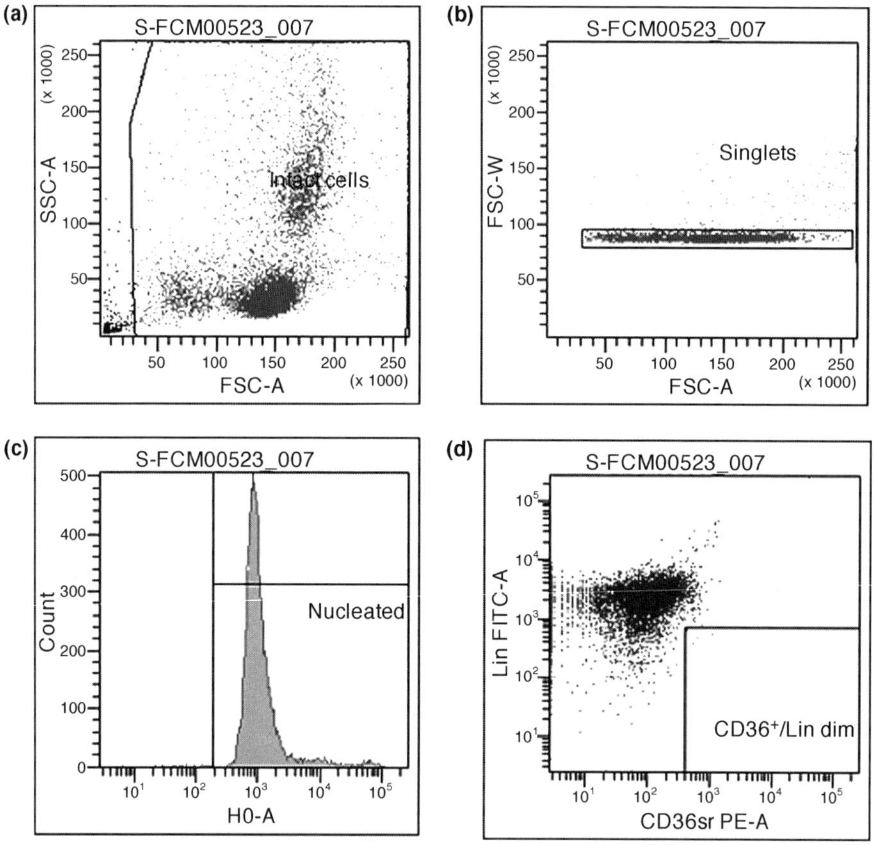

FIGURE 6.2 CEC flow cytometry gating strategy. (a) Debris is excluded from analysis based on low forward and side laser scatter. "Intact cells" are gated to (b). (b) Cell aggregates are excluded from analysis based on forward scatter pulse width. "Singlets" are gated to (c). (c) Only HO$^+$ cells are included in analysis. "Nucleated" cells are gated to (d). (d) CEC gate is tested on CD36$^-$ cells recovered from magnetic separation. This gate should be virtually devoid of events.

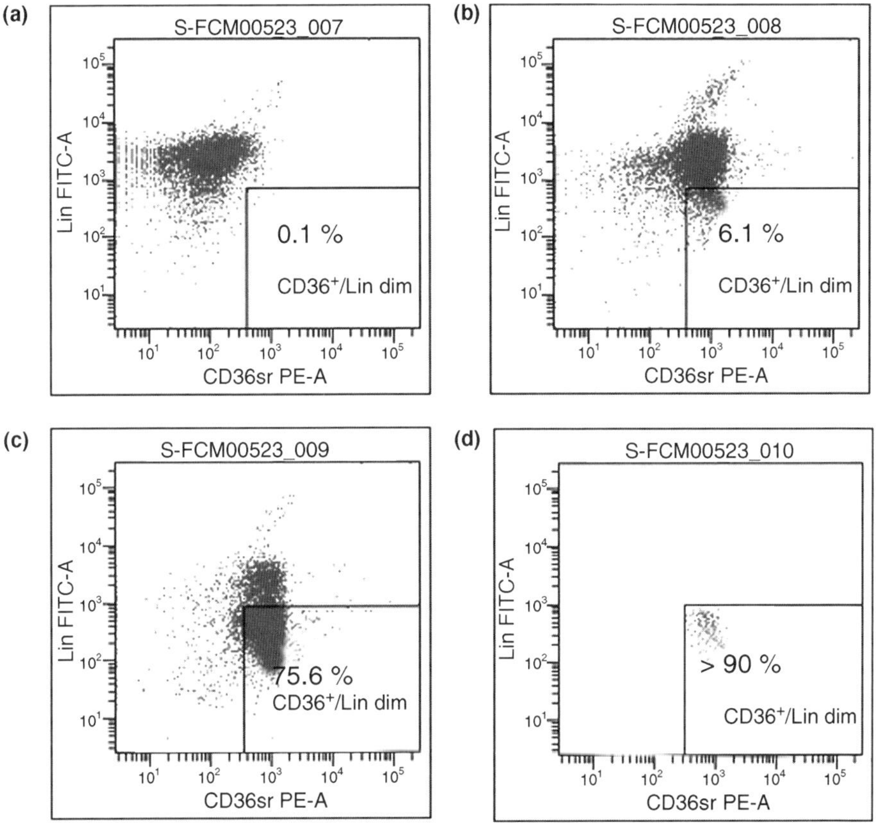

FIGURE 6.3 CEC sorting strategy. (a) Sort gate is applied to CD36$^-$ cells recovered from magnetic separation to confirm the absence of cellular events. (b) CD36$^+$ magnetically enriched cells with sort gate applied. (c) Postsort analysis after an enrichment sort. (d) Postsort analysis after second round of sorting for purity. All samples sorted for subsequent gene expression analysis were of >90% purity.

Other antigens investigated for CEC/CEP identification included CD31 (PECAM-1), CD54 (ICAM-1), CD105 (endoglin), CD106 (VCAM-1), CD133 (AC133), CD144 (VE-Cadherin), CD146 (MCAM), rat endothelial cell antigen (RECA)-1, vascular endothelial growth factor receptor (VEGFR)-1, VEGFR-2, VEGFR-3, and von Willebrand factor (vWF). It is worth mentioning that CD146 is the most widely used primary identifier in human CEC publications. Contrary to claims in early publications, CD146 expression is not specific to CEC but is found on some leukocyte subsets (e.g., activated T-cells) [15] and therefore negative selection of leukocytes is still necessary. It was hoped that some of these antigens (e.g., CD133) might be useful to distinguish CEC from CEP and provide information about activation status (e.g., CD144), but results were equivocal. Within the CD36$^+$/Lin$^-$ population, very few cells were positive for any of the other antigens examined, standard error for percent positive was very large in every dose group analyzed, and good correlation was not observed among any two antigens making robust statistical analysis problematic. In fact, VWF and CD31 staining, for instance, was virtually mutually exclusive. Differential immunophenotyping results led to the hypothesis that rat CECs/CEPs are a complex mixture that may include cells that are in different states of maturation or differentiation, originate from different sites, and/or are of hematopoietic and nonhematopoietic origin. This agrees well with published data describing many differences in growth, morphology, structural organization, protein expression, and function in endothelial cells from various sources throughout the vasculature [18, 21].

6.6.3 Secondary Validation

Despite the nonuniformity in surface protein expression, a cell population identified as $CD36^+$ and/or DiI-Ac-LDL^+ did contain an endothelial-like population as confirmed by gene expression analysis. $CD36^+/Lin^-$ and/or DiI-Ac-LDL^+/Lin^- populations were isolated by cell sorting and gene expression analysis (TaqMan) performed using a small panel of endothelial-related gene probes. This panel included TEK (TIE2, CD202B), MCAM (CD146, MUC18), EDN1 (endothelin1, ET-1), VWF, FLT1 (VEGFR-1), FLK1 (KDR, VEGFR-2), PECAM1 (PECAM-1, CD31), and CDH5 (VE-Cadherin, CD144). Gene expression paralleled flow cytometry results in that not all "endothelial-specific" genes were present and differential expression was observed across animals (control only). Only EDN1 was consistently expressed in sorted CEC/CEP fractions. Utility of probes was confirmed using rat heart homogenate. This supported the hypothesis of a complex mixture of cell types within the $CD36^+$/DiI-Ac-LDL^+ population. As with protein expression, heterogeneity in gene expression in endothelium has been reported and attributed to differences in functional or differentiation status that, in some cases, can be modulated by drug treatment [22, 23]. Heterogeneity in gene expression patterns has also been demonstrated across endothelial subtypes from distinct regions of the heart [24–26].

In addition to confirmatory gene expression experiments, cell culture of normal rat PBMC and sorted $CD36^+Lin^-$ and DiI-Ac-LDL^+/Lin^- cells was performed. First, a modified human colony forming unit (CFU)-Hill assay [27] using rat PBMC was successfully employed to confirm existence of CEP. Longer term culture resulted in disappearance of colonies and emergence of spindle-shaped cells that stained positively for CD31, von Willebrand factor, and rat endothelial antigen (RECA)-1 and negatively for smooth muscle α-actin (SMA), an immunophenotype consistent with mature endothelial cells. CFU-Hill assay confirmed existence of CEP in normal rat blood but required ProNectinF, as opposed to traditional fibronectin, as substrate [28]. The CFU-Hill assay begins with PBMC that are plated for 2 days, then nonadherent cells are removed and replated for the actual colony-forming assessment. This nonadherent fraction was examined by flow cytometry to confirm existence of $CD36^+/Lin^-$ cells. In addition, sorted $CD36^+/Lin^-$ cells were placed in culture directly, treating them as "nonadherent PBMC." Although colonies formed as with PBMC, longer term culture did not produce outgrowth of endothelial-like cells. This was likely due to the absence of other cell types that are necessary for outgrowth in culture. Monocytes and T cells have both been implicated as necessary for endothelial maturation and/or vascularization [29, 30].

6.6.4 Drug Toxicity Experiment

All experiments up to this point were performed in-house using freshly prepared lysed whole blood samples from just a few rats at a time. But the in-life portion of definitive studies, including dosing with tool compounds, were to be performed at remote locations. Each experiment consisted of four (vehicle control, low-, mid-, and high-) dose groups with six rats per group. This required that samples be fixed and/or frozen to survive transport and short-term storage (only a few rat samples could be sorted per day). In initial experiments, paraformaldehyde (PFA) fixation proved incompatible with mRNA analysis, so an alternative preservation method was sought that provided satisfactory flow cytometry results and sufficient amount of recoverable mRNA. In addition, all fixatives and buffers were made with DEPC-treated water. All vessels that came into contact with samples were rinsed with DEPC-treated water before use. The cell sorter was decontaminated with bleach and rinsed with DEPC-treated water before each sort. RNA integrity/yield was compared among sorted and unsorted leukocytes isolated from fresh lysed whole blood and preservation methods under consideration included PFA (4% and 0.2%, repeated with DEPC-treated water formulation), RNAlater™, Cyto-Chex®, "RNA-friendly" fixative (formalin/EtOH/AmOH), formaldehyde/MetOH, 90% EtOH, 90% MeOH, EtOH/acetone, and fresh-frozen in RPMI-10% DMSO. Further attempts to improve integrity included adding RNAsecure™ to PFA, formulating with chloride/citrate buffer instead of PBS, and trial of different commercially available RNA isolation and extraction kits (two each). In the end, only fresh-frozen samples gave both satisfactory flow cytometry and RNA integrity results. Samples could be stored up to 7 months without appreciable impact on immunophenotyping results. There was a slight reduction in RNA integrity/yield after four weeks.

For definitive studies, global gene expression profiling of $CD36^+/Lin^-$ cells by microarray was performed. The goals of this effort were to identify potential protein targets useful for more robust immunophenotypic characterization and to provide information on vascular injury pathogenesis in rats dosed with known vascular toxicants. Doses were chosen based on previous experience and/or literature review. One such study is outlined as follows.

6.6.5 Methods

6.6.5.1 Dosing/Sample Collections Male Sprague–Dawley rats (350–500 g; 6 per dose group) were given vehicle (sterile water), 1, 3, or 10 mg/kg/day serotonin (5HT) subcutaneously (2 mL/kg) for 4 days and blood was collected terminally approximately 24 h after the last dose. Postnecropsy pathologic examination of mesenteric rolls with lesion scoring and statistical analysis confirmed induction of dose-dependent mesenteric arterial vascular injury using this study design.

After carbon dioxide asphyxiation, rats were exsanguinated via the caudal vena cava by collecting blood (approximately 10 mL per rat) into a 10 mL syringe with an 18 gauge needle containing 200 μL of 10% w/v disodium EDTA. Blood was processed immediately; else syringes were rocked at room temperature until processed (<15 min). EDTA-anticoagulated blood was aliquoted, sans needle, into 50 mL conical tubes in 5 mL volumes. Ice-cold ammonium chloride RBC lysis buffer (40 mL/tube) was added and tubes were inverted gently several times to mix. Samples were incubated at RT for approximately 20 min to allow lysis of RBCs, centrifuged at 200 g for 5 min at RT, and supernatants removed, then washed with 20 mL PBS. Pellet was resuspended in 5 mL of 10% DMSO-RPMI, vortexed, and samples were aliquoted (1 mL) into 4.5 mL cryovials and placed on dry ice for transport. Upon arrival (same day) samples were transferred to freezer (−80°C) for storage (up to 7 days).

Enrichment and Cytometry. Cryovials were removed from freezer and thawed in 37°C water bath. PBS (2 ml) was added to each vial and vortexed. Replicate sample vials were combined and washed with PBS before addition of antibody cocktail containing CD3-, CD45R-, CD42d-, and CD11b/c-(Lin)-FITC and CD36-PE. Cells were incubated on ice in the dark for 20 min, washed again, and then resuspended in 2 μg/mL HO in AutoMACS running buffer (AB, 400 μL). Anti-PE microbeads (200 μL) were added to each sample and incubated at 4°C for 15 min with gentle tilting on Speci-Mix. AB (9.4 mL) was added (10 mL total volume) and samples were centrifuged at 300 g for 10 min at 4°C. Supernatants were removed and pellets resuspended in 500 μL AB and then separated using the POSSELDS AutoMACS program for positive selection of rare events. Positive ($CD36^+$ enriched) and negative (lineage) fractions were both collected into 12 × 75 mm polystyrene tubes and processed for cell sorting. Collected fractions were centrifuged at 200 g for 5 min at 4°C. $CD36^+$ enriched fraction was resuspended in 200 μL volume and lineage fraction was resuspended in 1 mL volume of PBS w/1% BSA and placed on wet ice until cell sorting. $CD36^+/Lin^-/HO^+$ cells were sorted and reanalyzed to assess purity (always $\geq$ 90%) and then processed for RNA isolation.

6.6.5.2 GeneChip Analysis Appropriate volume of RNA lysis buffer was added to sort sample and contents transferred to 1.5 mL microfuge tube and vortexed vigorously. Samples were freeze-thawed twice by placing them on dry ice followed by 37°C water bath and then stored at −80°C until RNA isolation using Absolutely RNA Microprep Kit (Stratagene). Samples were concentrated using RNA Clean-up Kit-5 (Zymo Research). Total RNA from each concentrate was quantified using a modified version of RiboGreen RNA Quantification Kit Low Range Assay (Molecular Probes). Quality was assessed by the presence of 18S and 28S ribosomal bands using Agilent 2100 Bioanalyzer and RNA Pico LabChip Kit (Agilent). RNA samples were stored at −80°C until GeneChip analysis.

All samples were randomized into two groups and assigned to two separate scientists for amplification along with one rat liver RNA, serving as a technical control, per group. All samples were amplified and biotin-labeled using the NuGEN

Ovation™ Biotin System. Size and integrity of amplified cDNA (SPIA) products and fragmented and biotin-labeled SPIA product were assessed using the Agilent 2100 BioAnalyzer RNA 6000 Nano LabChip® system. Fragmented and biotin-labeled SPIA yields and purities were determined by measuring O.D.260, O.D.280, and O.D.320 using NanoDrop. Targets containing 2.2 μg of fragmented and biotin-labeled SPIA were prepared for hybridization onto the Affymetrix GeneChip Rat Genome 230 + 2.0 arrays. Samples were hybridized for approximately 18 h at 45°C in a hybridization oven at 60 rpm rotation. Following hybridization, arrays were washed and stained with R-phycoerythrin streptavidin using the Affymetrix EukGE-WS2v4_450_OneFill protocol. Arrays were scanned using a GeneChip® Scanner 3000. Array quality was analyzed by principal component analysis; results indicated zero outliers due to overall array quality.

6.6.6 Results and Discussion

Statistical evaluation of linear dose response was used to identify toxicologically relevant genes, that is, genes whose expression changed linearly with increasing dose and hence increasing severity of vascular injury. For each gene, dose response was evaluated by application of a statistical trend test to expression values to obtain a *P*-value for linear response to increasing dose. Specifically, a one-way ANOVA analysis across dose groupings was followed by calculation of a *t*-statistic using a post hoc contrast for linear trend. A gene-specific reference distribution for the observed *t*-statistic was obtained using 100,000 permutations of the dose labels and recalculating the *t*-statistic for each permutation. A *P*-value for dose response was then obtained by comparing the observed *t*-statistic against the reference distribution. Over 1602 statistically significant dose-responsive genes were identified ($P < 0.01$). This is above the number expected by chance based on a chip of this size (31,099 genes). Representative data from four genes of interest is presented in Figure 6.4. These results demonstrated that gene expression analysis of sorted CECs in preclinical vascular toxicity studies is feasible. Many other tool compounds were investigated using a similar study design providing a wealth of data that is still being explored.

6.7 MOUSE BONE MARROW ASSESSMENT

Bone marrow differentials in preclinical toxicology studies are routinely assessed by manual microscopic evaluation. This method is tedious and labor intensive, with several hundred cells being counted per slide. Evidence suggests that flow cytometric analysis of rodent bone marrow can identify major hematopoietic subsets and reveal hematologic abnormalities [31–34]. Therefore, flow cytometric bone marrow differential analysis was investigated as a replacement to the manual cytologic technique. To this end, a multiparameter flow cytometric panel consisting of antibodies against erythroid (ter119), leukocyte (CD45), and myeloid (CD11b) antigens and the DNA binding dyes Hoechst 33342 (HO) and 7-aminoactinomycin D (7AAD) was devised

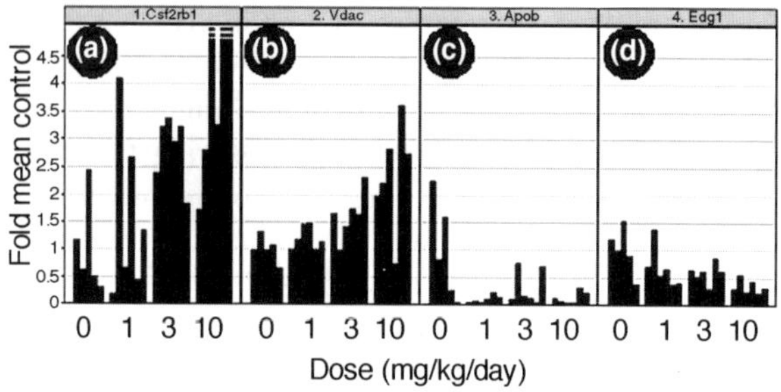

FIGURE 6.4 Rat CEC GeneChip statistical analysis revealed dose-responsive genes. Four representative genes are presented. (a) Colony stimulating factor 2 receptor, $\beta 1$ (Csf2rb1). (b) Voltage-dependent anion channel 1 (Vdac1). (c) Apolipoprotein B (Apob). (d) Endothelium differentiation sphingolipid G-protein-coupled receptor 1 (Edg1). (a) and (b) increased and (c) and (d) decreased in a dose-dependent manner relative to mean control expression levels.

to identify viable nucleated cells of erythroid, myeloid, and lymphoid lineages for comparison with manual cytologic analysis of mouse bone marrow.

6.7.1 Methods

6.7.1.1 Sample Collection Mice were killed by carbon dioxide asphyxiation and exsanguination. Both femurs were removed and dissected free of soft tissues. Right femurs were used for flow cytometric analysis and cytospin preparations and left femurs for bone marrow smear preparation. Previous experiments confirmed no statistically significant difference in bone marrow differentials among femurs in same animals.

6.7.1.2 Cytology Bone marrow smears were prepared using standard methods. Cytospin slides were prepared from collected whole bone marrow and sorted cells. Cells (2×10^5) were washed and resuspended in 250 μL PBS. Cytospins were prepared using Polysine™-coated slides (Erie Scientific Company, Portsmouth, NH) on a cytospin 2 (Shandon, Inc., Pittsburg, PA) and then fixed and stained using an Advia S60 automated slide-staining system (Bayer, Tarrytown, NY). Cells were visualized with a Nikon Eclipse E-800 microscope (Nikon, New York) using a Plan Apo100X oil objective (Nikon) and photomicrographs were taken with a Spot RT Slider digital camera (Diagnostic Instruments, Inc., Sterling Heights, MI). Images were prepared with Photoshop (Adobe Systems, San Jose, CA).

6.7.1.3 Flow Cytometry Myeloid, lymphoid, and nucleated erythroid populations were positively identified using fluorochrome-conjugated monoclonal antibodies. Antibody cocktail consisted of antimouse CD45-APCCy7, ter119-PE, and CD11b-APC. HO (2 μg/ml) and 7AAD (1 μg/ml) were used for nuclear recognition and dead cell exclusion, respectively. First, 7AAD staining was performed (10 min, RT) and then samples were washed, fixed (1 % PFA), and stored refrigerated overnight. Smears and cytospins were evaluated and scored by an experienced hematology technologist.

In order to confirm that manual cytologic characterization was in agreement with flow cytometric immunophenotype for the major lineage populations, nucleated erythroid, myeloid, and lymphoid lineage cells were sorted and reanalyzed to assess purity and cytospins were prepared for cytologic evaluation. Nucleated erythroid cells were identified as $HO^+/CD45^-/ter119^+$ events. Myeloid cells were identified as $HO^+/CD45^+/ter119^-/CD11b^+$ and included both granulocytes and monocytes. Lymphocytes were identified as $HO^+/CD45^+/ter119^-/CD11b^-$ events. 7-AAD$^+$, $ter119^+/CD45^+$ double-positive, and $ter119^-/CD45^-$ double-negative events were excluded as dead, aggregates, and debris, respectively. Sorting experiments revealed that flow cytometric identification of nucleated erythroid, lymphoid, and myeloid bone marrow subsets agreed very well with manual cytologic classification. The cell type sorted agreed with cytologic characterization in all cases and variation in stated purity of sorted populations by flow cytometric reanalysis versus cytologic assessment of cytospins was less than 3% for all populations.

In an effort to further investigate the utility of flow cytometric bone marrow analysis in toxicology studies, recombinant human erythropoietin (rhEpo) was used as a tool compound and administered to mice. A single subcutaneous injection of rhEpo at a dose of 3000 U/kg is sufficient to stimulate erythroblast production in mouse bone marrow with maximal effect observed 48 h postdose [35]. Therefore, bone marrow was collected approximately 48 h after a single injection. A dose of 1000 U/kg was included to investigate dose–response relationships. The effects on mouse bone marrow populations were monitored by flow cytometric analysis and cytologic assessment of bone marrow smears and results were compared.

6.7.2 Results and Discussion

A comparison of flow cytometry and cytology results produced similar trends in all monitored end points. Both methods revealed dose-responsive changes in bone marrow, although more instances of statistical significance were observed in flow cytometry data. This can be attributed to lower intragroup variation present in flow cytometry results versus cytology that is not surprising given that a much larger number of cells are interrogated in flow cytometric analysis. A comparison of methods using least square means was used to statistically confirm this assertion (Figure 6.5). Results strongly suggest greater sensitivity of flow cytometric methodologies to identify drug-induced bone marrow changes as demonstrated by dose-dependent changes seen in nucleated erythroid and lymphoid compartments on a percentage basis, nucleated erythroid on an absolute basis, and M:E (myeloid: erythroid) ratio whereas no dose-dependent changes of statistical significance were observed by cytology.

A trend for flow cytometric erythroid enumeration values to be below those determined cytologically was observed (Figure 6.6). Although this discrepancy in erythroid enumeration appeared striking at first glance, paired t-test comparison of methods revealed statistically significant difference among methods only in absolute

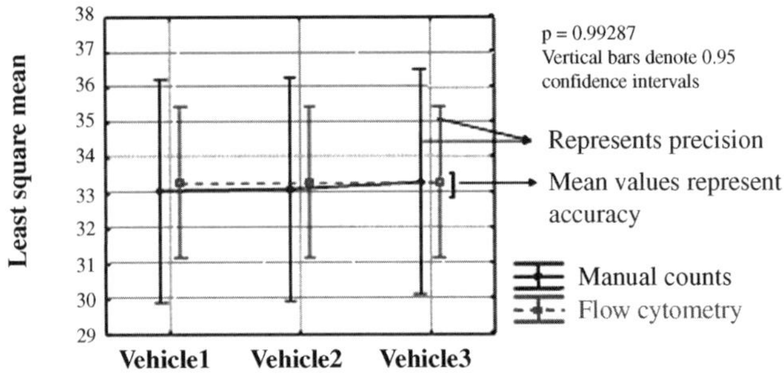

FIGURE 6.5 Statistical comparison of accuracy and precision between manual and flow cytometric bone marrow methods. Least square mean estimation was performed and 95% confidence intervals determined for flow cytometric and manual methods in control animals. Data collected from 15 vehicle control mice including measurements of nucleated erythroid, myeloid, and lymphoid were used for analysis. Precision and accuracy between flow cytometric and manual methods were compared. Mean values with associated errors indicate accuracy and confidence intervals are a representation of precision.

erythroid in the 1000 U/kg dose group. M:E ratio is a commonly used bone marrow assessment. Repeated measure analysis of variance (ANOVA) modeling of M:E ratios actually revealed very good correlation ($R = 0.99$) among methods (Figure 6.7). Regardless, this discrepancy of absolute values was of some concern.

A definite advantage of flow cytometry over manual cytologic assessment of bone marrow cells is less ambiguity in determination of cell types as monoclonal antibodies are employed allowing both positive and negative selection. For

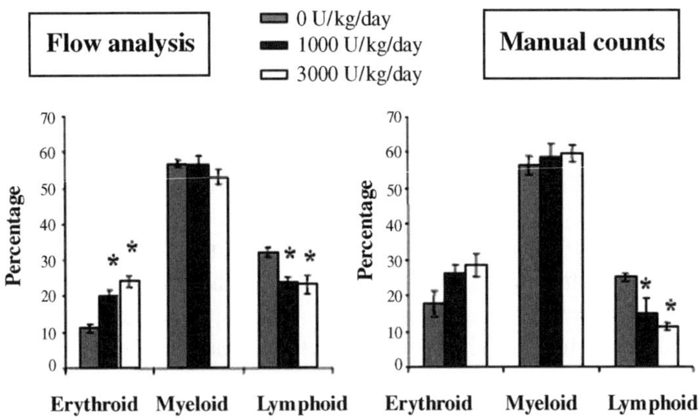

FIGURE 6.6 Comparison of total bone marrow analysis methods in mice dosed with recombinant human erythropoietin. Right femur was used for smear preparation. Smears were stained with Advia S60 automated slide staining system and analyzed manually with standard techniques. Left femur was flushed and prepared for flow cytometry. *Note*: *Denotes significant difference from control ($p \leq 0.05$).

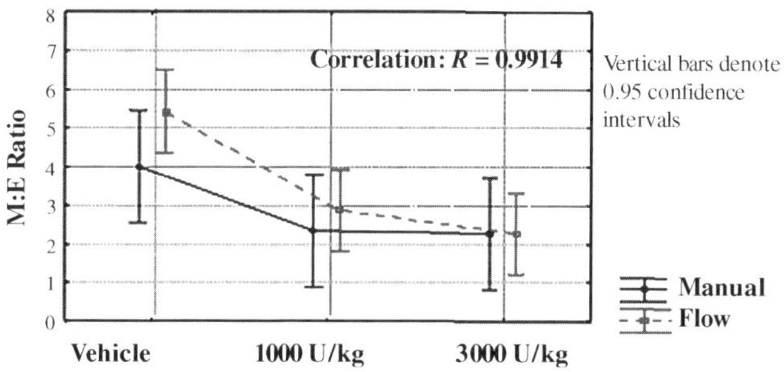

FIGURE 6.7 Comparison of M:E ratio methods in mice dosed with recombinant human erythropoietin. Mice were dosed with vehicle, 1000 U/kg or 3000 U/kg rhEpo. Bone marrow differentials and M:E ratios were measured by flow cytometric and cytologic methods. Graph presents results from a repeated measure analysis of variance (ANOVA) model.

example, in order to be characterized as a lymphocyte a cell must stain positively for CD45 but negatively for both CD11b and ter119. There is little chance for investigator bias to negatively impact cell determination on a cell-by-cell basis, whereas manual cytologic determination is based solely on a single medical technologist's judgment for each individual cell. In addition, flow cytometry has the advantage of decreased absolute counting errors due to the large number of cells interrogated (at least 10,000) versus only 100–500 for manual methods. Therefore, a case could easily be made as to why the flow cytometry results would be more accurate and reliable. However, the reality is that manual cytologic assessment is the "Gold Standard" and the incongruence observed simply cannot be dismissed. Possible explanations for this incongruence, other than stating that one method is right and the other one is wrong, do exist. The first is that manual and flow cytometric analyses were not performed on identical samples. Flow cytometric analysis was performed on suspensions of bone marrow cells from the left femur and cytology was performed on a smear created from the right femur. However, previous experiments did not reveal significant differences between femurs. Instead, differences may be attributed to the fact that smears are created from a small portion of the total femoral bone marrow, whereas flow cytometric preparation homogenizes the entire femoral bone marrow population, precluding potential sampling errors. A second possible explanation for lack of parity is the fact that some events are excluded from analysis during flow cytometric assessment. This is necessitated by the existence of minor populations of cell aggregates and nonstaining cells. Although the nonstaining population does not stain positively for any of the monoclonal antibodies used, it does stain with the nuclear dye, Hoechst 33342. This staining pattern argues that the events are not hematopoietic lineage cells and should, in fact, be excluded from analysis. The identity of these events could be primitive stem cells that do not yet express lineage markers, other resident bone marrow cell types (e.g., osteoblasts, osteoclasts, and osteocytes), or perhaps free nuclei. It is tempting to speculate that these events are free nuclei of erythroid cells that were in the process of nuclear extrusion upon harvest, but there is no direct evidence of this at present. In addition, there is a small percentage of coincident or aggregate events that also must be excluded in analysis of flow cytometry samples, but this seems unlikely to negatively impact the overall differential as cells excluded as aggregates are preferentially not one cell type. It should be noted that the exclusion of cell aggregates is a common practice in flow cytometric analysis and is performed to assure that data are from individual cells.

The possibility that preparation of mouse bone marrow samples for flow cytometric analysis leads to preferential removal and/or destruction of erythroid cells is another possibility that cannot be discounted. In order to have full confidence in flow cytometric results from mouse bone marrow, the lack of parity between flow cytometry and manual cytology would need to be resolved.

6.8 SUMMARY

The term high-content screening (HCS) or cell-based screening plays a major role in drug discovery and development in a pharmaceutical company and was originally employed in bioimaging-based systems. As flow cytometry technology has evolved to provide expanded color capabilities, the technology has become both a robust and an economic cell-based HCS system assaying thousands of cells per second. Flow cytometry provides several key advantages over bioimaging systems. First, this higher single-cell throughput technology allows researchers to analyze smaller population subsets with greater statistical accuracy. Second, the cell sorting can be used to collect these minor cell populations for secondary molecular testing, gaining additional data for understanding the mechanism and kinetics.

While flow cytometry provides advantages during the preclinical toxicity screening process (e.g., more information in less time), there are also challenges. First, the researchers must develop the assay. In the case of the oxidative stress model, a thorough titration of the labels was needed to optimize the assay. Any change in cell line or culture conditions would likely require similar titrations to optimize the assay. Biomarker assay development requires the availability of the appropriate antibodies and substrates to label key biomarkers. As was discussed with the CEC/CEP assay, the process may be one of elimination as well (gating out of the platelets, T and B cells that shared the CD36 marker). The development process requires a detailed knowledge of cell lineage markers, intracellular processes, and access to labeled antibodies/substrates for evaluation of these markers and processes. Functional substrates such as those employed in the oxidative stress study may have a fixed color or may be available in modified forms to allow selection of a "convenient" color during the assay development.

Second, the system must be tested with compounds that have a well-characterized toxicity profile (e.g., serotonin treatment with rats) and the outcomes must be validated with information from well-known secondary testing outcomes (e.g., gene or protein expression studies). The sorted $CD36^+/Lin^-$ cells were tested for both culture characteristics and gene expression and showed a less definitive outcome with complex and hetergeneous response in both assays.

The bone marrow study provided a more straightforward system for evaluation and validation. The flow data subset analysis information was developed using a well-characterized panel of mouse antibodies and was confirmed with a secondary test using microscopic cytology. The rhEpo treatment was known to trigger erythroblast production. The bone marrow subset data trended with similar end points between flow and cytology; however, more statistically significant variations were observed in the flow cytometry subsets.

Looking at multiple physiological end points in a single cell in the oxidative stress assay (lipid peroxidation, glutathione concentration, mitochrondial membrane potential, and finally apoptosis) provides a time-based look into a complex system. The observation propagation or abrogation of the stress effects to other compartments provides information on the mechanism of toxicity.

Ironically, the availability and complexitiy of the technology itself provide the least obstacle. Benchtop flow cytometers/analyzers are readily available in hospitals and clinics and used for a variety of routine diagnostic and monitoring applications. The platforms have evolved to provide robust, easy to operate systems with ever-expanding color capabilities. The more complex cell sorting systems do require a trained operator.

Proof of the value-added aspects of the implementation of flow cytometric biomarker-based analysis in preclinical toxicity testing will be the key. Reduced time, reagent, and animal costs are one form of value addition to the pharmaceutical company. Clinicians will require evidence of the robustness of the assay and assay outcomes. However, because there is a routine collection of a variety of tissue types amenable to flow cytometry during any study, once an assay design is establised, data could be generated from logged samples for statistical analysis within a fairly reduced time frame.

ACKNOWLEDGMENTS

The research described in this chapter was the result of the efforts of many scientists, technicians, and statisticians. The authors wish to acknowledge the following individuals for their contributions to this work: Rogely Boyce, Ryan Boyle, Yifeng Chen, Steve Clark, Harris Cohen, Molly Cool-Bainter, Deidre Dalmas, Cynthia Eberly, Kendall Frazier, Barry Gailliard, Tracy Gales, Vicki Gallagher, Jim Greenwood, Catherine Hu, Jim Huffnagle, Steven Hughes, Angela Hughes-Earle, Amy Landis, Michelle Lee, Nancy Lee Karen Lynch, Michal Magid-Slav, Beverly Maleeff, Chrisee Mignot, Rosanna Mirabile, Eric Moore, David Mullins, Padma Narayanan, Georgina Paulazzo, Dana Pietrzak, Robert Ray, Kate Rhodes, Kim Roland, Kelly Ross, Lester Schwartz, Marshall Scicchitano, Stacie Sink, Krista Stayer, Laura Storck, Kay Tatsuoka, Pattie Terreson, Bharath Thippireddy, Heath Thomas, Roberta Thomas, Doug Thudium, Patrick Warren, Mark Watkins, Gang Xu, Tiffany Yarnall, and Cindy Zhang.

REFERENCES

1. Szaniszlo P, Wang N, Sinha M, Reece LM, Van Hook JW, Luxon BA, Leary JF. Getting the right cells to the array: gene expression microarray analysis of cell mixtures and sorted cells. *Cytometry A* 2004;59:191–202.
2. Hofmann M, Zerwes H-G. Identification of organ-specific T cell populations by analysis of multiparameter flow cytometry data using DNA-chip analysis software. *Cytometry A* 2006;69A:533–540.
3. Achuthanandam R, Quinn J, Capocasale RJ, Bugelski PJ, Hrebien L, Kam M. Sequential univariate gating approach to study the effects of erythropoietin in murine bone marrow. *Cytometry A* 2008;73A:702–714.
4. Bosci J, Tarnok A. Toward automation of flow data analysis. *Cytometry A* 2008;73A:679–680.
5. Chan C, Feng F, Ottinger J, Foster D, West M, Kepler TB. Statistical mixture modeling for cell subtype identification in flow cytometry. *Cytometry A* 2008;73A:693–701.
6. Bus JS, Aust SD, Gibson JE. Paraquat toxicity: proposed mechanism of action involving lipid peroxidation. *Environ. Health Perspect.* 1976;16:139–146.
7. Shapiro, HM. Practical Flow Cytometry, 4th ed., John Wiley & Sons Inc., Hoboken, NJ, 2003.
8. Storozhevykh TP, Yuryavichyus AI, Sorokina EG, Pinelis VG. Induction of cyclosporin A-sensitive pore in mitochondria of intact neurons during uncoupling of oxidative phosphorylation. *Bull. Exp. Biol. Med.* 2001;131(5):440–443.
9. Park S-J, Wu C-H, Gordon JD, Zhong X, Emami A, Safa AR. Taxol induces caspase-10-dependent apoptosis. *J. Biol. Chem.* 2004;279(49):51057–51067.
10. Garcia-Calvo, M. Inhibition of human caspases by peptide-based and macromolecular inhibitors. *J. Biol. Chem.* 1998;273:32608–32613.
11. Kerns W, Schwartz L, Blanchard K, Burchiel S, Essayan D, Fung E, Johnson R, Lawton M, Louden C, MacGregor J, Miller F, Nagarkatti P, Robertson D, Snyder P, Thomas H, Wagner B, Ward A, Zhang J. Drug-induced vascular injury: a quest for biomarkers. *Toxicol. Appl. Pharmacol.* 2005;203:62–87.
12. Gerrity RG, Richardson M, Caplan BA, Cade JF, Hirsh J, Schwartz CJ. Endotoxin-induced vascular endothelial injury and repair. II. Focal injury, en face morphology, (3H) thymidine uptake and circulating endothelial cells in the dog. *Exp. Mol. Pathol.* 1976;24 (1):59–69.
13. Hladovec J. Circulating endothelial cells as a sign of vessel wall lesions. *Physiol. Bohemoslov.* 1978;27(2):140–144.
14. Blann AD, Woywodt A, Bertolini F, Bull TM, Buyon JP, Clancy RM, Haubitz M, Hebbel RP, Lip GY, Mancuso P, Sampol J, Solovey A, Dignat-George F. Circulating endothelial cells: biomarker of vascular disease. *Thromb. Haemost.* 2005;93:228–235.
15. Khan SS, Solomon MA, McCoy JP, Jr. Detection of circulating endothelial cells and endothelial progenitor cells by flow cytometry. *Cytometry B* 2005;64:1–8.
16. Smirnov DA, Foulk BW, Doyle GV, Connelly MC, Terstappen LW, O'Hara SM. Global gene expression profiling of circulating endothelial cells in patients with metastatic carcinomas. *Cancer Res.* 2006;66:2918–2922.

17. Craig LE, Spelman JP, Strandberg JD, Zink MC. Endothelial cells from diverse tissues exhibit differences in growth and morphology. *Microvasc. Res.* 1998;55:65–76.
18. Ewing P, Wilke A, Brockhoff G, Andreesen R, Eissner G, Holler E, Gerbitz A. Isolation and transplantation of allogeneic pulmonary endothelium derived from GFP transgenic mice. *J. Immunol. Methods* 2003;283:307–315.
19. Zhang X, Fitzsimmons RL, Cleland LG, Ey PL, Zannettino AC, Farmer EA, Sincock P, Mayrhofer G. CD36/Fatty acid translocase in rats: distribution, isolation from hepatocytes, and comparison with the scavenger receptor SR-B1. *Lab Invest.* 2003;83:317–332.
20. McFarland DC, Zhang C, Thomas HC. Confounding effects of platelets on flow cytometric analysis and cell-sorting experiments using blood-derived cells. *Cytometry A* 2006;69:86–94.
21. Garlanda C, Dejana E. Heterogeneity of endothelial cells: specific markers. *Arterioscler. Thromb. Vasc. Biol.* 1997;17:1193–1202.
22. Risau W. Differentiation of endothelium. *FASEB J.* 1995;9:926–933.
23. Thum T, Haverich A, Borlak J. Cellular dedifferentiation of endothelium is linked to activation and silencing of certain nuclear transcription factors: implications for endothelial dysfunction and vascular biology. *FASEB J.* 2000;14:740–751.
24. Chi JT, Chang HY, Haraldsen G, Jahnsen FL, Troyanskaya OG, Chang DS, Wang Z, Rockson SG, van de RM, Botstein D, Brown PO. Endothelial cell diversity revealed by global expression profiling. *Proc. Natl. Acad. Sci.* 2003;100:10623–10628.
25. Podgrabinska S, Braun P, Velasco P, Kloos B, Pepper MS, Skobe M. Molecular characterization of lymphatic endothelial cells. *Proc. Natl. Acad. Sci.* 2002;99: 16069–16074.
26. Thum T, Borlak J. Gene expression in distinct regions of the heart. *Lancet* 2000;355:979–983.
27. Hill JM, Zalos G, Halcox JP, Schenke WH, Waclawiw MA, Quyyumi AA, Finkel T. Circulating endothelial progenitor cells, vascular function, and cardiovascular risk. *N. Engl. J. Med.* 2003;348(7):593–600.
28. Thomas RA, Pietrzak DC, Scicchitano MS, Thomas HC, McFarland DC, Frazier KS. Detection and characterization of circulating endothelial progenitor cells in normal rat blood. *J. Pharmacol. Toxicol. Methods* 2009;60:263–274.
29. Rafii S, Lyden D. Therapeutic stem and progenitor cell transplantation for organ vascularization and regeneration. *Nat. Med.* 2003;9(6):702–712.
30. Hur J, Yang HM, Yoon CH, Lee CS, Park KW, Kim JH, Kim TY, Kim JY, Kang HJ, Chae IH, Oh BH, Park YB, Kim HS. Identification of a novel role of T cells in postnatal vasculogenesis: characterization of endothelial progenitor cell colonies. *Circulation* 2007;116(15):1671–1682.
31. Criswell KA, Bleavins MR, Zielinski D, Zandee JC. Comparison of flow cytometric and manual bone marrow differentials in Wistar rats. *Cytometry* 1998;32(1):9–17.
32. Criswell KA, Bleavins MR, Zielinski D, Zandee JC, Walsh KM. Flow cytometric evaluation of bone marrow differentials in rats with pharmacologically induced hematologic abnormalities. *Cytometry* 1998;32(1):18–27.
33. Saad A, Palm M, Widell S, Reiland S. Differential analysis of rat bone marrow by flow cytometry. *Comp. Haem. Int.* 2000;10:97–101.

34. Schlueter AJ, Bhatia SK, Li X, Tygrett LT, Yamashita Y, de Vries P, Waldschmidt TJ. Delineation among eight major hematopoietic subsets in murine bone marrow using a two-color flow cytometric technique. *Cytometry* 2001;43(4):297–307.
35. Capocasale R, Nesspor T, O'Brien J, Makropoulos D, Shamberger K, Bugelski P. Multi-parameter analysis of the *in vivo* effects of recombinant human erythropoietin on murine medullary late-stage erythroid precursors. International Society for Analytical Cytology Congress XXII Posters, 2004.

7

USE OF FLOW CYTOMETRY TO STUDY DRUG TARGET INHIBITION IN LABORATORY ANIMALS AND IN EARLY-PHASE CLINICAL TRIALS

DAVID W. HEDLEY

7.1 OVERVIEW

This chapter reviews the use of flow cytometry to study pharmacodynamics at the single-cell level during the transition from preclinical to early-phase testing of novel agents, including the potential for individual patient dose optimization based on pharmacodynamic monitoring. Although the emphasis is on cancer, particularly hematological malignancies where flow cytometry is uniquely suited, consideration is given to underlying principles that are applicable to inflammatory and autoimmune disorders and to solid tumors. Particular emphasis is placed on the use of panels of phosphospecific antibodies to monitor agents that act to inhibit signal transduction pathways, because of the importance of these agents and the unique capability of flow cytometry to monitor signaling networks at the single-cell level. However, this chapter touches on a broad range of applications to illustrate a field that is rich, and only partially explored.

Flow Cytometry in Drug Discovery and Development, Edited by Virginia Litwin and Philip Marder

7.2 STAGES IN THE DEVELOPMENT OF *IN VIVO* TESTING FOR NOVEL AGENTS

7.2.1 Developing Assays from First Principles

Most of the applications reviewed in this chapter can be defined in terms of cellular biochemistry. There is an unusual breadth in the range of applications used to monitor drug effects at the single-cell level, relative to other fields of flow cytometry. This reflects the scope of drug development in the era of molecular medicine. Although some applications can be regarded as "generic," for example, identifying and optimizing an appropriate antibody readout for drug effect, other techniques that have been published are completely novel and built up from first principles. It therefore follows that success in this field requires a multidisciplinary approach, involving flow cytometry technology, pharmacology, and cell biology.

In addition, for the biochemical changes to be measured, the type of material to be studied needs to be considered. This is particularly important when the intent is to use flow cytometry to monitor drug effects in patients rather than laboratory models. Obtaining sequential samples from solid tumor patients is clearly problematic. Although there are many recent reports where this has been successfully done using needle biopsies, very few studies have applied flow cytometry to these samples. However, after years of relative neglect, an appreciation of the analytical power of high dimensional flow cytometry to study solid tumors has resurfaced, in large measure due to interest in cancer stem cells [1, 2]. Further developmental work to study pharmacodynamic effects therefore appears warranted, perhaps initially based on primary human cancer xenografts rather than samples obtained directly from the patient.

7.2.2 Cell Signaling: Introduction

Cells respond to environmental cues, including growth factors and cell/cell or cell/extracellular matrix contact, by activating (or sometimes inhibiting) signal transduction pathways. These pathways involve protein interactions that result in posttranslational modifications, the commonest of which is phosphorylation at serine, threonine, or tyrosine sites. They are highly regulated by inhibitory and stimulatory feedback loops and through pathway interactions. In addition to regulating genes involved in cell growth, survival, and differentiation, signaling pathways also control cellular metabolism and motility. Cell signaling is disrupted in the early stages of cancer development, and it plays a major role in tumor progression toward more aggressive phenotypes. Therefore, there is intense interest in the modulation of these pathways as a novel approach to cancer treatment [3–10]. Most of the agents in current use are either monoclonal antibodies directed toward extracellular components of signaling pathways or small molecules that typically inhibit kinases through ATP competition. The Her2 inhibitor trastuzumab (Herceptin), used to target Her2 overexpressing breast cancers, and the bcr/abl kinase inhibitor imatinib (Gleevec), used to treat chronic myeloid leukemia (CML), are among the best known examples [11, 12].

However, large numbers of agents are currently under development or approved for routine use, and it is likely that drugs to target all the potentially oncogenic signaling pathways will become available within the next few years.

Although the major signaling pathways are well understood in terms of their key components and the fundamental biochemical processes involved, the regulation of cell signaling in normal and diseased tissues *in vivo* is highly complex and context sensitive. Complexity is particularly evident in cancers, where more than one signaling element is often affected and there is considerable cellular heterogeneity within the malignant cell population and its interactions with the surrounding stroma. This complexity explains why drugs that are highly effective in tissue culture are often much less active when used to treat cancer patients, and is a major challenge at the interface between the development of a drug and its effective use in the clinic.

The development of phosphospecific antibodies directed toward activated signaling proteins greatly facilitates the study of signal transduction in biological samples, including material obtained from cancer patients. Continuing improvement in antibody production techniques has resulted in the availability of a wide range of high-affinity and target-specific monoclonal antibodies that are suitable for flow cytometry applications. Single-cell measurement of activated signaling proteins using flow cytometry has a number of important advantages, relative to bulk assays:

1. It addresses cellular heterogeneity of signaling activity and allows measurements to be made in subpopulations of cells defined using standard phenotypic markers.
2. Activity can be measured using relatively small numbers of cells.
3. Signaling pathways function in highly regulated networks, and this can be appreciated with flow cytometry through the combined use of multiple phosphospecific antibodies.
4. Whole blood samples can be processed for signaling analysis by flow cytometry, which maintains the pharmacodynamic effect of agents that diffuse from the target during cell isolation procedures.

7.2.3 Cell Signaling: Assay Basics

As discussed in Chapter 15, these techniques have been successfully applied to patient samples and have identified novel and medically important alterations in cell signaling in both leukemia and inflammatory diseases. Although powerful, this approach has been relatively slow to gain widespread use due to the technical complexity of the assays, relative to more established flow cytometry methods such as immunophenotyping. The measurement of intracellular antigens requires fixation and permeabilization, and the optimum conditions can vary for different epitopes [13]. Importantly, the phosphorylation states of signaling proteins are necessarily highly dynamic, and can change within time frames of less than 1 min in response to pathway stimulation or by the action of phosphatases. These transient states can be stabilized by formaldehyde fixation, which is believed to destroy enzyme function and capture the phosphorylation state at the time when the fixative is added [14, 15]. It follows that in developing a flow cytometry protocol to study a particular signaling pathway, it is important to have a good appreciation of the kinetics and to recognize that the timing of fixation is critical.

Our laboratory was one of the first to describe the use of flow cytometry to study cell signaling based on panels of phosphospecific antibodies [14]. Although we gave consideration to the potential for more basic research on the study of signaling networks at the single-cell level, our primary goal was the development of a biomarker to allow pharmacodynamic monitoring of a novel Raf inhibitor, sorafenib (BAY 43-9006), that was entering phase I clinical trial. Through a series of steps, a flow cytometry protocol evolved that is robust, versatile, and relatively simple to learn [16]. It is therefore described in some detail.

Since the initial clinical trial of sorafenib involved solid tumor patients, where sequential cancer tissue sampling is problematic, the first-generation flow cytometry method used peripheral blood lymphocytes as a surrogate marker for pharmacodynamic monitoring. At that time (2000) it was believed that Raf kinase, a component of the ERK signaling pathway, was the major target of sorafenib; hence, the dose-limiting toxicity was expected to be due to the effects of Raf inhibition in normal tissues. Although the ERK pathway is normally latent in peripheral lymphocytes, it can be activated acutely via the T cell receptor, which is more physiological, or by phorbol ester that activates Raf kinase via protein kinase C [17]. Since the latter gave much more robust pathway activation, it was used for subsequent assay development [14]. Using isolated peripheral blood mononuclear cell (MNC) suspensions and standard flow cytometry techniques for intracellular antigen labeling, acute activation of the downstream elements MEK and ERK by phorbol ester can be readily detected by appropriate phosphospecific antibodies. Preincubation with a pathway inhibitor blunts this effect in a dose-dependent manner, and the IC_{50} value for sorafenib obtained in our initial experiments agreed with other cell-based assays for drug effect. However, drug target inhibition is rapidly lost when the pretreated samples are resuspended in drug-free medium, as it occurs when blood samples from patients are processed for MNC isolation. This is likely to be a problem with most other ATP-competitive kinase inhibitors. It can be solved by adding fixative directly to the blood sample, thereby maintaining the drug:target equilibrium. Red cells can then be removed by density gradient centrifugation, which is laborious, or more conveniently by lysis using 0.1% Triton X-100 for 30 min [16, 18]. Treatment with 4% formaldehyde for 10–15 min produces excellent preservation of forward and orthogonal light scatter, and has relatively little detrimental effect on cellular DNA content analysis. Red cell lysis using Triton X-100 also permeabilizes the nucleated cells, allowing access for the phosphospecific antibodies. This simple protocol is adequate for labeling many phosphorylated epitopes, but others such as P-ERK and P-STAT5 are masked by formaldehyde fixation and require antigen retrieval by alcohol treatment for optimum staining [16]. Most surface antigens are preserved at methanol concentrations up to 50%, but the higher concentrations needed for optimum labeling of the STAT signaling proteins can severely compromise important markers, including CD13, CD14, and CD19. This problem can be overcome if needed by adding the appropriate immunophenotypic antibody to the whole blood sample prior to the fixation step, although the possibility that antigen binding might activate cell signaling, as it occurs with the T cell receptor, needs to be considered.

7.2.4 Cell Signaling: Use in Pharmacodynamic Monitoring

7.2.4.1 Surrogate Marker Assays Particularly during the early phase of clinical testing, where the goal is to establish safe treatment schedules for subsequent efficacy testing, pharmacodynamic monitoring of drug effects on normal tissues can be informative. In phase I clinical trials of anticancer agents, this is often done using buccal mucosa or skin biopsies as a source of normal epithelium. However, quantitative analysis of this material can be problematic, and successful detection of pharmacodynamic effects depends on the relevant pathway being constitutively active, which is not always the case. The use of flow cytometry to measure effects in normal peripheral blood cells avoids these problems due to the inherently quantitative nature of flow cytometry and the ability to acquire large numbers of events for statistical analysis. Although most signal transduction pathways are quiescent in normal peripheral blood cells, many can be acutely activated *ex vivo*, and the extent of activation relative to an unstimulated control sample can be used as a measure of pathway inhibition.

Using phorbol ester to activate the ERK pathway in blood samples, we found that although target inhibition occurred at sorafenib concentrations in the tens of micromole range when isolated cells were resuspended in tissue culture medium, drug concentrations of several hundred micromoles were needed to produce an equivalent effect in whole blood [18]. This is explained by the high levels of protein binding that occur with this agent, and suggests the potential to compare signaling inhibition in serum-free medium versus whole blood as a rapid screen for the plasma or red cell binding of a novel compound. In our original phase I trial of sorafenib in solid tumor patients, we were unable to detect a pharmacodynamic effect at the maximum tolerated dose, and this result now appears to be most likely explained by the inability to achieve sufficient free drug to inhibit Raf kinase in peripheral blood due to protein binding rather than failure of the assay. In addition to its potential to monitor pharmacodynamic effects of ERK pathway inhibitors, phorbol ester stimulation in whole blood has also been adapted to monitor effects of the protein kinase C inhibitor enzastaurin, using an antibody that recognizes PKC phosphorylation sites in general [19].

The ERK pathway can also be acutely activated in normal peripheral blood monocytes by bacterial lipopolysaccharide (LPS). Although this acts through Toll-like receptors rather than growth factor receptors, downstream signaling proceeds via MEK and is more physiological than phorbol ester. The other MAP kinase pathways SAP/JNK and p38 are also acutely activated by LPS, suggesting the potential to use this as an assay for monitoring anti-inflammatory agents that target these pathways [20, 21]. Other signaling pathways that can be acutely activated in whole blood include JAK/STAT, which responds to GM-CSF or G-CSF (Figure 7.1a).

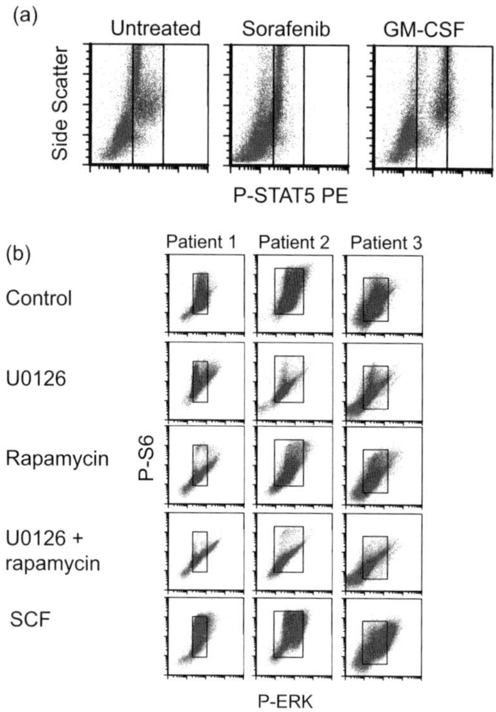

FIGURE 7.1 (a) Constitutive STAT5 phosphorylation in peripheral blasts of a patient with acute myeloid leukemia carrying a Flt-3 internal tandem duplication. Aliquots of the sample were incubated with the Flt-3 inhibitor sorafenib, or activated using GM-CSF, and then fixed and processed as described in the text. Blast cells were identified using anti-CD34 and color gated. The blast cells have a side scatter that is intermediate between lymphocytes and granulocytes, and their P-STAT5 fluorescence is relatively high in the untreated control sample. The intensity decreases to that of the granulocytes in the presence of sorafenib, consistent with the ITD mutation activating P-STAT5, whereas GM-CSF activates STAT5 in the granulocytes via JAK, so that their intensity is similar to that of the blast cells. This effect can be used as a positive control. (b) Constitutive activation of S6 ribosomal protein in peripheral blast cells (color gated) from three AML patients. Note the considerable intercellular heterogeneity of P-S6, which ranges over two decades in all three cases. The MEK inhibitor U0126 decreases P-S6 in patients 2 and 3 and P-ERK in all three cases. The mTOR inhibitor rapamycin decreases P-S6 in patient 1, but not the other cases, whereas the combination of U0126 and rapamycin reduces P-S6 to baseline in all three cases. A treatment protocol incorporating a rapamycin analogue might therefore be more effective in patient 1, and is predicted to cause a decrease in the constitutive P-S6 levels that could be monitored using this assay.

There is currently interest in agents that inhibit the PI3 kinase/Akt pathway, since this is aberrantly activated in many cancers, and a need for effective pharmacodynamic monitoring of these agents, given the importance of this pathway in the maintenance of normal tissues. Having been unsuccessful in attempts to develop a surrogate blood marker for PI3 kinase inhibition based on the use of phosphospecific antibodies to Akt, Bowers et al. devised an ingenious assay that makes use of the PI3 kinase dependence of platelet activation in response to fibrinogen. This assay uses the expression of the platelet activation marker CD41a in response to a synthetic peptide as a readout for PI3 kinase inhibition in peripheral blood samples [22]. When applied to blood samples obtained from tumor-bearing mice treated with the experimental PI3 kinase inhibitor wortmannin, the pharmacodynamic effects detected with this assay were found to correlate closely with the extent of Akt phosphorylation seen in tumor tissue, suggesting that this will prove to be a useful assay applied to cancer patients during early-phase clinical trials.

7.2.4.2 Leukemia In most leukemia patients, the malignant cells are present in the peripheral blood and are therefore readily sampled. However, in the case of acute myeloid leukemia (AML), which is the common form of aggressive leukemia in adults, cell proliferation takes place predominantly in the bone marrow where it is regulated by complex interactions with the stroma. The leukemic stem cell population is also believed to reside in the bone marrow. Bone marrow aspiration samples, therefore, probably give a better appreciation of signaling traffic in the leukemic cell population, and are also needed in cases where the leukemic blasts do not enter the circulation. However, they are less readily obtained than peripheral blood.

In contrast to normal blood cells, constitutively activated signaling proteins can be detected in the peripheral blast cells in the majority of AML patients, indicating some degree of autonomy from external sources of pathway activation. For example, STAT5 is activated via bcr/abl in cases of chronic myeloid leukemia, and by the Flt-3 receptor tyrosine kinase in cases of AML carrying the Flt-3 internal tandem duplex mutation (Figure 7.1a). The S6 ribosomal protein, which is involved in the regulation of protein translation, can be activated by several upstream signaling pathways, including ERK and Akt/mTOR, and we have shown that peripheral AML blast cells can show great heterogeneity in the extent to which S6 is constitutively activated by these pathways (Figure 7.1b) [23]. Using isolated peripheral blast cells from AML patients, Irish and Nolan observed significant differences in the extent to which signaling pathways could be acutely activated by different growth factors, with some of these patterns carrying important prognostic information [24]. Based on the above observations, it seems likely that complex cell signaling assays could be adapted for individual leukemia patient treatment selection by identifying the critical aberrant pathways in the blast population. However, that is outside the scope of this chapter.

The application of flow cytometry to the pharmacodynamic monitoring of signal transduction inhibitors in leukemia patients is still in its infancy, but appears to have major potential. Chronic myeloid leukemia, which is driven by the bcr/abl fusion protein, is the most obvious candidate for this since there is a consistent pattern of activated signaling elements present in nearly all cases. Flow cytometry has been used to measure dose-dependent loss of constitutively activated STAT5 and CrkL in CML cell lines treated with the bcr/abl inhibitor imatinib, and subsequently demonstrated in CML patients, although to date this has not developed into a routine clinical test [25, 26].

Our own laboratory has been focused on AML patients, linked to a large clinical and translational research program testing new approaches to treatment. In a multicenter phase I clinical trial conducted by the National Cancer Institute of Canada Clinical Trials Group, patients with chemotherapy refractory AML were treated with sorafenib, which at that time was believed to be a selective Raf kinase inhibitor, as a single agent [27]. Blood samples were obtained for pharmacodynamic monitoring predose and at weekly intervals during the first treatment cycle. The flow cytometry protocol originally developed for pharmacodynamic monitoring of sorafenib in peripheral blood lymphocytes was modified to include activation via the c-Kit receptor using its ligand stem cell factor (SCF), in addition to phorbol ester as used in the earlier version [18]. This was intended to provide a more physiological stimulus to the ERK pathway, but unexpectedly it was found that although no pharmacodynamic effect was detectable at any dose level using phorbol ester to drive the ERK pathway, similar to our finding in the solid tumor trial, there was an obvious dose-dependent inhibition of ERK activation when SCF was added to the blood samples, with almost 100% inhibition after 1 week treatment at the highest dose level (Figure 7.2). In patients who were randomized to sorafenib treatment on an intermittent schedule, drug target inhibition was lost after 2 weeks off treatment. Furthermore, the extent to which SCF activation of ERK signaling was inhibited in patients during treatment was found to correlate with the plasma drug concentration [27]. The apparent paradox that a pharmacodynamic effect could be detected using SCF to activate ERK, but not with phorbol ester, is explained by the fact that sorafenib has multiple signaling targets in addition to Raf kinase. Using normal donor-mobilized peripheral stem cells, we subsequently confirmed that ERK activation by SCF in the CD34 positive cells was inhibited at low micromolar concentrations of sorafenib that are clinically achievable, whereas several hundred micromoles were required to inhibit stimulation by phorbol ester [28].

Although this early sorafenib trial established the potential to monitor pharmacodynamic effects in sequential blood samples obtained from leukemia patients, several weaknesses in the flow cytometry protocol were identified, including poor sensitivity and difficulty in identifying the blast cells in some samples due to the degradation of light scatter. It was also felt that the technique was technically too difficult for point of care application. A major overhaul of the method was therefore undertaken, as described above, and applied to a new trial that also targeted the c-Kit receptor in AML patients, this time using imatinib in combination with second-line chemotherapy to treat patients whose disease was considered refractory to the first-line drugs, but who were still candidates for intensive myelo-ablative chemotherapy. An important feature of the optimized flow cytometry protocol is the ability to identify leukemic blast cells at frequencies as low as 0.1%, an important consideration for early-phase clinical trials in AML since patients enrolled onto these studies often have fairly slow tempo, low count disease [16]. The study design included a 4-day lead in period when the patients were treated with imatinib alone, and more intensive pharmacodynamic monitoring during this period. Although results from this study are still being analyzed, it is evident that when patients are treated with a fixed dose of imatinib, there is considerable heterogeneity in the level of c-Kit inhibition achieved,

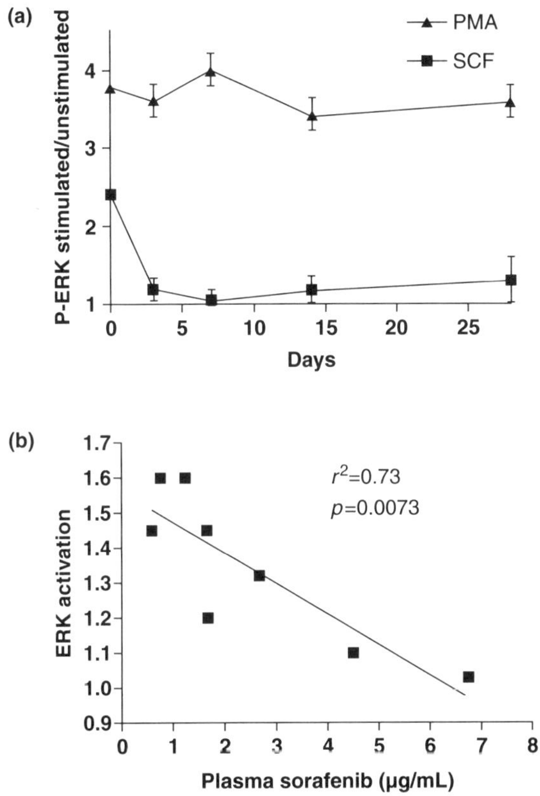

FIGURE 7.2 (a) Pharmacodynamic effects of sorafenib 400 mg given twice daily to a cohort of patients with acute myeloid leukemia. Whole blood samples were obtained at the indicated time points, and activated *ex vivo* using either PMA or stem cell factor prior to fixation. P-ERK was measured in the blast cells and the results plotted as the mean fluorescence intensities of the stimulated/unstimulated. Note that almost 100% inhibition is seen when the blasts are activated using SCF, whereas there is no significant inhibition using PMA. This is explained by the fact that sorafenib inhibits the SCF receptor c-Kit at the drug levels achieved, but not the Raf kinase. (b) Inhibition of P-ERK activation by SCF versus mean plasma sorafenib concentrations in the phase I clinical trial of sorafenib in acute myeloid leukemia. Each data point is the mean value for a patient cohort.

with some showing almost complete target inhibition, as we previously noted using sorafenib, and others showing no detectable pharmacodynamic effect. Furthermore, patients who showed a larger pharmacodynamic effect were significantly more likely to respond to chemotherapy, which is consistent with the hypothesis that increased signaling through c-Kit contributes to drug resistance in AML patients.

7.2.4.3 High-Complexity Cell Signaling Protocols There are many potential signaling targets expressed in leukemia, and there is increasing evidence for considerable heterogeneity in signaling activity between individual patients. It is

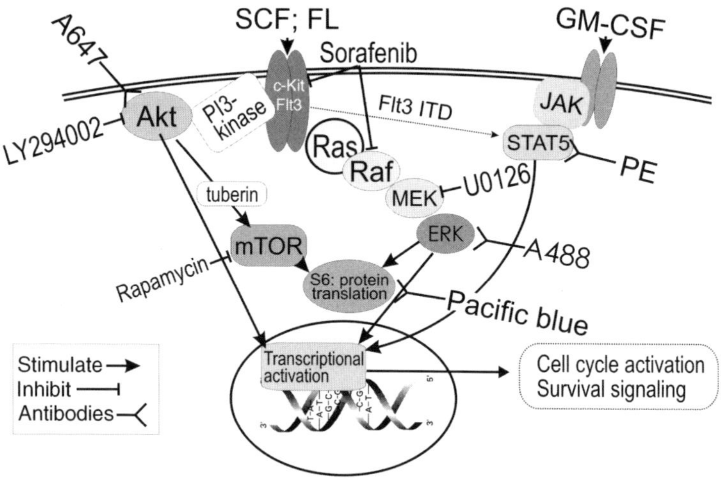

FIGURE 7.3 Schematic for a four phosphospecific antibody panel designed to test multiple potential signaling targets in AML patients. The assay has a modular design, and individual pathways can be tested by the addition of appropriate stimulants or inhibitors, as indicated. (See the color version of the figure in the Color Plates section.)

therefore important to extend the repertoire of flow cytometry applications to include multiple signaling pathways. With the widespread availability of multilaser flow cytometers and steady progress in reagent development, it is becoming feasible to develop high-complexity protocols for pharmacodynamic monitoring in leukemia patients. Figure 7.3 illustrates one that is currently being used in our own laboratory to monitor a multikinase inhibitor ENMD2076, based on a combination of four phosphospecific antibodies plus surface markers. The data shown in Figure 7.1 were produced using this protocol. Although for this particular drug the main interest is in Flt-3 and c-Kit inhibition, as illustrated in the figure, the assay is sensitive to other agents targeting a wide range of potential targets relevant to AML. It therefore has considerable flexibility and can be rapidly redeployed to clinical trials involving other signaling pathways, including PI3-kinase, Raf/MEK/ERK, and JAK/STAT.

7.2.4.4 Future Directions for Monitoring Signal Transduction Inhibitors in Leukemia This field is still evolving, and to some extent remains dependent on progress introducing new and effective targeted agents to treat leukemia patients. Future directions are likely to include the use of high-complexity flow cytometry protocols first to select appropriate treatment, including the potential for combining multiple targeted agents, and then to monitor effects during treatment (individualized medicine). To move this into mainstream clinical practice, there is a need for standardized reagents including reagent kits and eventually for automated sample preparation and data analysis. Our results from the recent imatinib trial, where the extent of pharmacodynamic effect was predictive of patient outcome, support the idea that eventually these assays might be done in real time, with the drug dose being adjusted to achieve optimum target inhibition. At least in principle, this appears to be technically achievable.

7.2.4.5 ***Monitoring Signal Transduction Inhibitors in Solid Tumors*** This is much more challenging due to the difficulties in obtaining multiple tissue samples and then disaggregating these for flow cytometry measurements. With accessible lesions, fine needle aspiration is an alternative to core or incisional biopsy that is less traumatic and therefore easier to repeat at intervals during treatment. Tumor samples obtained by fine needle aspiration are dispersed into single cells or small aggregates, and therefore easier to process for flow cytometry. Alternatively, they can be analyzed using slide-based cytometry, which has the advantages of allowing morphological assessment of the cell population and requiring smaller samples. Schwock et al. recently reported that this approach is technically feasible using human tumor xenografts, and therefore likely also applicable to human clinical trials [29].

7.3 OTHER TARGETS AND APPLICATIONS

7.3.1 Drug Resistance Mechanisms

The biochemical basis for cellular resistance to anticancer chemotherapy is fairly well understood in terms of alterations in drug uptake, metabolism, and the initiation of DNA damage, although the subsequent responses that determine if cell death or damage repair occurs remain an area of intense research. Although it is clear that drug resistance is multifactorial and that the relative contribution of the various mechanisms differs among tumors, it is likely that some predominate in individual tumors. Furthermore, the expression levels of some drug resistance mechanisms often increase during drug exposure resulting in acquired drug resistance. Probably the best characterized of these in the multidrug resistance (MDR) efflux is pump P-glycoprotein (Pgp). This is a member of a large family of ATP-dependent membrane transporters, and is able to actively transport several classes of chemotherapy agents out of the cell [30, 31]. Examples include anthracyclines, vinca alkaloids, and taxanes. Increased levels of Pgp occur during exposure to these agents *in vitro*, and are implicated in drug resistance in the clinic.

7.3.1.1 ***P-Glycoprotein Inhibitors*** In the 1980s it was recognized that Pgp-mediated drug resistance could be overcome using pharmacological inhibitors, and several of these were developed for testing in patients. Given that Pgp also appears to play a role in protecting normal hematopoietic stem cells from chemotherapy effects and that Pgp-like molecules are involved in the biliary excretion of some drugs, combining Pgp inhibitors with chemotherapy has the potential to increase normal host toxicity; hence, the need for functional markers to monitor pharmacodynamic effects in patients.

Rubin et al. described an optimized flow cytometry protocol to monitor PD effects of the potent Pgp inhibitor LY335979 (zosuquidar) that illustrates some important general principles [32]. These authors exploited the high levels of Pgp expression seen in natural killer (NK) cells to provide a surrogate biomarker, since it is easier to obtain sequential samples from normal blood cells rather than tumor tissue. In this case, it is reasonable to expect that the levels of target inhibition in NK cells reflect those in the cancer. They used rhodamine 123 as the Pgp substrate rather than a fluorescent anticancer drug, since this has a much greater quantum efficiency and is therefore detectable using low concentrations (50 nM). This limits the rate of bulk flow into the cells, so that its efflux via Pgp is more evident. Because LY335979 tightly binds to Pgp, target inhibition is maintained during mononuclear cell isolation, whereas a weakly binding inhibitor is expected to dissociate. To standardize the assay, they compared rhodamine uptake in the absence or presence of excess LY335979 added *ex vivo* to completely inhibit Pgp and expressed the results for NK cells (identified by dual staining with anti-CD56) using a formula based on these values. This assay was then applied to a large number of clinical trial patients, and it clearly identified a wide range of drug target inhibition during treatment, with almost 100% inhibition in some patients. Importantly, Pgp inhibition measured by the flow cytometry assay was found to correlate closely with the plasma drug concentration in the same blood samples (Figure 7.4) [32]. This paper therefore provides an excellent example of the rational development and standardization of a flow cytometry assay to monitor a complex process in a clinical trial and the correlation between the pharmacodynamics and pharmacokinetics of a novel agent.

7.3.2 Cell Cycle Regulation

Flow cytometry is well suited to the study of normal and abnormal cell cycle regulation. This is fundamental to understanding the action of cancer chemotherapy

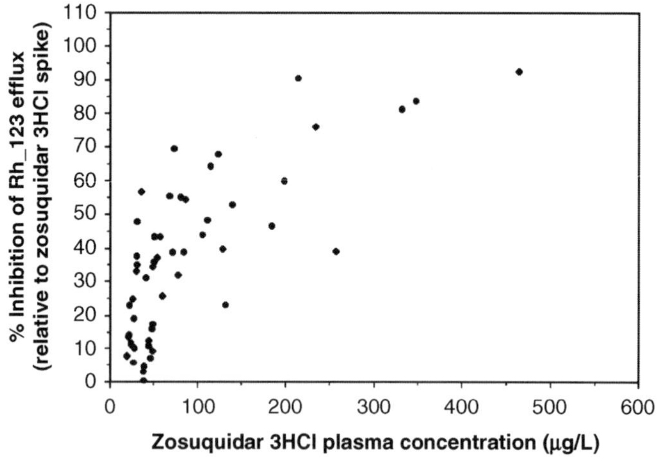

FIGURE 7.4 Results from Rubin et al. [32], showing the close correlation between a flow cytometry assay for P-glycoprotein functional activity in peripheral blood NK cells, and the corresponding plasma concentrations of the P-glycoprotein inhibitor zosuquidar obtained in a phase I clinical trial.

and was therefore one of the earliest applications in cancer research. Although this work has so far not achieved the original goal of transforming the science of combination chemotherapy in the clinic, there has been a recent convergence involving detailed understanding of the basic mechanisms of cell cycle regulation and the development of novel anticancer agents that selectively target these mechanisms. Importantly, specific and high-affinity antibodies directed toward the key molecular regulators are becoming available, opening up the potential to develop a new generation of flow cytometry protocols to study cell cycle regulation in experimental cancer models and in patients [33]. Combined with the capabilities of modern data analysis software and multilaser flow cytometers, this is likely to become a very productive area for research over the next few years.

7.3.2.1 ***Aurora Kinase Inhibitors*** Aurora kinases play an important role in mitotic progression and are frequently overexpressed in cancers. There is therefore interest in the development of anticancer agents that target the two main forms, Aurora A and Aurora B, and these can be considered representative of a broader range of potential drug targets that is emerging from recent research into cell cycle regulation [34, 35]. Wilkinson et al. recently tested a novel Aurora B inhibitor, AZD1152, in a human tumor xenograft models [36]. To monitor *in vivo* pharmacodynamic effects, they combined a standard DNA content protocol with an antibody to phosphorylated histone H3, which is an important Aurora B target. As expected, in control tumors the P-H3 positive cells were confined to the G_2/M peak. Drug treatment produced a striking decrease in P-H3, accompanied by the emergence of a population of cells with a 4C DNA content consistent with Aurora B inhibition (Figure 7.5). These results suggest the potential to develop this application for the study of other cell cycle disrupting agents in animal models, as well as in samples obtained from patients during drug treatment.

7.3.3 Histone Modification

In addition to mutational loss of tumor suppressors, their expression can be silenced epigenetically by histone modifications that involve deacetylation. Histone deacetylase (HDAC) inhibitors are therefore novel and potentially effective anticancer agents, and several have now been introduced into clinical trial [37, 38]. As with other novel agents discussed in this chapter, although HDAC inhibitors differ in their effects from classical chemotherapy agents, they have the potential for major side effects [39]. Furthermore, there appears to be considerable interpatient variability in their effects, indicating a need for laboratory methods able to monitor drug effects in patients. An interesting paper by Kummar et al. describes the use of a flow cytometry technique for surrogate measurement of HDAC inhibition in peripheral blood samples, based on an antiacetylated lysine antibody combined with surface immunophenotypic markers [40]. This was applied to a phase I clinical trial of the novel HDAC inhibitor MS-275, treating patients with refractory cancers. The results show a clear increase in acetylated proteins in subpopulations of normal blood cells that was dependent on the dose of drug given [40]. Although the antibody used was not specific

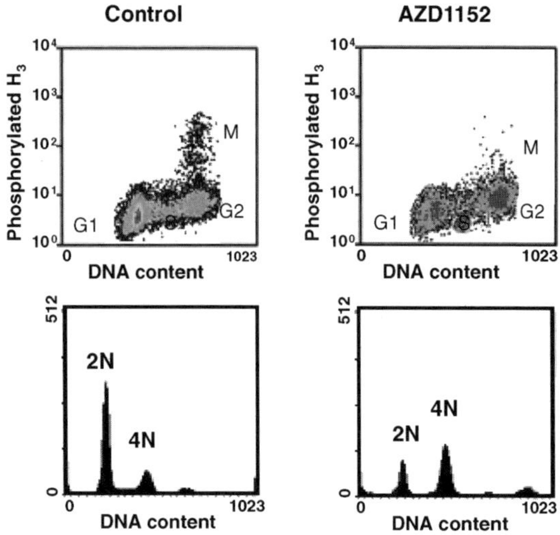

FIGURE 7.5 Results from Wilkinson et al. [36], showing *in vivo* inhibition of Aurora B kinase using the novel agent AZD1152. Nude mice bearing SW620 colon cancer xenografts were treated with drug for 48 h and the tumors excised, disaggregated, and dual stained for DNA content and phosphorylated histone H5. Note the loss of P-H3 in the treated tumor, which is associated with a polyploidy DNA content, consistent with Aurora B kinase inhibition.

for acetylated histones, this paper does illustrate the future potential for flow cytometry techniques to study histone modifications and the effects of newer drugs that selectively modify them.

7.3.4 Rare Event Detection and Analysis: Cancer Stem Cells

The ability to make measurements in rare subpopulations of cells has long been recognized as one of the unique strengths of flow cytometry. This has been reemphasized in recent years with the renewed interest in cancer stem cells, which are believed to represent a small but critical subpopulation in many types of cancer. Although there is some controversy as to whether they are true stem cells, there is now an extensive literature showing that human cancers can be grown from a limited number of cancer cells, viably flow sorted based on the surface expression of putative stem cell markers such as CD133, and that the tumors formed recapitulate the phenotypic heterogeneity seen in the original specimen, whereas the bulk cancer cell population is not able to initiate cancers when implanted into immune-deprived mice [1, 2, 41]. At the present time, cancer-initiating (or stem) cells are defined by their ability to grow tumors at limiting dilution rather than by an unambiguous immunophenotype. However, further progress in this field should allow the development of more sophisticated flow cytometry protocols combining intracellular markers to study their biology with respect to drug treatment. Importantly, several reports suggest that this subpopulation of cancer-initiating cells can be resistant to standard cancer treatments and that developmental pathways that regulate cell fate determination, such as hedgehog or Notch, are involved in their maintenance [41–44]. If this were confirmed in cancer patient samples, then drugs that target these pathways might potently enhance the effectiveness of standard cancer treatments. This question is therefore of great interest to cancer stem cell researchers, and there is potential for flow cytometry to make a major contribution to the field because of its ability to study complex processes in rare subpopulations of cells.

7.4 FUTURE PROSPECTS

The development of molecular medicine and the introduction of personalized treatments have gathered momentum over the last few years, and this trend is expected to continue. It is increasingly recognized that new therapeutic agents need to be developed in parallel with laboratory methods able to identify patients likely to benefit (theranostics) and also to monitor the effects during treatment. Eventually some of these tests are expected to move from the preclinical and early-phase clinical trial setting into routine patient care. A common theme running through the flow cytometry methods discussed in this chapter is their potential for development from a research study to standardized clinical tests. Clinical flow cytometry is a mature discipline within laboratory medicine that undergoes a process of continuing professional, technical, and scientific development. It is expected that over the next few years, we will see increasing numbers of flow applications for studying novel drugs moving from research methods to large-scale clinical trials and eventually into use for routine patient care.

REFERENCES

1. O'Brien CA, Pollett A, Gallinger S, Dick JE. A human colon cancer cell capable of initiating tumour growth in immunodeficient mice. *Nature* 2007;445(7123):106–110.
2. Quintana E, Shackleton M, Sabel MS, Fullen DR, Johnson TM, Morrison SJ. Efficient tumour formation by single human melanoma cells. *Nature* 2008;456(7222):593–598.
3. Bjornsti MA, Houghton PJ. The TOR pathway: a target for cancer therapy. *Nat. Rev. Cancer* 2004;4(5):335–348.
4. Brandts CH, Berdel WE, Serve H. Oncogenic signaling in acute myeloid leukemia. *Curr. Drug Targets* 2007;8(2):237–246.
5. Hennessy BT, Smith DL, Ram PT, Lu Y, Mills GB. Exploiting the PI3K/AKT pathway for cancer drug discovery. *Nat. Rev. Drug Discov.* 2005;4(12):988–1004.
6. Jiang J, Hui CC. Hedgehog signaling in development and cancer. *Dev. Cell* 2008;15 (6):801–812.
7. Kota J, Caceres N, Constantinescu SN. Aberrant signal transduction pathways in myeloproliferative neoplasms. *Leukemia* 2008;22(10):1828–1840.
8. Levis M. Recent advances in the development of small-molecule inhibitors for the treatment of acute myeloid leukemia. *Curr. Opin. Hematol.* 2005;12(1):55–61.

9. Paez J, Sellers WR. PI3K/PTEN/AKT pathway. A critical mediator of oncogenic signaling. *Cancer Treat. Res.* 2003;115: 145–167.

10. Sawyers C. Targeted cancer therapy. *Nature* 2004;432(7015):294–297.

11. Druker BJ, Talpaz M, Resta DJ, Peng B, Buchdunger E, Ford JM, et al. Efficacy and safety of a specific inhibitor of the BCR-ABL tyrosine kinase in chronic myeloid leukemia. *N. Engl. J. Med.* 2001;344(14):1031–1037.

12. Hortobagyi GN. Overview of treatment results with trastuzumab (Herceptin) in metastatic breast cancer. *Semin. Oncol.* 2001;28 (6 Suppl. 18): 43–47.

13. Bauer KD, Jacobberger JW. Analysis of intracellular proteins. *Methods Cell Biol.* 1994;41: 351–376.

14. Chow S, Patel H, Hedley DW. Measurement of MAP kinase activation by flow cytometry using phospho-specific antibodies to MEK and ERK: potential for pharmacodynamic monitoring of signal transduction inhibitors. *Cytometry* 2001;46(2):72–78.

15. Perez OD, Nolan GP. Simultaneous measurement of multiple active kinase states using polychromatic flow cytometry. *Nat. Biotechnol.* 2002;20(2):155–162.

16. Chow S, Hedley D, Grom P, Magari R, Jacobberger JW, Shankey TV. Whole blood fixation and permeabilization protocol with red blood cell lysis for flow cytometry of intracellular phosphorylated epitopes in leukocyte subpopulations. *Cytometry A* 2005;67(1):4–17.

17. Barnard D, Diaz B, Clawson D, Marshall M. Oncogenes, growth factors and phorbol esters regulate Raf-1 through common mechanisms. *Oncogene* 1998;17(12):1539–1547.

18. Tong FK, Chow S, Hedley DW. Pharmacodynamic monitoring of BAY 43-9006 (Sorafenib) in phase I clinical trials involving solid tumor and AML/MDS patients, using flow cytometry to monitor activation of the ERK pathway in peripheral blood cells. *Cytometry B* 2006;70: 107–114.

19. Green LJ, Marder P, Ray C, Cook CA, Jaken S, Musib LC, et al. Development and validation of a drug activity biomarker that shows target inhibition in cancer patients receiving enzastaurin, a novel protein kinase C-beta inhibitor. *Clin. Cancer Res.* 2006;12 (11 Part 1): 3408–3415.

20. Rowlett RM, Chrestensen CA, Nyce M, Harp MG, Pelo JW, Cominelli F, et al. MNK kinases regulate multiple TLR pathways and innate proinflammatory cytokines in macrophages. *Am. J. Physiol. Gastrointest. Liver Physiol.* 2008;294(2):G452–G459.

21. Tudhope SJ, Finney-Hayward TK, Nicholson AG, Mayer RJ, Barnette MS, Barnes PJ, et al. Different mitogen-activated protein kinase-dependent cytokine responses in cells of the monocyte lineage. *J. Pharmacol. Exp. Ther.* 2008;324(1):306–312.

22. Bowers RK, Marder P, Green LJ, Horn CL, Faber AL, Thomas JE. A platelet biomarker for assessing phosphoinositide 3-kinase inhibition during cancer chemotherapy. *Mol. Cancer Ther.* 2007;6(9):2600–2607.

23. Chow S, Minden MD, Hedley DW. Constitutive phosphorylation of the S6 ribosomal protein via mTOR and ERK signaling in the peripheral blasts of acute leukemia patients. *Exp. Hematol.* 2006;34(9):1183–1191.

24. Irish JM, Hovland R, Krutzik PO, Perez OD, Bruserud O, Gjertsen BT, et al. Single cell profiling of potentiated phospho-protein networks in cancer cells. *Cell* 2004;118 (2):217–228.

25. Hamilton A, Elrick L, Myssina S, Copland M, Jorgensen H, Melo JV, et al. BCR-ABL activity and its response to drugs can be determined in CD34+ CML stem cells by CrkL phosphorylation status using flow cytometry. *Leukemia* 2006;20(6):1035–1039.

26. Hedley DW, Chow S, Goolsby C, Shankey TV. Pharmacodynamic monitoring of molecular-targeted agents in the peripheral blood of leukemia patients using flow cytometry. *Toxicol. Pathol.* 2008;36(1):133–139.

27. Crump M, Hedley D, Kamel-Reid S, Leber B, Wells R, Brandwein J, et al. A randomized phase I clinical and biologic study of two schedules of sorafenib in patients with myelodysplastic syndrome or acute myeloid leukemia: a NCIC (National Cancer Institute of Canada) Clinical Trials Group Study. *Leuk. Lymphoma* 2010;51 (2):252–260.

28. Tong FK, Chow S, Hedley D. Pharmacodynamic monitoring of BAY 43-9006 (Sorafenib) in phase I clinical trials involving solid tumor and AML/MDS patients, using flow cytometry to monitor activation of the ERK pathway in peripheral blood cells. *Cytometry B Clin. Cytom.* 2006;70(3):107–114.

29. Schwock J, Ho JC, Luther E, Hedley DW, Geddie WR. Measurement of signaling pathway activities in solid tumor fine-needle biopsies by slide-based cytometry. *Diagn. Mol. Pathol.* 2007;16(3):130–140.

30. Sheps JA, Ling V. Preface: the concept and consequences of multidrug resistance. *Pflugers Arch.* 2007;453(5):545–553.

31. Ling V, Charles F. Kettering Prize. P-glycoprotein and resistance to anticancer drugs. *Cancer* 1992;69(10):2603–2609.

32. Rubin EH, De Alwis DP, Pouliquen I, Green L, Marder P, Lin Y, et al. A phase I trial of a potent P-glycoprotein inhibitor, Zosuquidar trihydrochloride (LY335979), administered orally in combination with doxorubicin in patients with advanced malignancies. *Clin. Cancer Res.* 2002;8(12):3710–3717.

33. Darzynkiewicz Z, Crissman H, Jacobberger JW. Cytometry of the cell cycle: cycling through history. *Cytometry A* 2004;58(1):21–32.

34. Gautschi O, Heighway J, Mack PC, Purnell PR, Lara PN, Jr., Gandara DR. Aurora kinases as anticancer drug targets. *Clin. Cancer Res.* 2008;14(6):1639–1648.

35. Vader G, Lens SM. The Aurora kinase family in cell division and cancer. *Biochim. Biophys. Acta* 2008;1786(1):60–72.

36. Wilkinson RW, Odedra R, Heaton SP, Wedge SR, Keen NJ, Crafter C, et al. AZD1152, a selective inhibitor of Aurora B kinase, inhibits human tumor xenograft growth by inducing apoptosis. *Clin. Cancer Res.* 2007;13(12):3682–3688.

37. Blum KA, Advani A, Fernandez L, Van Der JR, Brandwein J, Kambhampati S, et al. Phase II study of the histone deacetylase inhibitor MGCD0103 in patients with previously treated chronic lymphocytic leukaemia. *Br. J. Haematol.* 2009;147(4):507–514.

38. Byrd JC, Marcucci G, Parthun MR, Xiao JJ, Klisovic RB, Moran M, et al. A phase 1 and pharmacodynamic study of depsipeptide (FK228) in chronic lymphocytic leukemia and acute myeloid leukemia. *Blood* 2005;105(3):959–967.

39. Vanhaecke T, Papeleu P, Elaut G, Rogiers V. Trichostatin A-like hydroxamate histone deacetylase inhibitors as therapeutic agents: toxicological point of view. *Curr. Med. Chem.* 2004;11(12):1629–43.

40. Kummar S, Gutierrez M, Gardner ER, Donovan E, Hwang K, Chung EJ, et al. Phase I trial of MS-275, a histone deacetylase inhibitor, administered weekly in refractory solid tumors and lymphoid malignancies. *Clin. Cancer Res.* 2007;13 (18 Part 1): 5411–5417.

41. Lee CJ, Dosch J, Simeone DM. Pancreatic cancer stem cells. *J. Clin. Oncol.* 2008;26 (17):2806–2812.

42. Donnenberg VS, Donnenberg AD. Multiple drug resistance in cancer revisited: the cancer stem cell hypothesis. *J. Clin. Pharmacol.* 2005;45(8):872–877.
43. Harrison H, Farnie G, Howell SJ, Rock RE, Stylianou S, Brennan KR, et al. Regulation of breast cancer stem cell activity by signaling through the Notch4 receptor. *Cancer Res.* 2010;70(2):709–718.
44. Ischenko I, Seeliger H, Schaffer M, Jauch KW, Bruns CJ. Cancer stem cells: how can we target them? *Curr. Med. Chem.* 2008;15(30):3171–3184.

8

CD4 T CELL ASSESSMENTS IN EVALUATION OF HIV THERAPEUTICS

Thomas N. Denny, Raul Louzao, John Wong, and Brooke Walker

8.1 INTRODUCTION

Recent figures from the Joint United Nations Programme on HIV/AIDS (UNAIDS) estimate that 33 million people are living with human immunodeficiency virus (HIV) worldwide and an estimated 3 million are on antiretroviral (ARV) therapy (Figures 8.1 and 8.2) [1]. As global efforts continue to make life-saving HIV therapeutics more accessible and affordable to patients, there is a growing need to ensure that routine monitoring of CD4 T cells is affordable and available for patients on ARV therapy. CD4 has proven to be a fundamental immunological marker in HIV infection that can indicate disease progression, drug adherence or resistance, and pathogen susceptibility (e.g., opportunistic diseases such as *Pneumocystis carinii* pneumonia (PCP)). The first clinical trials to assess the efficacy and safety of AZT validated CD4 as a critical prognostic marker of the progression to AIDS and demonstrated its value to accurately predict treatment success or failure. These findings led to the current use of HIV-RNA levels and CD4 levels as end points for drug approval by the U.S. Food and Drug Administration (FDA) in clinical trials for antiretroviral drugs. International and national clinical guidelines on ARV therapy and opportunistic infections (OIs) rely on CD4 T cell counts to recommend when to initiate therapy and prophylaxis against OIs. Timely monitoring of CD4 T cells in HIV-infected patients is critical to assess the efficacy of any ARV drug regimen. It is, therefore, of vital importance that CD4 measurements are evaluated accurately and precisely to ensure that patients receive

Flow Cytometry in Drug Discovery and Development, Edited by Virginia Litwin and Philip Marder

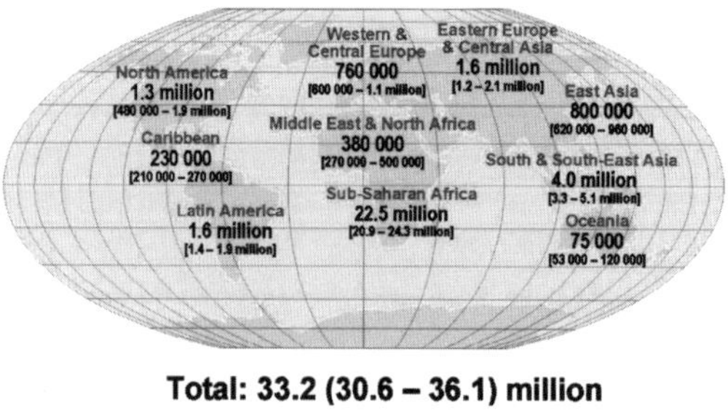

FIGURE 8.1 Adults and children estimated to be living with HIV, 2007 (from UNAIDS, Core slides: global summary of the HIV and AIDS epidemic, 2008).

the most effective treatment. Flow cytometry is regarded as the predicate technique in the evaluation of CD4 T cells. Since the discovery of the HIV virus, flow cytometric methods have evolved to incorporate improved gating strategies coupled with enhanced multicolor analysis techniques that have improved the precision of measuring CD4 counts. CD4 T cell enumeration strategies have also evolved to make testing more accessible and affordable in resource-limited settings [2]. Testing can now be conducted with low-cost point-of-care technologies that can be used in nontertiary settings, which allows more routine monitoring. Patients may remain close to their village or community and need not travel long distances under difficult environmental conditions to have their CD4 levels checked.

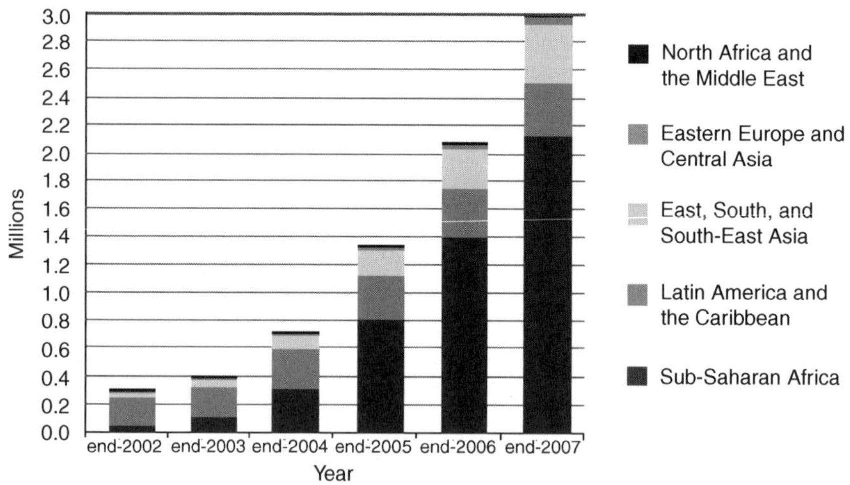

FIGURE 8.2 Number of people receiving antiretroviral drugs in low- and middle-income countries, 2002–2007 (from UNAIDS, Graphics from 2008 Report on the Global AIDS Epidemic).

This chapter will illustrate the prognostic value of CD4 as an immunological marker in the first AZT trials, which led to its current use as an immunological end point in clinical trials of ARV drugs and in patient management to detect its adherence and resistance to ARV therapy. It will also examine the role of CD4 T cell counts in guidelines on ARV therapy and prophylaxis against opportunistic infections. Finally, the chapter will discuss the evolution of CD4 enumeration strategies since the discovery of the HIV virus. This discussion will include the future of CD4 assessment methodology and technology such as developing testing alternatives aimed at getting low-cost monitoring to patients in resource-limited settings.

8.2 HISTORICAL BACKGROUND OF THE HUMAN IMMUNODEFICIENCY VIRUS

Between 1979 and 1981, health clinics throughout the United States saw an unexpected increase in the treatment of men who have sex with men who presented with *Pneumocystis carinii* pneumonia and a relatively rare skin cancer, Kaposi's sarcoma [3–6]. Most of these same patients also presented with a series of other opportunistic infections, including oral mucosal candidiasis and *cytomegalovirus* (CMV) infection. These opportunistic infections had rarely been seen before, but independent research groups in New York and California began to report high rates of these infections among men who have sex with men who had previously been healthy. Subsequent immunological testing revealed that the patients had a reduced percentage and absolute number of T cells and a significant decrease in the helper T cell (CD4 cell) population [3–10]. Scientists hypothesized that this severe immunodeficiency and the resulting opportunistic infections could be the result of a newly acquired syndrome caused by an unknown infectious agent. In May 1983, the human immunodeficiency virus was finally isolated from an AIDS patient and was characterized as a retrovirus belonging to the human T cell leukemia viruses (HTLV) group [8, 9].

HIV was found to selectively target and destroy CD4 T cells by binding to the CD4 antigen using the chemokine receptor 5 (CCR5) or CXC chemokine receptor 4 (CXCR4) to replicate in the body [10]. HIV causes a virtual complete depletion of CD4 T cells as levels can drop to an average of 60% of their original number within 12–18 months after seroconversion [11]. This depletion has a devastating effect on host immunity because it ultimately reduces the production of B cells, monocytes, and natural killer cells [10]. The compromised immune system is not capable of producing a robust enough response to pathogens, which ultimately results in morbidity and mortality in the patient without therapeutic intervention. The destruction of the CD4 T cells correlates with the onset of AIDS [12]. AIDS is clinically defined as having HIV infection and a CD4 count less than 200 cells/μL; however, studies suggest that HIV-infected patients can progress to AIDS with CD4 counts ranging from 150 to 300 cells/mm^3 during 12–36 months [13, 14]. When CD4 counts are less than 200 cells/μL, patients with HIV infection are more susceptible to developing the most common opportunistic infections, *Pneumocystis carinii* pneumonia, *cytomegalovirus*, and *Mycobacterium avium* complex (MAC) [14]. They are also susceptible to develop cancers such as Kaposi's sarcoma and lymphomas [15]. HIV-infected patients with Kaposi's sarcoma are more likely to develop PCP within 24 months if their CD4 counts are below 200 cells/mm^3 [14].

8.3 FIRST AZT CLINICAL TRIALS

After the discovery of the HIV virus, there was an urgent need to develop therapeutics that would either eliminate the virus from the body or keep it from replicating. The first viable candidate as an HIV therapeutic was zidovudine (AZT), a nucleoside reverse transcriptase inhibitor (NRTI). AZT is a thymidine analogue that was initially designed as an anticancer drug, but was found to inhibit the DNA polymerase enzyme reverse transcriptase, required by HIV to replicate in the host system [16]. After initial safety evaluations showed that AZT was well tolerated by patients, a double-blind, placebo-controlled trial of the efficacy was conducted [16, 17]. The trial enrolled 282 patients with AIDS manifested by *Pneumocystis carinii* pneumonia and patients with advanced AIDS-related complex, a term used to describe symptoms typically experienced before the progression to AIDS. The criteria for response were based on clinical end points such as survival, occurrence and frequency of opportunistic infections, occurrence of AIDS-associated neoplasms, Karnofsky performance status, body weight, and the number and severity of symptoms. Patients were stratified into two groups based on absolute number of CD4 T cells, one group with <100 cells/mm^3 and the other group with 101–499 cells/mm^3. Over the course of 24 weeks, 145 subjects received AZT, while 137 received a placebo. At the end of 24 weeks, patients who received AZT had increased CD4 counts, while patients who received the placebo had decreased counts. Patients with AIDS-related complex who received AZT showed the largest increase in CD4 counts through the end of the study. This group had median baseline levels of 190 cells/mm^3 that increased to 340 cells/mm^3 after 20 weeks. Patients with AIDS had CD4 counts that initially increased, but then gradually decreased to below baseline levels after 20 weeks. This group had baseline levels of 54 cells/mm^3 that initially increased to 133 cells/mm^3 after 4 weeks, but then declined to 49 cells/mm^3 at 20 weeks. This trial first demonstrated the importance of using CD4 counts to measure the efficacy of treatment during stage of disease. While clinical end points were the preferred criteria for response, CD4 counts had vital prognostic value as biomarkers for disease progression because a higher baseline count correlated with a higher increase in CD4 T cells and a decrease in frequency of opportunistic infections. Patients who were clinically defined as having AIDS and who had a low baseline CD4 count did not see the same immunological response from the drug.

After this single clinical trial that demonstrated its efficacy, AZT was approved by the FDA on March 19, 1987 with accelerated approval. This fast track approval system was designed to push viable candidates quickly through FDA approval to make them accessible to patients. AZT was approved while clinical trials continued testing the efficacy and safety of the drug. There was mounting criticism that the drugs had been approved too quickly without sufficient testing of the toxicity of the drug. In 1990, the NIH–NIAID AIDS Clinical Trials Group conducted a large, multi-institutional trial to test the efficacy of AZT in asymptomatic adults with CD4 counts below 500 cells/mm^3 [18]. The primary criteria for response were clinical end points, including survival and the development of AIDS or AIDS-related complex defined as two successive CD4 counts <200 cells/mm^3 and the presence of at least two symptoms indicative of AIDS-related complex. The secondary criteria for response included an increase in CD4 cell counts or a reduced rate of decline, a decrease in serum HIV p24 antigen levels, or both. Of the study participants, 428 subjects received placebo, 453 received 500 mg of zidovudine, and 457 received 1500 mg of ziduvudine; however, the placebo arm was terminated by the independent board monitoring data and safety because treatment of AZT at a lower dosage was shown to delay the onset of AIDS. Patients receiving the 500 mg of AZT saw a median annual increase of 39 CD4 cells/mm^3, while the group receiving 1500 mg saw a median annual increase of 26 CD4 cells/mm^3. AZT had the greatest effect on patients with 200–499 CD4 cells/mm^3. This finding suggested that initiating therapy at earlier stages of HIV infection would have a greater impact than during the late stage of the disease.

Although the first AZT trial indicated that the drug was effective at increasing CD4 counts and preventing opportunistic infections, further trials were needed to assess whether the duration of treatment impacted CD4 counts and whether drug resistance could be detected. Another trial testing the safety and efficacy of AZT during the early stage of infection investigated the relationship between stage of infection and the effect on CD4 counts [19]. Based on the finding that AZT generated a larger increase in CD4 T cells in patients in early-stage disease as seen with asymptomatic HIV infection and counts less than 500 cells/mm^3, this trial tested the safety and efficacy in patients with mildly symptomatic HIV infection to see if patients would experience sustained increases in CD4 T cells and less toxicity. This hypothesis was based on the understanding that a more robust immune system would not be as susceptible to potential drug toxicity.

In this large, multi-institutional trial of 711 subjects with mildly symptomatic HIV infection, patients were enrolled with CD4 T cell counts ranging from 200–800 cells/mm^3. Criteria for response were clinical end points, including the prevention of or delay to AIDS or advanced AIDS-related complex, and immunological end points such as a sustained increase in the absolute number of CD4 T cells. Patients were stratified into two groups – one group with CD4 T cell counts ranging 200–500 cells/mm^3 and the other group with 500–799 cells/mm^3. Again, the treatment group with baseline CD4 counts of 200–500 CD4 cells/mm^3 saw the greatest increase in CD4 T cell counts by the end of the study. It was determined that patients with CD4 T cell counts above 500 cells/mm^3 might not benefit from AZT because the risk of toxicity is too great and there were not enough data to support a positive immunological response. Recommendations from this study included the monitoring of CD4 T cell count in patients with relatively normal counts as a method of anticipating potential progression to AIDS.

In another trial that tested whether stage of disease affected CD4 counts, patients were enrolled as late-stage participants with CD4 counts <200 cells/mm^3 and the patients enrolled as early-stage participants had few symptoms and CD4 counts >200 cells/mm^3 [20]. The study found that 89% of patients given AZT during the late stage of the disease developed drug resistance, while 31% of patients given AZT during the early stage of the disease developed drug resistance. This study found that a low CD4 count was predictive of increased probability of resistance. The findings that AZT had a greater impact on patients with early-stage HIV infection led the U.S. Department of Health and Human Services to revise their guidelines regarding the initiation of treatment with AZT [20].

The AZT trials first demonstrated the prognostic value of CD4 as a biomarker to evaluate the efficacy of antiretroviral drugs based on the strong correlation between CD4 T cell counts and disease staging [21–23]. These studies showed that CD4 clearly has value as an indicator of immune dysfunction and disease progression in HIV-infected patients. The recommendation that CD4 counts be used to monitor patients with normal counts underlines CD4's significance in patient management as demonstrated by the Center for Disease Control's 1993 Revised Classification System and Expanded Surveillance Case Definition [13]. In 1997, the FDA changed approval guidelines of antiretroviral drugs to use CD4 count and HIV-RNA levels in clinical trials assessing the efficacy of new candidates [24, 25]. Prior to1997, traditional approval relied on clinical end points in trials of ARV drugs as seen in the first AZT trials. Studies continued to show that CD4 levels and plasma HIV-RNA levels correlated with the onset of certain clinical conditions and, subsequently, an increase in CD4 count and a decrease in viral load indicated that the risk for AIDS-related illness and death declined. Based on these findings, the FDA's Division of Antiviral Drug Products convened a meeting of the Antivirals Advisory Committee in July 1997. The committee concluded that HIV-RNA is a suitable end point for both accelerated approval and traditional approval and recommended that increases in CD4 counts must also be consistent with changes in viral load. Based on the finding that CD4 was a suitable surrogate marker, clinical end points were no longer necessary, but remained an option. While studies have shown that CD4 counts are appropriate surrogate end points in clinical trials, there is conflicting evidence that other factors may contribute to the clinical benefit of the drug independent of the CD4 count [26].

Since 1992, there have been 32 different drugs developed to attack the HIV virus utilizing a variety of different mechanisms, which have shown to enhance CD4 levels and decrease HIV-RNA (Table 8.1). The classes of drugs are as follows:

- Multiclass combination products
- Nucleoside reverse transcriptase inhibitors
- Nonnucleoside reverse transcriptase inhibitors (NNRTIs)
- Protease inhibitors (PIs)
- Fusion inhibitors
- Entry inhibitors – CCR5 coreceptor antagonist
- HIV integrase strand transfer inhibitors

TABLE 8.1 Antiretroviral Drugs Used in the Treatment of HIV Infection

Brand Name	Generic Name	Manufacturer Name	Approval Date	Time to Approval (Months)
		Multiclass combination products		
Atripla	Efavirenz, emtricitabine and tenofovir disoproxil fumarate	Bristol-Myers Squibb and Gilead Sciences	12-July-06	2.5
		Nucleoside reverse transcriptase inhibitors (NRTIs)		
Combivir	Lamivudine and zidovudine	GlaxoSmithKline	27-Sep-97	3.9
Emtriva	Emtricitabine, FTC	Gilead Sciences	02-Jul-03	10.0
Epivir	Lamivudine, 3TC	GlaxoSmithKline	17-Nov-95	4.4
Epzicom	Abacavir and lamivudine	GlaxoSmithKline	02-Aug-04	10.0
Hivid	Zalcitabine, dideoxycytidine, ddC	Hoffmann-La Roche	19-Jun-92	7.6
Retrovir	Zidovudine, azidothymidine, AZT, ZDV	GlaxoSmithKline	19-Mar-87	3.5
Trizivir	Abacavir, zidovudine, and lamivudine	GlaxoSmithKline	14-Nov-00	10.9
Truvada	Tenofovir disoproxil fumarate and emtricitabine	Gilead Sciences, Inc.	02-Aug-04	5.0
Videx EC	Enteric coated didanosine, ddI EC	Bristol-Myers Squibb	31-Oct-00	9.0
Videx	Didanosine, dideoxyinosine, ddI	Bristol-Myers Squibb	9-Oct-91	6.0
Viread	Tenofovir disoproxil fumarate, TDF	Gilead	26-Oct-01	5.9
Zerit	Stavudine, d4T	Bristol-Myers Squibb	24-Jun-94	5.9
Ziagen	Abacavir sulfate, ABC	GlaxoSmithKline	17-Dec-98	5.8
		Nonnucleoside reverse transcriptase inhibitors (NNRTIs)		
Intelence	Etravirine	Tibotec Therapeutics	18-Jan-08	6.0
Rescriptor	Delavirdine, DLV	Pfizer	4-Apr-97	8.7
Sustiva	Efavirenz, EFV	Bristol-Myers Squibb	17-Sep-98	3.2
Viramune	Nevirapine, NVP	Boehringer Ingelheim	21-Jun-96	3.9
		Protease inhibitors (PIs)		
Agenerase	Amprenavir, APV	GlaxoSmithKline	15-Apr-99	6.0
Aptivus	Tipranavir, TPV	Boehringer Ingelheim	22-Jun-05	6.0

(*continued*)

TABLE 8.1 (*Continued*)

Brand Name	Generic Name	Manufacturer Name	Approval Date	Time to Approval (Months)
Crixivan	Indinavir, IDV,	Merck	13-Mar-96	1.4
Fortovase	Saquinavir (no longer marketed)	Hoffmann-La Roche	7-Nov-97	5.9
Invirase	Saquinavir mesylate, SQV	Hoffmann-La Roche	6-Dec-95	3.2
Kaletra	Lopinavir and ritonavir, LPV/RTV	Abbott Laboratories	15-Sep-00	3.5
Lexiva	Fosamprenavir Calcium, FOS-APV	GlaxoSmithKline	20-Oct-03	10.0
Norvir	Ritonavir, RTV	Abbott Laboratories	1-Mar-96	2.3
Prezista	Darunavir	Tibotec, Inc.	23-Jun-06	6.0
Reyataz	Atazanavir sulfate, ATV	Bristol-Myers Squibb	20-Jun-03	6.0
Viracept	Nelfinavir mesylate, NFV	Agouron Pharmaceuticals	14-Mar-97	2.6
		Fusion inhibitors		
Fuzeon	Enfuvirtide, T-20	Hoffmann-La Roche & Trimeris	13-Mar-03	6.0
		Entry inhibitors: CCR5 coreceptor antagonist		
Selzentry	Maraviroc	Pfizer	06-August-07	8.0
		HIV integrase strand transfer inhibitors		
Isentress	Raltegravir	Merck & Co., Inc.	12-Oct-07	6.0

Approved by the U.S. Food and Drug Administration.

8.4 IMPORTANCE OF CD4 MONITORING IN ANTIRETROVIRAL THERAPY

HIV is ultimately fatal in most cases without therapeutic intervention. The majority of deaths are due to the opportunistic infections that manifest in a susceptible immune system. In the absence of treatment, most infected persons will progress to AIDS within 7–10 years [27]. While increasing numbers of patients are on antiretroviral therapy, it is necessary to monitor the efficacy of these drugs. Viral load measurements have proven to be a better predictive marker of progression to AIDS; however, CD4 is a better predictor of death compared to HIV-RNA levels [22, 24, 28]. In addition, CD4 T cell counts are a more cost-effective monitoring alternative to viral load measurements and, based on current technologies, easier to establish in laboratories. This is especially applicable in resource-limited settings where viral load testing may not be available or may not be affordable to patients. The monitoring of CD4 T cell counts is vital to determine when to initiate antiretroviral therapy and when to initiate prophylaxis against opportunistic infections. The Centers for Disease Control 1993 Revised Classification System for HIV infection emphasizes the clinical importance of the CD4 T cell count in the categorization of HIV-related clinical conditions [13]. Although ARV therapy and prophylactic therapies for different types of OIs increase survival rate and decrease the risk of developing coinfections in HIV-infected patients, they can be toxic, invasive, and expensive; therefore, it is important to evaluate the risk of progression to AIDS and developing OIs using CD4 counts to avoid giving unnecessary treatment.

Normal CD4 levels for adults are around 1000 cells/μL, but can range from 450 to over 1500 cells/μL (Table 8.2) [29, 30]. As stated previously, the clinical definition of AIDS is having CD4 levels at less than 200 cells/μL [12, 13]. At this point, the risk of developing opportunistic infections increases. Prophylaxis against opportunistic infections should be started at various CD4 counts depending on the type of infection. The three most common AIDS-related opportunistic infections associated with increased mortality include *Pneumocystis carinii* pneumonia, *cytomegalovirus*, and *Mycobacterium avium* complex [14]. Primary prophylaxis for *Pneumocystis carinii* pneumonia should be started at a CD4 count of <200 cells/μL or if there is a history of *oropharyngeal Candida*, and prophylaxis should be discontinued if the CD4 count stays greater than 200 cells/μL for at least 3 months [31, 32]. For toxoplasmosis, criteria for starting primary prophylaxis include the presence of immunoglobulin G (IgG) antibody to *Toxoplasma* and a CD4 count of <100 cells/μL [31]. Treatment may be discontinued with a CD4 count greater than <200 cells/μL for at least 3 months [31, 32]. Primary prophylaxis against disseminated *Mycobacterium avium* complex should be started with a CD4 count <50 cells/μL and can be stopped if the CD4 count stays greater than <100 cells/μL for at least 3 months [31, 32]. For relatively rare OIs such as Cryptococcosis, Histoplasmosis, and *Cytomegalovirus* Retinitis, routine testing in asymptomatic individuals is recommended only for patients living in regions where these OIs are endemic. Antifungal prophylactic therapy is not usually recommended because of the lack of demonstrated survival benefits and the possibility of drug interactions, drug resistance, and cost [32].

Highly active antiretroviral therapy (HAART) works by suppressing viral replication, which allows CD4 T cells to reconstitute. There is no universal or global consensus as to when to initiate antiretroviral therapy. Reference ranges on the initiation of ARV and prophylaxis against OIs vary from more developed countries to less developed countries based on disease prevalence, financial burden, and the capacity and infrastructure of the health systems [2]. The U.S. Health and Human Services recommends starting ARV when counts are <200 cells/μL and when counts are <350 cells/μL, patient comorbidities and individual scenarios should be taken into account before starting ARV [33]. ARV should also be initiated in pregnant women, those having a history of AIDS-defining illness, persons with HIV-associated nephropathy, and in individuals coinfected with the Hepatitis B virus (HBV) (Table 8.3) [33]. The World Health Organization (WHO) recommends that ARV be considered at CD4 counts <350 cells/μL and that ARV should be initiated in all patients who have CD4 counts <200 cells/μL. Where CD4 testing is unavailable, the WHO clinical staging criteria can be used (Table 8.4). It

TABLE 8.2 Normal Human Blood Lymphocyte Subpopulations at Various Ages

	Age Groups						
Subpopulations (Cells/μL)	Cord Blood	2–3 Months	4–8 Months	12–23 Months	2–5 Years	7–17 Years	Adult
Total lymphocytes							
Median (%)	5400 (41%)	5680 (66%)	5990 (64%)	5160 (59%)	4060 (50%)	2400 (40%)	2100 (32%)
Confidence intervals (%)	4200 (35%\)–6900 (47%)*	2920 (55%\)–8840 (78%)	3610 (45%\)–8840 (79%)	2180 (44%\)–8270 (72%)	2400 (38%\)–5810 (64%)	2000 (36%\)–2700 (43%)*	1600 (28%\)–2400 (39%)*
CD3 T cells							
Median (%)	3100 (55%)	4030 (72%)	4270 (71%)	3300 (66%)	3040 (72%)	1800 (70%)	1600 (73%)
Confidence intervals (%)	2400 (49%\)–3700 (62%)*	2070 (55%\)–6540 (78%)	2280 (45%\)–6450 (79%)	1460 (53%\)–5440 (81%)	1610 (62%\)–4230 (80%)	1400 (66%\)–2000 (76%)*	960 (61%\)–2600 (84%)*
CD4 T cells							
Median (%)	1900 (35%)	2830 (52%)	2950 (49%)	2070 (43%)	1800 (42%)	800 (37%)	940 (46%)
Confidence intervals (%)	1500 (28%\)–2400 (42%)*	1460 (41%\)–5116 (64%)	1690 (36%\)–4600 (61%)	1020 (31%\)–3600 (54%)	900 (35%\)–2860 (51%)	700 (33%\)–1100 (41%)*	540 (32%\)–1660 (60%)*
CD8 T cells							
Median (%)	1500 (29%)	1410 (25%)	1450 (24%)	1320 (25%)	1180	800 (30%)	520 (27%)
Confidence intervals (%)	1200 (26%\)–2000 (33%)*	650 (16%\)–2450 (35%)	720 (16%\)–2490 (34%)	570 (16%\)–2230 (38%)	630 (22%\)–1910 (38%)	600 (27%\)–900 (35%)*	270 (13%\)–930 (40%)*
B cells (CD19 or CD20)†							
Median (%)	1000 (20%)†	900 (23%)	900 (23%)	900 (23%)	900 (24%)	400 (16%)†	246 (13%)†
Confidence intervals (%)	200 (14%\)–1500 (23%)*	500 (19%\)–1500 (31%)	500 (19%\)–1500 (31%)	500 (19%\)–1500 (31%)	700 (21%\)–1300 (28%)	300 (12%\)–500 (22%)*	122 (10%\)–632 (31%)
CD4:CD8 ratio							
Median	1.2	2.2	2.1	1.6	1.4	1.3	1.7
Confidence intervals	0.8–1.8	1.3–3.5	1.2–3.5	1.0–3.0	1.0–2.1	1.1–1.4	0.9–4.5

Source: Stiehm ER, editor. *Immunologic Disorders in Infants and Children*, 4th ed., 1996.

Confidence intervals given are the 5th to 95th percentiles except where indicated (*); these are the 25th to 75th percentiles. B cells use the CD20 antigen except where indicated (†); these use the CD19 antigen. The lymphocyte % is the percentage of total leukocytes. The CD3, CD4, CD8, and B cell (CD19 or CD20) % is the percentage of total lymphocytes. Each subgroup contains at least 22 normal subjects. Combined data from Erkellor-Yuksel FM. et al. *J. Pediatr*. 1992;120:216–222 (cord blood, 7–17 years; B cells); Denny T. et al. *JAMA* 1992;267:1484–1488 (2–3 months to 5 years); Fahey JL, cited in Giorgi, JV et al. in Rose NR et al. *Manual of Clinical Laboratory Immunology*, 4th ed., American Society for Microbiology, Washington, DC, 1992, pp. 174–181 (adults).

TABLE 8.3 Indications for Initiating Antiretroviral Therapy for the Chronically HIV-1 Infected Patient[a]

Clinical Condition and/or CD4 Count	Recommendations
History of AIDS-defining illness (AI) CD4 count <200 cells/mm^3 (AI) CD4 count 200–350 cells/mm^3 (AII) Pregnant women[b] (AI) Persons with HIV-associated nephropathy (AI)Persons coinfected with hepatitis B virus (HBV), when HBV treatment is indicated (treatment with fully suppressive antiviral drugs active against both HIV and HBV is recommended.) (BIII)	Antiretroviral therapy should be initiated
Patients with CD4 count >350 cells/mm^3 who do not meet any of the specific conditions listed above	The optimal time to initiate therapy in asymptomatic patients with CD4 count >350 cells/mm^3 is not well defined. Patient scenarios and comorbidities should be taken into consideration

[a]Updated December 1, 2007.
[b]Panel on Antiretroviral Guidelines for Adults and Adolescents. Guidelines for the use of antiretroviral agents in HIV-1-infected adults and adolescents, Department of Health and Human Services, November 3, 2008; 1–139. Available at http://www.aidsinfo.nih.gov/ContentFiles/AdultandAdolescentGL.pdf. Accessed Jun 16, 2008, p. 7, Table 5a.

TABLE 8.4 WHO Clinical Staging of HIV/AIDS for Adults and Adolescents

Primary HIV infection
Asymptomatic
Acute retroviral syndrome
Clinical stage 1
Asymptomatic
Persistent generalized lymphadenopathy
Clinical stage 2
Moderate unexplained weight loss (<10% of presumed or measured body weight)
Recurrent respiratory infections (sinusitis, tonsillitis, otitis media, and pharyngitis)
Herpes zoster
Angular cheilitis
Recurrent oral ulceration
Papular pruritic eruptions
Seborrheic dermatitis
Fungal nail infections

(*continued*)

TABLE 8.4 (*Continued*)

Clinical stage 3

- Unexplained severe weight loss (>10% of presumed or measured body weight)
- Unexplained chronic diarrhea for >1 month
- Unexplained persistent fever for >1 month (>37.6°C, intermittent or constant)
- Persistent oral candidiasis (thrush)
- Oral hairy leukoplakia
- Pulmonary tuberculosis (current)
- Severe presumed bacterial infections (e.g., pneumonia, empyema, pyomyositis, bone or joint infection, meningitis, bacteremia)
- Acute necrotizing ulcerative stomatitis, gingivitis, or periodontitis
- Unexplained anemia (hemoglobin <8 g/dL)
- Neutropenia (neutrophils <500 cells/μL)
- Chronic thrombocytopenia (platelets <50,000 cells/μL)

Clinical stage 4

- HIV wasting syndrome, as defined by the CDC (see Table 8.3)
- Pneumocystis pneumonia
- Recurrent severe bacterial pneumonia
- Chronic herpes simplex infection (orolabial, genital, or anorectal site for >1 month or visceral herpes at any site)
- Esophageal candidiasis (or candidiasis of trachea, bronchi, or lungs)
- Extrapulmonary tuberculosis
- Kaposi sarcoma
- *Cytomegalovirus* infection (retinitis or infection of other organs)
- Central nervous system toxoplasmosis
- HIV encephalopathy
- Cryptococcosis, extrapulmonary (including meningitis)
- Disseminated nontuberculosis mycobacteria infection
- Progressive multifocal leukoencephalopathy
- Candida of the trachea, bronchi, or lungs
- Chronic cryptosporidiosis (with diarrhea)
- Chronic isosporiasis
- Disseminated mycosis (e.g., histoplasmosis, coccidioidomycosis, penicilliosis)
- Recurrent nontyphoidal Salmonella bacteremia
- Lymphoma (cerebral or B cell non-Hodgkin)
- Invasive cervical carcinoma
- Atypical disseminated leishmaniasis
- Symptomatic HIV-associated nephropathy
- Symptomatic HIV-associated cardiomyopathy
- Reactivation of American trypanosomiasis (meningoencephalitis or myocarditis)

Source: WHO case definitions of HIV for surveillance and revised clinical staging and immunological classification of HIV-related disease in adults and children, August 7, 2006.

is also recommended to use successive counts to determine when to initiate ARV rather than one count that could have a higher degree of variability due to technical bias or biological reasons [34, 35]. The frequency of counts also plays a role in accuracy [34].

HAART has shown not to be as effective in patients with advanced disease; therefore, stage of infection has a critical role in how well ARV therapy will work [36]. Studies have shown that patients who have immune systems that are capable of controlling the virus and preventing opportunistic infections may exhibit a healthier prognosis when on HAART [25]. Therapy must be initiated before the patient's immune system is too compromised to effectively reconstitute CD4 T cells [36]. Routine monitoring of CD4 counts can anticipate progression to AIDS and can also detect drug resistance allowing the initiation of therapy or changing to alternative ARV therapeutics to combat resistance. HIV-infected patients should have CD4 levels routinely checked one to two times per year. In resource-limited settings where CD4 testing is not readily available or is too expensive, clearer evidence of immunosuppression may be needed before treatment can be recommended. Complications from long-term use of HAART are still unknown; however, studies have shown that delaying treatment has long-term disadvantages in terms of patient survival [37]. Studies have shown that patients who were severely immunodeficient at the start of therapy experienced higher rates of AIDS and death up to 6 years later [37]. In addition, there is increasing evidence to suggest that there is an increased risk of resistance to HAART if treatment is started with a low CD4 count. A recent study found that the frequency of HIV resistance mutations was 50% among patients who started HAART with a CD4 T cell count between 0 and 349 cells/mm^3 compared to a frequency of 22% among patients who started HAART at a CD4 T cell count $\geq$350 cells/mm^3 [38]. This finding suggests that initiating HAART at higher CD4 cell counts may decrease the risk of developing resistance to therapy.

8.5 EVOLUTION OF CD4 TESTING AND METHODOLOGY

Over the past 25 years, there have been major technological advances in the way that CD4$^+$ T cells are enumerated. The utility of CD4 measurements as a prognostic marker for HIV infection was an early finding in the understanding of HIV/AIDS [39]. The technology available to perform this testing included flow cytometry as well as manual counting by light and fluorescent microscopy. These experiments rely on the use of monoclonal antibodies targeted to the CD4 antigen on T lymphocytes. The CD4 molecule is a transmembrane glycoprotein approximately 55 kDa in size that is involved in cell activation upon its recognition to target antigen-presenting cells [40]. CD4 positive T cells are selected during their maturation from hematopoietic stem cells in the bone marrow to peripheral blood circulation via the thymus, by their affinity to MHC II complexes [41, 42]. Production of antibody to the CD4 molecule in the late 1970s led to further identification of the involvement of the T helper cell population in cellular immunity. Variability in antibody specificity led to the strategy of using antibody combinations to exclude non-T cells in the analysis. Prior to antigen nomenclature conventions, the OKT4 antibody clone developed by Ortho Diagnostics (Ortho Clinical Diagnostics, Rochester, NY), T4 clone by Coulter Corporation (Beckman Coulter, Inc., Miami, FL) and Leu-3 clone by Becton Dickinson {BD Biosciences, San Jose, CA] were the first antibodies widely used for CD4 cell enumerations. Hybridoma technology led the way to refining the production of antibodies with higher specificity. The use of highly specified monoclononal antibodies complements the rapid throughput of flow cytometry. While manual counting methods are still an option, flow cytometry for its reliability and ease of operation has become the predicate technique for the quantitative analysis of CD4 T-lymphocytes.

Absolute values of CD4 positive cells can be calculated using a two-system approach, combining the lymphocyte counts obtained from a hematology analyzer and the percentage value from a flow cytometer. However, statistical error inherent in this dual platform method can contribute to variability in the determination of CD4 absolute values. Alternatively, the use of standardization beads or a fixed volume of specimen in a single platform approach allows a more direct measurement of the absolute marker cell count in blood samples. The single platform approach has gained wide acceptance as the preferred method in performing CD4 enumeration when monitoring HIV/AIDS patients. The manual method for CD4 enumeration, however overshadowed by automation and instrumentation development, has maintained its utility particularly in resource-limited settings. The Manual CD4 Kit from Beckman Coulter consists of CD14 labeled cytospheres, CD4 labeled cytospheres, and an erythrocyte-lysing reagent. Whole blood collected in EDTA is initially incubated with CD14 reagent to block monocyte binding and then labeled with CD4 conjugated cytospheres. Lysing solution is added to this mixture and this preparation is loaded onto a hemocytometer for counting by light microscopy. CD4 positive cells are identified by the presence of three or more large latex spheres attached to the cell surface (Figure 8.3). Monocytes react with the CD14 reagent to yield a brown coloration and are excluded from the analysis. The $CD4^+$ cells are counted and reported as cells/μL.

To date, there have been two flow cytometry-based systems marketed that are dedicated to single-platform CD4 absolute counting, Becton Dickinson (San Jose, CA) and Guava Technologies (Millipore Corporation, Billerica, MA). The Becton Dickinson FACSCount system introduced in 1994 uses counting beads to directly measure the absolute values of CD3, CD4, and CD8 populations. Recent revisions to

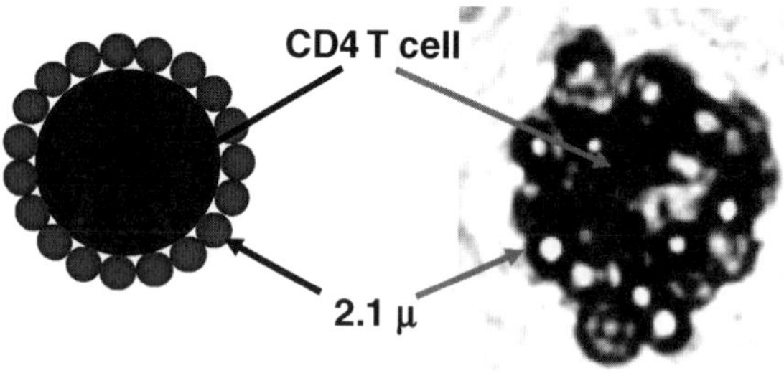

FIGURE 8.3 Rosetting cytospheres (from Frank Mandy, National Laboratory for HIV Immunology (NHLI), Public Health Agency of Canada, Ottawa, ON). (See the color version of the figure in the Color Plates section.)

this FACSCount system provides for the testing of CD4 percentage as well as CD4 absolute values. The Guava EZCD4 system measures the absolute CD4 values directly by using a microcapillary fixed volume method. In the Guava system, separate two-color panels of CD3/CD4 and CD3/CD8 are available for testing. Becton Dickinson FACSCount and Guava EZCD4 systems are FDA approved. In addition to the dedicated CD4 flow cytometry systems, there are single and dual platform methods available to perform CD4 absolute counts. Among the manufacturers of these flow cytometry systems are Accuri Cytometers (Accuri Cytometers Inc., Ann Arbor, MI), Beckman Coulter, Becton Dickinson, Partec (Partec GmbH, Muester, Germany), and PointCare (PointCare Technologies, Marlborough, MA). A flow cytometric method called Pan-Leukogate or PLG CD4 was recently introduced by investigators at the National Health Laboratory Systems in the Republic of South Africa. This counting approach uses a two-color antibody panel to report percentage and absolute CD4 values. The reagent used in this method is licensed through Beckman Coulter (Beckman Coulter, Inc., Miami, FL) and is compatible with most cytometry systems. Several new nonflow cytometric systems are in development to perform CD4 testing. LabNow (LabNow, Inc., Austin, TX) and Zyomyx (Zyomyx, Inc., Hayward, CA) use microchip-based technology to test for both percent and absolute values. Point-of-care technology developed by Beckman Coulter, Burnet Institute, and Millennium Biotechnology use a rapid low-technology dipstick-like approach for semiquantitative evaluation of CD4 values.

8.6 FUTURE OF CD4 T CELL ASSESSMENT METHODOLOGIES AND TECHNOLOGIES

Strategies to develop low-cost laboratory assays with high degree of accuracy and precision are more pronounced in the management and treatment of HIV than with any other disease. This is particularly relevant to sustaining the level of care of this disease in countries with limited resources. The challenge is to deliver a system of testing that is accurate, simple, low cost, and readily available in local settings. Initiatives supported by private industry as well as government and private funding are underway to develop such a system. There are several fronts emerging from these initiatives. In the area of instrumentation, the development of a compact and simple flow cytometer is the focus of manufacturers, old and new alike. In the area of reagent development, the reintroduction of lyophilization technology is advancing the field in reducing the cost related to reagent transport and storage. The removal of refrigeration needs in the handling of these lyophilized reagents is a considerable cost savings factor. In the area of system innovations, the development of a point-of-care method for CD4 testing is gaining the most fervor. The point-of-care system will provide accurate and reliable testing in remote as well as well-populated areas where decisions to start drug treatment depend on quick and easily obtained CD4 counts.

CD4 T cell assessments are useful prognostic markers in HIV infection because they can indicate disease progression and drug adherence or resistance and susceptibility to OIs. They are invaluable for use in clinical trials to assess the efficacy and safety of HIV therapeutics and in patient management to assess the efficacy of an ARV drug regimen. As HIV therapeutics become more affordable and accessible to patients, routine monitoring of CD4 T cells should be equally as accessible to patients particularly in resource-limited settings where HIV infection is most prevalent. As new methods and instrumentation are developed for CD4 testing, it will be increasingly important to evaluate these new approaches against previously established methods. Correlation of methods should be evaluated in experiments involving within-laboratory comparisons of instrumentation as well as good performance in external quality assessment (EQA) programs. These measures will ensure that whatever be the technique used to assess the $CD4^+$ T cells, the patient receives the most appropriate and timely treatment.

REFERENCES

1. UNAIDS. 2008 Report on the Global AIDS Epidemic. Available at http://www.unaids.org/en/KnowledgeCentre/HIVData/GlobalReport/2008/2008_Global_report.asp.
2. Barnett D, Walker B, Landay A, Denny TN. CD4 immunophenotyping in HIV infection. *Nat. Rev. Microbiol.* 2008;6(Suppl. 11):S7–S15.
3. Gottlieb MS et al. *Pneumocystis carinii* pneumonia and mucosal candidiasis in previously healthy homosexual men: evidence of a new acquired cellular immunodeficiency. *N. Engl. J. Med.* 1981;305:1425–1431.
4. Masur H et al. An outbreak of community-acquired *Pneumocystis carinii* pneumonia: initial manifestation of cellular immune dysfunction. *N. Engl. J. Med.* 1981;305: 1431–1438.
5. Siegal FP, Lopez C, Hammer GS et al. Severe acquired immunodeficiency in male homosexuals, manifested by chronic perianal ulcerative herpes simplex lesions. *N. Engl. J. Med.* 1981;305:1439–1444.
6. Schroff RW, Gottlieb MS, Prince HE, Chai LL, Fahey JL. Immunological studies of homosexual men with immunodeficiency and Kaposi's sarcoma. *Clin. Immunol. Immunopathol.* 1983;27:300–314.
7. Ammann AJ et al. Acquired immune dysfunction in homosexual men: immunologic profiles. *Clin. Immunol. Immunopathol.* 1983;27:315–325.
8. Barré-Sinoussi F et al. Isolation of a T-lymphotropic retrovirus from a patient at risk for acquired immune deficiency syndrome (AIDS). *Science* 1983;220:868–871.
9. Popovic M, Sarngadharan MG, Read E, Gallo RC. Detection, isolation, and continuous production of cytopathic retroviruses (HTLV-III) from patients with AIDS and pre-AIDS. *Science* 1984;224:497–500.
10. Bowen DL, Lane HC, Fauci AS. Immunopathogenesis of the acquired immunodeficiency syndrome. *Ann. Intern. Med.* 1985;103(5):704–709.
11. Giorgi JV, Detels R. T-cell subset alterations in HIV-infected homosexual men: NIAID multicenter AIDS cohort study. *Clin. Immunol. Immunopathol.* 1989;52:10–18.
12. Margolick JB et al. Decline in total T cell count is associated with onset of AIDS, independent of $CD4^+$ lymphocyte count: implications for AIDS pathogenesis. *Clin. Immunol. Immunopathol.* 1998;88:256–263.

13. Centers for Disease, Control. 1993 revised classification system for HIV infection and expanded surveillance case definition for AIDS among adolescents and adults. *MMWR Recomm Rep.* 1992;41(RR-17):1–19.

14. Masur H, Ognibene FP, Yarchoan R, Shelhamer JH, Baird BF, Travis W, Suffredini AF, Deyton L, Kovacs JA, Falloon J, et al. CD4 counts as predictors of opportunistic pneumonias in human immunodeficiency virus (HIV) infection. *Ann. Intern. Med.* 1989;111(3):223–231.

15. Longo DL, Steis RG, Lane HC, Lotze MT, Rosenberg SA, Preble O, Masur H, Rook AH, Fauci AS, Jacob J, et al. Malignancies in the AIDS patient: natural history, treatment strategies, and preliminary results. *Ann. N. Y. Acad. Sci.* 1984;437:421–430.

16. Yarchoan R, Klecker RW, Weinhold KJ, Markham PD, Lyerly HK, Durack DT, Gelmann E, Lehrman SN, Blum RM, Barry DW, et al. Administration of 3′-azido-3′-deoxythymidine, an inhibitor of HTLV-III/LAV replication, to patients with AIDS or AIDS-related complex. *Lancet* 1986;8481:575–580.

17. Fischl MA, Richman DD, Grieco MH, Gottlieb MS, Volberding PA, Laskin OL, Leedom JM, Groopman JE, Mildvan D, Schooley RT, et al. The efficacy of azidothymidine (AZT) in the treatment of patients with AIDS and AIDS-related complex. A double-blind, placebo-controlled trial. *N. Engl. J. Med.* 1987;317(4):185–191.

18. Volberding PA, Lagakos SW, Koch MA, Pettinelli C, Myers MW, Booth DK, Balfour HH, Jr., Reichman RC, Bartlett JA, Hirsch MS, et al. Zidovudine in asymptomatic human immunodeficiency virus infection. A controlled trial in persons with fewer than 500 CD4-positive cells per cubic millimeter. The AIDS Clinical Trials Group of the National Institute of Allergy and Infectious Diseases. *N. Engl. J. Med.* 1990;322(14):941–949.

19. Fischl MA, Richman DD, Hansen N, Collier AC, Carey JT, Para MF, Hardy WD, Dolin R, Powderly WG, Allan JD, et al. The safety and efficacy of zidovudine (AZT) in the treatment of subjects with mildly symptomatic human immunodeficiency virus type 1 (HIV) infection. A double-blind, placebo-controlled trial. The AIDS Clinical Trials Group. *Ann. Intern. Med.* 1990;112(10):727–737.

20. Richman DD, Grimes JM, Lagakos SW. Effect of stage of disease and drug dose on zidovudine susceptibilities of isolates of human immunodeficiency virus. *J. Acquir. Immune Defic. Syndr.* 1990;3(8):743–746.

21. Mellors JW et al. Plasma viral load and CD4 + lymphocytes as prognostic markers of HIV-1 infection. *Ann. Intern. Med.* 1997;126:946–954.

22. Fahey JL et al. The prognostic value of cellular and serologic markers in infection with human immunodeficiency virus type 1. *N. Engl. J. Med.* 1990;322:166–172.

23. Jacobson LP, Li R, Phair J, Margolick JB, Rinaldo CR, Detels R, et al. Evaluation of the effectiveness of highly active antiretroviral therapy in persons with human immunodeficiency virus using biomarker-based equivalence of disease progression. *Am. J. Epidemiol.* 2002;155:760–770.

24. Olsen, CH, Gatell, J, Ledergerber, B, Katlama, C, Friis-Møller, N, Weber, J, Horban, A, Staszewski, S, Lundgren, JD, Phillips, AN, EuroSIDA Study Group. Risk of AIDS and death at given HIV-RNA and CD4 cell count, in relation to specific antiretroviral drugs in the regimen. *AIDS* 2005;19(3):319–330.

25. U.S. Department of Health and Human Services Food and Drug Administration Center for Drug Evaluation and Research (CDER). Antiretroviral drugs using plasma HIV RNA measurements: clinical considerations for accelerated and traditional approval. Available

at http://www.fda.gov/downloads/Drugs/GuidanceComplianceRegulatoryInformation/Guidances/ucm070968.pdf. Accessed October 2002.

26. HIV Surrogate Marker Collaborative, Group. Human immunodeficiency virus type 1 RNA level and CD4 count as prognostic markers and surrogate end points: a meta-analysis. HIV Surrogate Marker Collaborative Group. *AIDS Res. Hum. Retroviruses* 2000;16: 1123–1133.
27. Moss AR, Bacchetti P. Natural history of HIV infection. *AIDS* 1989;3(2):55–61.
28. Ledergerber B, Egger M, Erard V, Weber R, Hirschel B, Furrer H, Battegay M, Vernazza P, Bernasconi E, Opravil M, Kaufmann D, Sudre P, Francioli P, Telenti A. AIDS-related opportunistic illnesses occurring after initiation of potent antiretroviral therapy: the Swiss HIV Cohort Study. *JAMA* 1999;282(23):2220–2226.
29. Stiehm ER, editor. *Immunologic Disorders in Infants and Children*, 4th ed., W.B. Saunders Company, Philadelphia, PA, 1996, p.217.
30. Denny T. et al. Lymphocyte subsets in healthy children during the first 5 years of life. *JAMA* 1992;267:1484–1488.
31. U.S., Public Health Service and the Infectious Diseases Society of, America. Criteria for starting, discontinuing, and restarting opportunistic infection prophylaxis for adults with human immunodeficiency virus infection, 2002.
32. Kaplan JE, Benson C, Holmes KH, Brooks JT, Pau A, Masur H, Centers for Disease Control and Prevention (CDC), National Institutes of Health, HIV Medicine Association of the Infectious Diseases Society of America. Guidelines for prevention and treatment of opportunistic infections in HIV-infected adults and adolescents: recommendations from CDC, the National Institutes of Health, and the HIV Medicine Association of the Infectious Diseases Society of America. *MMWR Recomm. Rep.* 2009;58(RR-4):1–207.
33. Panel on Antiretroviral Guidelines for Adults and, Adolescents. Guidelines for the use of antiretroviral agents in HIV-1-infected adults and adolescents, Department of Health and Human Services, November 3, 2008, pp.1–139. Available at http://www.aidsinfo.nih.gov/ContentFiles/AdultandAdolescentGL.pdf. Accessed Jun. 16, 2008, p. 7, Table 5a.
34. Phillips AN, Lee CA, Elford J, Janossy G, Kernoff PB. The cumulative risk of AIDS as the CD4 lymphocyte count declines. *JAIDS* 1992;5:148–152.
35. Malone JL, Simms TE, Gray GC, Wagner KF, Burge JR, Burke DS. Sources of variability in repeated T-helper lymphocyte counts from human immunodeficiency virus type 1-infected patients: total lymphocyte count fluctuations and diurnal cycle are important. *J. Acquir. Immune. Defic. Syndr.* 1990;3(2):144–151.
36. Dyrhol-Riise AM, Voltersvik P, Rosok BI, Olofsson J, Asjo B. Normalization of CD4 + cell numbers and reduced levels of memory CD8 + cells in blood and tonsillar tissue after highly active antiretroviral therapy in early HIV type-1 infection. *AIDS Res. Hum. Retroviruses* 2000;16:191–201.
37. Antiretroviral Therapy Cohort, Collaboration. Importance of baseline prognostic factors with increasing time since initiation of highly active antiretroviral therapy: collaborative analysis of cohorts of HIV-1-infected patients. *J. Acquir. Immune. Defic. Syndr.* 2007;46:607–615.
38. Uy J, Armon C, Buchacz K, Wood K, Brooks JT; the HOPS Investigators. Initiation of HAART at higher CD4 cell counts is associated with a lower frequency of antiretroviral drug resistance mutations at virologic failure. *J. Acquir. Immune. Defic. Syndr.* 2009;51 (4):450–453.

39. Klatzmann D, Champagne E, Charmaret S, Gruest J, Guetard D, Hercend T, Gluckman JC, Montagnier L. T-lymphocyte T4 molecule behaves as the receptor for the human retrovirus LAV. *Nature* 1984;312:767–768.

40. Bierer BE, Sleckman BP, Ratnofsky SE, Burakoff SJ. The biological roles of CD2, CD4, and CD8 in T-cell activation. *Annu. Rev. Immunol.* 1989;7:579–599.

41. Biddison WE, Rao PE, Talle MA, Goldstein G, Shaw S. Possible involvement of the OKT4 molecule in the T cell recognition of class II HLA antigens. Evidence from studies of cytotoxic lymphocytes specific for SB antigens. *J. Exp. Med.* 1982;156:1065–1076.

42. Bowman MR, MacFerrin KD, Schreiber SL, Burakoff SJ. Identification and structural analysis of residues in the V1 region of CD_4 involved in interaction with human immunodeficiency virus envelope glycoprotein gp120 and class II major histocompatibility complex molecules. *Proc. Natl. Acad. Sci. USA* 1990;87:9052–9056.

9

MONITORING THE CELLULAR COMPONENTS OF THE IMMUNE SYSTEM DURING CLINICAL TRIALS: A TRANSLATIONAL MEDICINE APPROACH

Virginia Litwin and James Andahazy

9.1 TRANSLATIONAL MEDICINE IN DRUG DEVELOPMENT

Historically, research scientists, clinical scientists, and pharmaceutical scientists operated in isolation, rarely exchanging information. However, the emergence of the translational medicine approach has facilitated a bidirectional flow of information from "bench to bedside." The importance of this paradigm shift is highlighted in the NIH Roadmap for Clinical and Translational Science that describes a pathway to shepherd biomedical discoveries into clinical application and advance both basic and clinical research (Figure 9.1) [1, 2]. Moreover, translational research is emerging as a field in its own right [3].

The translational medicine approach is now routinely applied in the drug development process in order to improve the success rate of lead compounds as safe and efficacious treatments [4]. The major stages of the drug development process are target identification, target validation, pharmaceutical formulation, toxicity evaluation, and clinical development (Figure 9.2). During the early stages of drug development (target identification and validation), pharmaceutical scientists utilize information generated by research scientists defining the molecular pathways and mechanistic origins of diseases. In the later stages of drug development (toxicity evaluation and clinical development), pharmaceutical scientists utilize information

Flow Cytometry in Drug Discovery and Development, Edited by Virginia Litwin and Philip Marder

FIGURE 9.1 Translational medicine: accelerating the progression of scientific advances from the bench to the bedside.

generated by clinical scientists characterizing the pathology and pathophysiology of diseases.

Early clinical trials primarily assess pharmacokinetics (PK) and safety in the human population, but they now also evaluate efficacy through the inclusion of pharmacodynamic (PD) biomarkers. The most accepted definition of a biomarker is a "biological substance or features that can be used to indicate normal biological process, disease process, or responses to therapy" [5]. PD biomarkers are included in clinical trials under the supposition that they may substitute for, or augment, defined clinical end points. By utilizing PD biomarkers that represent the disease pathology, indications of efficacy can be obtained early on in the clinical development process in smaller clinical trials and thus potentially "buy down the risk" of costly, large, late-stage trials that focus on clinical outcomes [6]. In addition, PD biomarkers can potentially help bridge the gap between drug discovery and clinical development that is especially important for novel drug targets that have not yet been tested in the human population [7].

For an effective use of PD biomarker data, the following questions will be addressed during analysis: Is there a correlation between the PK data and the PD biomarker? Is there a relationship between the drug activity and the PD biomarker? Is there a correlation between the PD biomarker and the clinical outcome?

The translational medicine approach provides details to these questions not in a unidirectional manner, but rather provides a dynamic cyclical flow of information. In the translational medicine cycle, the PD biomarker data are used by scientists, moving through each stage from basic research to drug and compound development. By completing the translational medicine cycle, data from the clinical trials can provide

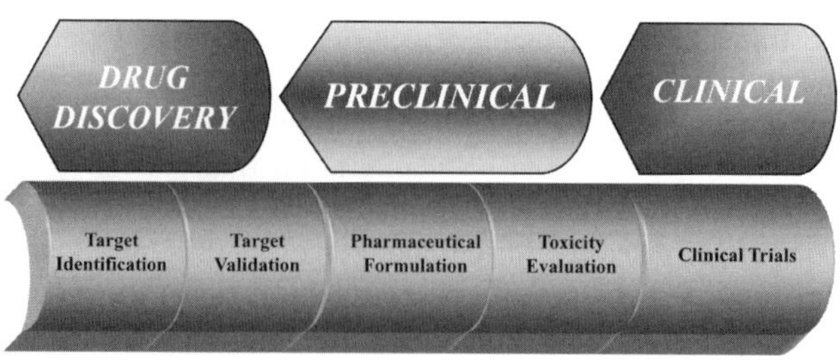

FIGURE 9.2 Drug development lifecycle.

information back to both the research and the clinical scientists, such as additional details regarding mechanism of action, or disease-specific diagnostic or prognostic biomarkers.

Examples of the translational medicine approach are evident in the development of therapeutic compounds that target components of the immune system in diseases that encompass an inflammatory component. A recent review of the development of anti-TNF treatment provides an elegant illustration of this approach [8]. The process begins in the laboratories conducting basic research in immunology where the increasingly more complex mysteries of immune system continue to be elucidated: new cytokines, receptors, signaling pathways, transcription factors, cellular subsets, and functional activities are identified and characterized at a rapid rate (Figure 9.3). There are three major phases to translational medicine: basic science, clinical research, and drug development.

Historically, in the first phase, discoveries in immunology were achieved using elegant *in vivo* and *in vitro* murine models that first employed inbred mouse strains and adoptive transfer experiments, and later molecular manipulations involving gene knockouts and gene transfer systems. *In vitro* cellular systems and molecular tools were then used to verify and refine the findings in the human system. Even though these highly controlled experimental approaches fall short of fully translating to the complex realities of the whole, the basic premises are valid. Critical information such as defining the key cellular components, soluble factors, molecular pathways, and rate-limiting steps in the immune system has been gained through the use of these models. In addition, these experimental models provide insight into potential new drug targets.

In the next phase, clinical research scientists help to characterize the pathogenesis of immune abnormalities by conducting extensive case studies in which known immune parameters are observed and analyzed in patients with immunological disorders. This extensive immunological characterization in patient populations provides insights into potential disease-specific biomarkers.

In the last phase, pharmaceutical scientists translate all of this information into drug development. Information on the need for new compounds comes from clinical research. Direction for selection of PD biomarker would most likely be obtained from the case study reports. Potential PD biomarkers will also come from basic research and experimental modeling.

If the PD biomarker is a cellular component of the immune system, flow cytometry represents the most useful analytical tool for both extensive phenotypic and functional analyses. Discrimination and enumeration of the cellular components of the immune system, and the ever-increasing number of distinct subsets, require simultaneous evaluation of multiple surface and/or intracellular characteristics. Multiparametric flow cytometry is the ideal (and perhaps the only) technology to obtain these measurements because of the number of data points that can be simultaneously captured. Data can be collected with relatively high throughput and good precision. Moreover, functional assays such as cytokine production, antigen-specific proliferation, apoptosis, and even cytotoxicity can be assessed with flow cytometry.

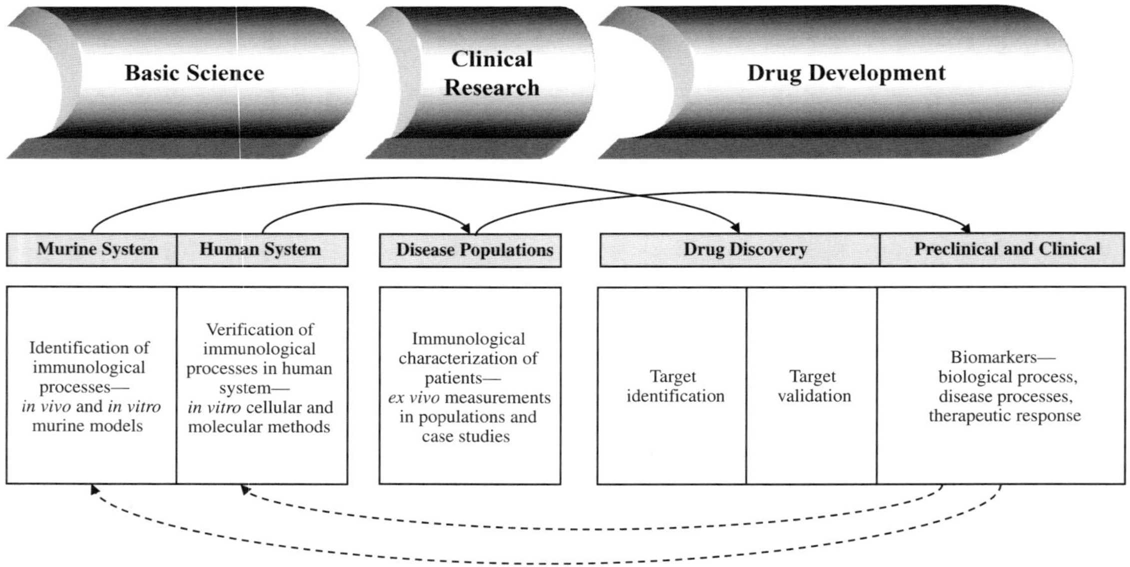

FIGURE 9.3 Translational medicine application in drug development. The development of new drug entities depends on the progression of scientific advances from basic science laboratories to clinical research studies and to drug discovery laboratories. One example would be in the area of immunology where discoveries made in the basic science laboratories are applied when characterizing immunological abnormalities in autoimmune populations. In addition, the information would be used to identify new molecular targets for therapeutic intervention in the treatment of autoimmunity. Conversely, biomarker results obtained during preclinical or clinical studies may assist in elucidating new immunological processes and feed information back to the basic science laboratories.

9.2 BASIC RESEARCH: CELLULAR COMPONENTS OF THE IMMUNE SYSTEM

The immune system has traditionally been subdivided into two distinct arms, the innate immune system and the adaptive immune system. In recent years, the interaction and interdependence between the two systems have come to light and thus the distinctions between the two have been blurred. The major effector cells of the innate immune system include neutrophils, monocytes, dendritic cells (DCs), and natural killer (NK) cells. The innate immune system provides the first line of defense against pathogens that are identified by pattern recognition-type cellular receptors such as the Toll-like receptors (TLR). These evolutionarily conserved receptors have limited diversity and require no antigenic priming. The innate immune cells bearing these receptors have no capacity for memory responses. In contrast, in the adaptive immune system foreign antigen is detected by highly specific, membrane-bound receptors on the surface of T and B lymphocytes. The T-cell antigen receptor (TCR) and the B-cell antigen receptor (BCR or membrane immunoglobulin) are highly specific and have the capacity for unlimited diversity. Although the adaptive immune system requires antigen priming, once activated it is capable of long-term immunity by mounting vigorous memory responses even decades after the initial exposure. Functional responses of the adaptive immune system are either cellular or humoral (antibody-driven) responses. T-lymphocyte cells represent 50–90% of the peripheral blood lymphocytes and are often referred to as "directors" of the adaptive immune response. T-lymphocyte cells are divided into two major subsets: T-helper cells (Th) expressing CD4 and cytotoxic T cells (CTLs) expressing CD8. The immunoglobulin (Ig) producing B lymphocytes are the principal cellular mediators of humoral immune responses. Most of the adult B cells (80%) are found in the lymph nodes and thus they represent only 4–25% of the peripheral blood lymphocytes.

In the initial stages of an antigen-specific immune response, the interaction of pathogens (intracellular, parasitic, and extracellular) or allergens with the cellular components of the innate immune system results in cytokine production and antigen presentation to the cellular components of the adaptive immune system (Figure 9.4). Next naïve, or antigen-inexperienced, T-lymphocyte cells proliferate and generate a variety of antigen-specific effector and memory cells with distinct functions [9]. After the immune response has subsided, central memory cells patrol lymphoid tissue while effector memory cells actively launch immune responses to subsequent assaults. It is not yet clear which cells are actually most important for maintaining immunological memory. CD4 T-lymphocyte cells are the crucial component of the adaptive immune system. Effector CD4 T-lymphocyte cells are heterogeneous in terms of their surface phenotype, cytokine production, and homing potential; they differentiate into Th1, Th2, Th17, and regulatory T cell (Treg) (Figure 9.5). Interferon gamma (IFN-γ) and interleukin-12 (IL-12) initiate the differentiation of Th1 cells that are characterized by IFN-γ secretion and induction of inflammatory immune responses, and are required for cell-mediated immunity against intracellular pathogens. In contrast, IL-4 drives Th2 cell differentiation that results in IL-4 production, induction of humoral immune responses, and clearance of extracellular pathogens. More recently, Th17, a distinct

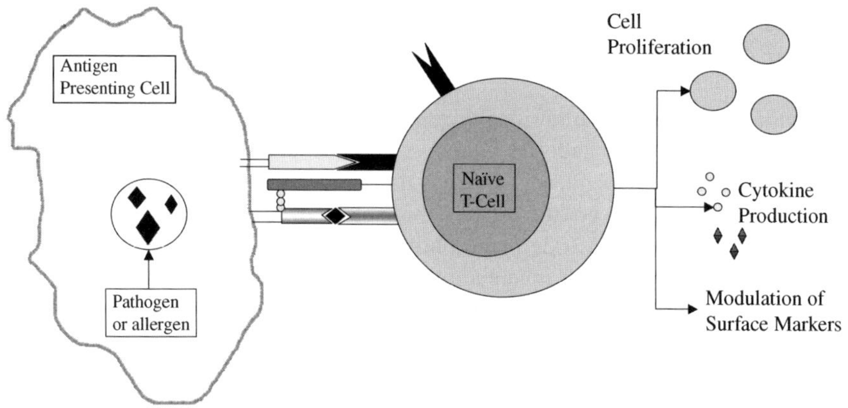

FIGURE 9.4 Initiation of an immune response. A pathogen or allergen is presented by an antigen-presenting cell (innate immune system) to a naïve T lymphocyte (adaptive immune system) resulting in the initiation of an antigen-specific immune response. The T-cell responses—proliferation, cytokine production, and modulation of cell surface marker—can be monitored using flow cytometric methods.

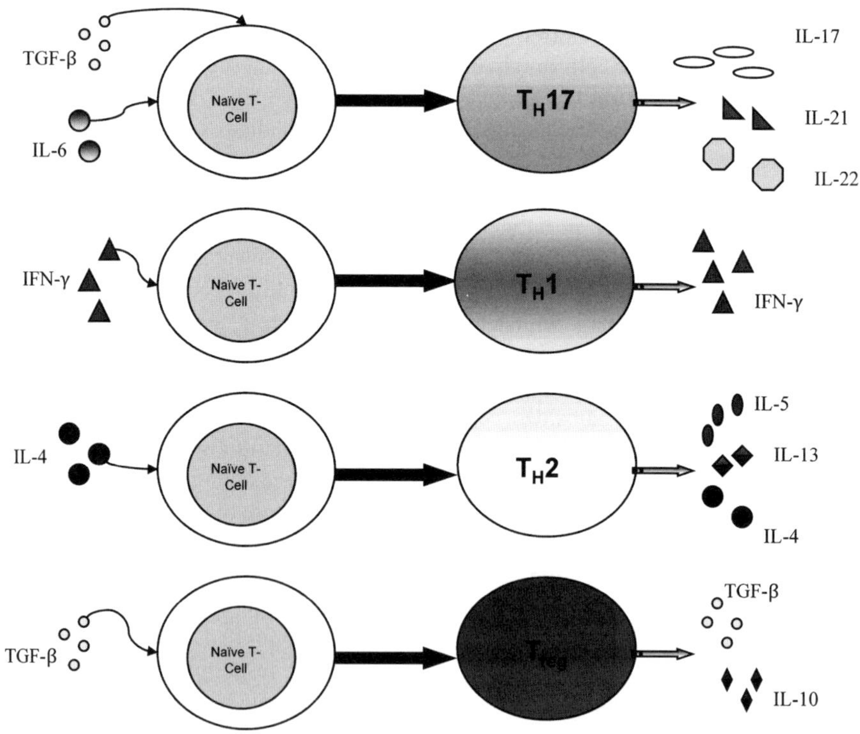

FIGURE 9.5 CD4 T-lymphocyte differentiation and subsets.

T-helper lineage, has been described. Th17 differentiation is induced by TGF-β, IL-6, and maintained IL-23 [10]. Th17 cells secrete IL-17 (a potent proinflammatory cytokine), IL-21, and IL22 and are potent inducers of tissue inflammation and the clearance of pathogens not adequately controlled by Th1 and Th2 cells [11]. They play a key role in driving autoimmune pathologies [12]. Treg cells are identified phenotypically as $CD4^+$, CD25bright, and intracellular $Foxp3^+$, and prevent immune attack against normal and neoplastic cells by directly suppressing the activation of effector $CD4^+$ and $CD8^+$ T cells [13].

Antigen activation of naïve B cells ($CD19^+$, $sIgM^+$, $sIgD^+$) results in generation of memory B cells ($CD19^+$, $CD27^+$, $sIgD^-$) and terminally differentiated plasma cells that secrete large quantities of antigen-specific antibody (Ab).

9.3 CLINICAL RESEARCH: CHARACTERIZATION OF ABNORMALITIES OF THE CELLULAR COMPONENTS OF THE IMMUNE SYSTEM

Many diseases and syndromes fall under the umbrella of inflammatory autoimmune disorders such as systemic lupus erythematosus (SLE), rheumatoid arthritis (RA), psoriasis, Crohn's disease, Sjögren's syndrome, and systemic sclerosis. Most of these rheumatic disorders are heterogeneous diseases with a clinical spectrum, ranging from mild to life threatening, and variability in secondary organ system involvement (e.g., kidney and heart). Clinical research scientists are actively studying patients with autoimmune disorders with the goal of elaborating the pathophysiology and identifying disease-specific molecular signatures. There is a growing body of literature suggesting that hyper-responsive B cells, abnormalities in T-lymphocyte cell signaling, and abnormalities in Treg function contribute to the development of autoimmune inflammation. Nevertheless, the cause of these disorders and their pathogenesis remains poorly understood [14–16]. To illustrate the importance and utility of the research conducted by clinical scientists, three examples will be focused on.

SLE is primarily a disease of B-cell origin characterized by the loss of immune tolerance to multiple self-antigens resulting in the production of numerous autoantibodies, in particular, antinuclear autoantibodies [14, 17]. There is a correlation between SLE flare-ups and systemic B-cell activation, defined as an expansion of plasma blasts/plasma cells ($CD19^+$, $CD20^-$) and pregerminal center cells ($CD19^+$, IgD^+, CD38high) combined with a decrease in circulating naïve B cells. In particular, an increase in a homogeneous subset of $CD27^-$, IgD^-, and $CD95^+$ memory B cells shows a strong correlation with disease activity [18]. Thus, chronic B-cell activation is gaining acceptance as a pathogenic biomarker for SLE [19].

Several T-cell abnormalities have also been characterized in SLE patients. Spontaneous apoptosis of circulating T cells is thought to contribute to chronic lymphopenia and the generation of autoantigens [20]. Double-negative (DN) T cells ($CD3^+$, $CD4^-$, and $CD8^-$) are also thought to contribute to SLE pathogenesis as they are expanded in the peripheral blood of SLE patients and have been shown to be potent producers of the inflammatory cytokines IL-17 and IFN-γ [21]. Likewise, extensive

T-cell infiltration has been observed in chronically inflamed tissues of SLE patients. Although the mechanisms underlying the infiltration process from the blood stream to inflamed sites are not fully characterized, an upregulation of adhesion molecules, such as CD44, is associated with T-cell tissue migration [22]. Consequently, it has been suggested that CD44 may represent a legitimate therapeutic target for SLE [10]. Thus, CD44 upregulation on T cells and an expansion of peripheral DN T cells may also represent disease-related biomarkers for SLE [11]. Numerous, other surface markers and subsets have been reported to correlate with SLE disease activity such as the involvement of dendritic cell function (DC) and elevated numbers of $CXCR3^+$ and $CD4^+$ T cells in urine [23, 24].

RA is another example of an autoimmune disease well characterized by clinical research. RA is a chronic inflammatory autoimmune disease occurring in approximately 1% of adults that begins with swollen joints and ultimately leads to debilitating joint destruction. Inflammatory cells such as monocytes and neutrophils, together with T and B cells, infiltrate the synovium in RA patients. Given that lymphocyte homing and migration are controlled in part by interactions between chemokines and their receptors, it is not surprising that chemokine receptor imbalances on peripheral lymphocytes have been reported in RA [25]. The percentage of B cells expressing the chemokine receptor CXCR5 that is expressed on naïve B cells and plays a role in homing the secondary lymphoid organs is decreased in RA patients compared to control groups. In addition, CXCR3, a chemokine receptor for inflammatory chemokines that is upregulated during inflammation and plays a role in cellular recruitment to inflamed tissue, was found to be upregulated on peripheral B cells from RA patients. Although there was a considerable intersubject variability, this phenotype (decreased percentage of $CXCR5^+$ B cells accompanied by an increase in the expression of CXCR3) is a good candidate for a disease-specific biomarker and merits more robust evaluation. Correlation between the disease activity and the level of memory B cells (IgD^-, $CD27^+$) in the synovium and in the peripheral blood has also been described for RA patients [26].

Psoriasis is another well-characterized autoimmune disease; it is primarily a T-cell-mediated inflammatory autoimmune condition. Unlike SLE and RA, psoriasis research has not found any links to B-cell abnormalities. Psoriasis occurs in approximately 2–3% of adults and is primarily localized in the skin and less often in the joints [27, 28]. The pathogenesis includes DC and T-cell activation, differentiation, and expansion. CD2 is a known costimulatory molecule for T-lymphocyte activation in psoriasis patients that has made anti-CD2 drugs good targeted therapies for the treatment of psoriasis. There is a correlation between an anti-CD2 drug treatment and a decrease in T lymphocytes. Specifically, a marked decrease in effector memory T cells ($CCR7^-$, $CD45RA^-$) in both $CD4^+$ and $CD8^+$ T lymphocyte cell populations is observed. $CD4^+$ and $CD8^+$ T-lymphocyte cells are responsible for IFN-γ production, which also decreases along with the effector memory T cells. Naïve T cell ($CCR7^+$, $CD45RA^+$) and central memory T cell ($CCR7^+$, $CD45RA^-$) populations had no sustained decrease in the presence of therapeutics [29]. The latter makes a strong case for subset-specific targeted therapeutics that do not affect the overall populations of cells that will reduce the potential for serious adverse events.

In addition to autoimmunity, other therapeutic areas such as oncology, cardiovascular disease, or metabolic disease are now also known to involve aspects of the immune system or an inflammatory component and can be elucidated using PD biomarkers. For example, type I diabetes is now considered a T-cell-mediated immune disease [30]. Cardiovascular disease is thought to have an important inflammatory component. Mechanistically, atherosclerosis is primarily an inflammatory disease mediated by cellular components of both the innate and the adaptive immune systems [31]. While cancer is primarily a proliferative disorder, it can be characterized using flow cytometric assays.

9.4 CLINICAL DRUG DEVELOPMENT I: MONITORING THE CELLULAR COMPONENTS OF THE IMMUNE SYSTEM TO ESTABLISH THE EFFECT OF THERAPEUTIC INTERVENTION

As discussed above, abnormalities in peripheral blood leucocytes have been described in various therapeutic areas: autoimmunity, oncology, cardiovascular, and metabolic disease. The nature of the abnormalities—be it a change in the abundance of a given cell type or a change in the receptor expression on a particular cell type—continues to be dissected and appreciated. Whether these cellular biomarkers are broad signatures of chronic, systemic inflammation or disease-specific biomarkers remains to be determined. Monitoring these cellular abnormalities during clinical testing of immune modulating treatments will provide a unique opportunity to validate the cellular abnormalities with regard to their utility as biomarkers and significance in disease pathogenesis.

For example, there are several therapeutic compounds in development targeting $CD20^+$ B cells for the treatment of SLE and RA. CD20 is expressed on most B-cell lymphomas and all normal B cells with the exception of very early progenitors and terminally differentiated plasma cells. Anti-CD20 treatment that results in a prolonged but transient depletion of peripheral B cells and was first approved for treatment of B-cell tumors in non-Hodgkin's lymphoma and chronic lymphocytic leukemia is now emerging as a supplementary treatment for autoimmune diseases with B-cell involvement. The hypothesis is that treatment with CD20 targeting compounds will result in depletion of autoreactive B cells while leaving the rest of the immune system intact.

In preliminary studies with rituximab, a cytotoxic chimeric monoclonal antibody (mAb) directed against CD20, improvement in the clinical manifestations of SLE and RA has been reported [32, 33]. Data supporting a primary study objective—the degree of B-cell depletion—were generated with high-sensitivity flow cytometric assays able to measure B-cell levels in the range of 0.1 cells/μL whole blood [33, 34]. Some studies suggested that the degree of B-cell depletion after the primary treatment was predictive of clinical outcome [34]. In addition to monitoring B-cell depletion, flow cytometric assays were used to monitor the status of the immune system by measuring other cell types of both the adaptive and the innate immune systems and to assess functionally distinct B-cell subsets (naïve, memory) during posttreatment B-cell replenishment. The B-cell repopulation was reported to be dominated by naïve B cells ($CD27^-$, IgD^+). Moreover, the sustained low number of circulating memory B cells ($CD27^+$, IgD^-) correlated with a reduction in serum rheumatoid factor and good clinical outcomes. In contrast, an accumulation of memory B cells in the synovium correlated with disease flare-ups [22].

Compounds in development targeting $CD2^+$ include Alefacept, a CD2 binding chimeric protein. CD2 is expressed on all thymocytes, T cells, NK cells, and a small subset of DC. The heterotypic interaction between CD2 and its major ligand CD58 (leukocyte function antigen (LFA)-3) enhances TCR-mediated signaling. CD2:CD58 interaction can also directly stimulate T-cell proliferation and differentiation independent of TCR engagement. During clinical evaluation of Alefacept for psoriasis vulgaris, flow cytometric assays were used to support primary efficacy objectives [29]. Analysis of total circulating T-cell levels as well as naïve, central memory, and effector memory CD4 and CD8 T cells during treatment provided critical support to the proof of concept for Alefacept. Selective decreases in circulating effector memory T cells were observed after Alefacept treatment while the naïve T-cell population was left intact.

Several monoclonal antibody-based cancer therapeutics are available on the market (Ritxuan, Avastin, Herceptin, Erbitux, Vectibix, and Campath) with a greater number in clinical development. Additional immunotherapeutics include cytokines, adjuvants, and vaccines against oncogenic viruses. As such, the immune biomarkers utilized for autoimmune conditions will also likely be applicable to oncology.

9.5 CLINICAL DRUG DEVELOPMENT II: RISK/BENEFIT OF MONITORING CELLULAR COMPONENTS OF THE IMMUNE SYSTEM DURING CLINICAL TRIALS

Quite a bit of information is available about lymphocyte subsets and the generation of an immune response, including intracellular biochemical and molecular details, yet much remains unknown particularly with regard to the whole animal model and the clinical significance of specific immune subsets with respect to the pathophysiology of immune-related disorders. Monitoring components of the immune system during clinical testing has the potential to provide critical information on basic biology of the immune system and insight into disease pathogenesis in addition to the primary objective of gathering pharmacodynamic information. There are examples of bidirectionality in translation medicine where information from drug development studies feeds back to basic and clinical researchers. The success of B-cell depletion in treatment of RA and SLE has increased the understanding of the role of B-cell functions in rheumatic conditions. Furthermore, tracking B-cell replenishment has provided valuable insight into the basic biology of B-cell ontogeny and trafficking. Results of the clinical studies where selective depletion of memory T cells was achieved were valuable in underscoring the relative contribution of different cell types to the pathogenesis of psoriasis. Moreover, much remains unresolved regarding the T-cell differentiation pathway [9].

A major challenge of translating information from the research community into useful biomarker targets and biomarker assays is that all too often publications contain little or no validation data and shockingly sometimes no actual flow cytometry data. Without the context of method validation, in particular, precision assessments, it is not possible to establish if the changes in cellular subsets reported are valid or an artifact of method variability. Even though pharmaceutical scientists will perform the appropriate "Fit-for-Purpose" method validation prior to testing clinical specimens, there is an associated risk when the initial decision to pursue a biomarker is based on information generated with poorly validated methods [35]. Without a validated method, it is not possible to know if the method is capable of distinguishing changes that are statistically significant. Moreover, proper assay development and validation of flow cytometric methods is more time consuming and expensive compared to the more commonly used biomarker platforms such as immunoassays.

Another challenge is that with the exception of the major lymphocyte subsets (T cells, CD4 T cells, CD8 T cells, NK cells, and B cells), there are no reference intervals for the lymphocyte subsets commonly evaluated. Reference intervals can be used to assess clinical improvement. For example, LDL cholesterol has been established as a possible marker for cardiovascular risk. LDL levels can be monitored both to assess medical treatment of atherosclerosis and to serve as an indicator of heart attack or stroke risk; when values are within optimal range, treatment is considered successful. Similarly in HIV, CD4 counts are routinely monitored to assess treatment efficacy. With a validated method, it is possible to statistically determine if changes are correlated with treatment; however, without reference values in healthy and diseased populations, it is difficult to understand if the changes are clinically meaningful. Analysis of the more exploratory PD biomarker data raises many questions that contribute to the difficulty in interpreting the data. What does a change indicate? What does the lack of a change indicate? How should this information affect decision making in the clinical development of the compound?

Sampling is also a challenge when monitoring the cellular components of the human immune system. By and large, sampling is limited to the peripheral blood that is a more accessible, less invasive sampling site than the secondary lymphoid tissues (bone marrow, lymph node, and spleen) or inflamed tissue (intestine, synovium) where specific immune responses are generated. Testing is conducted under the assumption that the peripheral blood represents a true snapshot of the immune activation and inflammatory responses elsewhere. For example, B-cell depletion is monitored in the periphery, yet the goal is to deplete the B-cell population in the lymph nodes where most of the cells reside. Similarly, Treg cells act in tissues to control established inflammation and it is unclear if their levels in the peripheral blood are reflective of overall levels and activity.

Perhaps, the most challenging aspect of measuring cellular components of the immune system is the lack of a consensus phenotype for most commonly evaluated cell types [36]. In many studies where flow cytometric analysis was "limited" to four- or five-color detection, CD4 and CD8 T-cell subsets were evaluated by two additional parameters, CD45RA (or CD45RO) and CD62L or CD27 or CCR7 in order to identify naïve and memory T cells. More recently, labs equipped with the newer instruments with 6–18 color capabilities have used additional markers to define these T-cell subsets. The questions that arise now are how comparable are the data? Can results from studies using one phenotype (CD45RA and CD27) be compared to others in which a different phenotype was used (CD45RA and CCR7, or more advanced phenotypes)? Would the trends be the same? Are we comparing apples and oranges or merely tangerines and oranges? If a newer, more extensive, phenotype has been described, should that new convention be adopted even though earlier studies were conducted with a different phenotype? If so, can the data be compared?

A similar lack of definitive phenotypes exists for Th1 and Th2. Initial reports in the literature phenotypically identified Th1 and Th2 on the basis of CD45RA, CXCR3, and CCR4. More recently, Th1 and Th2 have been described as $CCR7^-$, $CCR5^+$, $CXCR3^+$ and $CCR7^-$, $CCR3^+$, $CCR4^+$ [36]. Many questions arise. Is the first phenotype still legitimate? Are CD45RA and CCR7 interchangeable? Is it sufficient to look at CXCR3 only without further stratifying into CCR5 positive cells? Can one evaluate the phenotype directly from whole blood without the *ex vivo* step to induce cytokine secretion? Does this approach have the advantage of possibly providing a more accurate approach to assess the immune status *in vivo*?

At present, the phenotype for Treg is well established as $CD4^+$, CD25bright, and intracellular $Foxp3^+$ or $CD4^+$, CD25bright, and CD127dim/- in cases where intracellular staining is not feasible. This was not the case as recently as 4–5 years ago, before Foxp3 was identified as the definitive additional marker for human Treg and good-quality reagents were available. In those early days, enumeration of human Treg by immunophenotyping without sorting and functional studies was not possible. In such cases, being ahead of the curve is not to one's advantage.

9.6 SUMMARY

The translational medicine approach in drug development is still fairly new and the benefit of this approach has yet to be demonstrated. To date, very few of the many biomarkers evaluated in clinical trials have translated into useful disease-specific diagnostic or prognostic biomarkers. Nonetheless, the enthusiasm for this approach lies in its vast potential. Similarly, this is an exciting but challenging time in the evolution of flow cytometry. We have technologically reached the point where we can observe as many subsets as we have cellular markers leaving us with the conundrum that the more information we gather, the more questions we raise. Thus, translating basic scientific findings into useful cellular PD biomarkers and translating the information back is an enormous, albeit intriguing, challenge.

ABBREVIATIONS

Ab	antibody
BCR	B-cell receptor
CD	cluster of differentiation

CTL	cytotoxic T lymphocyte
DC	dendritic cell
DN	double negative
HIV	human immunodeficiency virus
IFN-γ	interferon gamma
Ig	immunoglobulin
IL	interleukin
LFA	leukocyte function antigen
mAb	monoclonal antibody
NIH	National Institutes of Health
NK	natural killer cell
PD	pharmacodynamics
PK	pharmacokinetics
RA	rheumatoid arthritis
SLE	systemic lupus erythratomas
TCR	T-cell receptor
Th	T-helper cell
Treg	regulatory T cell
TLR	Toll-like receptor

ACKNOWLEDGMENT

The authors thank Mr. Viq Pervaaz and Dr. Sophie Connetable for critical review of this chapter.

REFERENCES

1. Lean MEJ, Mann JI, Hoek JA, Elliot RM, Schofield G. Translational research: from evidence-based medicine to sustainable solutions for public health problems. *Br. Med. J.* 2008;337:a863.
2. (http://nihroadmap.nih.gov/clinicalresearch/overview-translational.asp)
3. Carpenter S. Carving a career in translational research. *Science* 2007;317(5840):966–967.
4. Feuerstein GZ, Rutkowski JL, Walsh FS, Stiles GL, Ruffolo RR, Jr., The role of translational medicine and biomarker research in drug discovery and development. *Am. Drug Discov.* 2007;2(1):23–28.
5. Atkinson AJ, Jr., Colburn WA, DeGruttola VG, DeMets DL, Downing GJ, Hoth DF, Oates JA, Peck CC, Schooley RT, Spilker BA, Woodcock J, Zeger SL. Biomarkers and surrogate endpoints: preferred definitions and conceptual framework. Biomarkers Definitions Working Group. *Clin. Pharmacol. Ther.* 2001;69:89–95.
6. Wan HI, Day M, Hurko O, Rutowski JL. Biomarkers for Alzheimer's disease: translation medicine approaches in development of disease modifying therapeutic. *Am. Pharmaceut. Rev.* 2008;11:83–92.
7. Vesterqvist OE, Park CH, Rutkowski JL. From biological measures to valid biomarker: predicting success of novel therapeutic. *Am. Drug. Discov.* 2007;3(3):56–59.

8. Feldmann, M. Translating molecular insights in autoimmunity into effective therapy. *Ann. Rev. Immunol.* 2009;27:1–27.
9. Reiner SL, Sallusto F, Lanzavecchia A. Division of labor with a workforce of one: challenges in specifying effector and memory T cell fate. *Science* 2007;317(5838): 622–625.
10. Awasthi A, Kuchroo VK. Th17 cells: from precursors to players in inflammation and infection. *Int. Immunol.* 2009;21(5):489–498.
11. Korn T, Bettelli E, Oukka M, Kuchroo VK. IL-17 and Th17 cells. *Annu. Rev. Immunol.* 2009;27:485–517.
12. Nakae S, Nambu A, Sudo K, Iwakura Y. Suppression of immune induction of collagen-induced arthritis in IL-17-deficient mice. *J. Immunol.* 2003;171:6173–6177.
13. Sakaguchi S, Powrie F. Reviewed in emerging challenges in regulatory T cell function and biology. *Science* 2007;317(5838):627–629.
14. Zikherman J, Weiss A. Antigen receptor signaling in the rheumatic diseases. *Arthritis Res. Ther.* 2009;11:202.
15. Skapenko A, Leipe J, Lipsky PE, Schulze-Koops H. The role of the T cell in autoimmune inflammation. *Arthritis Res. Ther.* 2005;7 (Suppl. 2): S4–S14.
16. Leipe J, Skapenko A, Lipsky PE, Schulze-Koops H. Regulatory T cells in rheumatoid arthritis. *Arthritis Res. Ther.* 2005;7:93–99.
17. Looney RJ, Anolik J, Sanz I. B lymphocytes in systemic lupus erythematosus: lessons from therapy targeting B cells. *Lupus* 2004;13:381–390.
18. Jacobi AM, Reiter K, Mackay M, Aranow C, Hiepe F, Radbruch A, Hansen A, Burmester GR, Diamond B, Lipsky PE, Dörner T. Activated memory B cell subsets correlate with disease activity in systemic lupus erythematosus: delineation by expression of CD27, IgD, and CD95. *Arthritis Rheum.* 2008;58(6):1762–1773.
19. Enghard P, Riemekasten G. Immunology and the diagnosis of lupus nephritis. *Lupus* 2009;18:287.
20. Fernandez DR, Telarico T, Bonilla E, Li Q, Banerjee S, Middleton FA, Phillips PE, Crow MK, Oess S, Muller-Esterl W, Perl A. Activation of mammalian target of rapamycin controls the loss of TCRζ in lupus T cells through HRES-1/Rab4-regulated lysosomal degradation. *J. Immunol.* 2009;182:2063–2073.
21. Li Y, Harada T, Juang YT, Kyttaris VC, Wang Y, Zidanic M, Tung K, Tsokos GC. Phosphorylated ERM is responsible for increased T cell polarization, adhesion, and migration in patients with systemic lupus erythematosus. *J. Immunol.* 2007;178: 1938–1947.
22. Crispín JC, Oukka M, Bayliss G, Cohen RA, Van Beek CA, Stillman IE, Kyttaris VC, Juang YT, Tsokos GC. Expanded double negative T cells in patients with systemic lupus erythematosus produce IL-17 and infiltrate the kidneys. *J. Immunol.* 2008;181: 8761–8766.
23. Kwok SK, Lee JY, Park SH, Cho ML, Min SY, Park SH, Kim HY, Cho YG. Dysfunctional interferon-α production by peripheral plasmacytoid dendritic cells upon Toll-like receptor-9 stimulation in patients with systemic lupus erythematosus. *Arthritis Res. Ther.* 2008;10:R29.
24. Enghard P, Humrich JY, Rudolph B, Rosenberger S, Biesen R, Kuhn A, Manz R, Hiepe F, Radbruch A, Burmester GR, Riemekasten G. CXCR3$^+$CD4$^+$ T cells are enriched in

inflamed kidneys and urine and provide a new biomarker for acute nephritis flares in SLE patients. *Arthritis Rheum.* 2009;60(1):199–206.

25. Henneken M, Dörner T, Burmester GR, Berek C. Differential expression of chemokine receptors on peripheral blood B cells from patients with rheumatoid arthritis and systemic lupus erythematosus. *Arthritis Res. Ther.* 2005;7:R1001.
26. Souto-Carneiro MM, Mahadevan V, Takada K, Fritsch-Stork R, Nanki T, Brown M, Fleisher TA, Wilson M, Goldbach-Mansky R, Lipsky PE. Alterations in peripheral blood memory B cells in patients with active rheumatoid arthritis are dependent on the action of tumour necrosis factor. *Arthritis Res. Ther.* 2009;11:R84.
27. Krueger JG. The immunologic basis for the treatment of psoriasis with new biologic agents. *J. Am. Acad. Dermatol.* 2002;46:1–23.
28. Lew W, Bowcock AM, Krueger JG. Psoriasis vulgaris: cutaneous lymphoid tissue supports T-cell activation and 'type 1' inflammatory gene expression. *Trends Immunol.* 2004;25 (6):295–305.
29. Chamian F, Lin SL, Lee E, Kikuchi T, Gilleaudeau P, Sullivan-Whalen M, Cardinale I, Khatcherian A, Novitskaya I, Wittkowski KM, Krueger JG, Lowes MA. Alefacept (anti-CD2) causes a selective reduction in circulating effector memory T cells (Tem) and relative preservation of central memory T cells (Tcm) in psoriasis. *J. Transl. Med.* 2007;5:27.
30. Lawson JM, Tremble J, Dayan C, Beyan H, Leslie RDG, Peakman M, Tree TIM. Increased resistance to $CD4^+$CD25hi regulatory T cell-mediated suppression in patients with type 1 diabetes. *Clin. Exp. Immunol.* 2008;154:353–359.
31. Ward JR, Wilson HL, Francis SE, Crossman DC, Sabroe I. Translational mini-review series on immunology of vascular disease: inflammation, infections and Toll-like receptors in cardiovascular disease. *Clin. Exp. Immunol.* 2009;156:386–394.
32. Abdwani R, Mani R. Anti-CD20 monoclonal antibody in acute life threatening haemolytic anaemia complicating childhood onset SLE. *Lupus* 2008;18:460–464.
33. Reynolds JA, Toescu1 V, Yee1 CS, Prabu1 A, Situnayake D, Gordon C. Effects of rituximab on resistant SLE disease including lung involvement. *Lupus* 2009;18:67–73.
34. Dass S, Rawstron AC, Vital EM, Henshaw K, McGonagle D, Emery P. Highly sensitive B cell analysis predicts response to rituximab therapy in rheumatoid arthritis. *Arthritis Rheumat.* 2008;58(10):2993–2999.
35. Lee JW, Devanarayan V, Barrett YC, Weiner R, Allinson J, Fountain S, Keller S, Weinryb I, Green M, Duan L, Rogers JA, Millham R, O'Brien PJ, Sailstad J, Khan M, Ray C, Wagner JA. Fit-for-purpose method development and validation for successful biomarker measurement. *Pharm. Res.* 2006;23(2).
36. Appay V, van Lier RAW, Sallusto F, Roederer M. Phenotype and function of human T lymphocyte subsets: consensus and issues. *Cytometry A* 2008;73(11):975–983.

10

IMMUNOGENICITY TESTING USING FLOW CYTOMETRY

DENISE M. O'HARA AND VALERIE THEOBALD

10.1 INTRODUCTION

It is widely recognized that clinical administration of biotherapeutic molecules in human subjects has the potential to elicit unwanted immune responses to the biotherapeutic product. Consequently, a fundamental component of the development and commercialization of biotherapeutics is to have the capability to monitor patients for the development of antidrug antibody (ADA) responses. Following detection of antibodies to the biotherapeutic, the ADA responses are characterized, for example, in terms of both specificity and neutralizing capacity and the outcomes are correlated to any clinical sequelae. Typically, immunogenicity testing utilizes bioanalytical methods specifically designed to detect antidrug binding and neutralizing antibodies. Flow cytometric cell-based immunogenicity assays can be employed when testing for neutralizing activity of antibody positive samples and included in a tiered design approach to detect and characterize specific antibody responses in clinical (and sometimes nonclinical) samples [1, 2].

The analytical power of the flow cytometer instrument lies in its ability to simultaneously quantify the intensity of fluorescent dyes on the surface of cells or intracellularly in addition to measuring light-scatter signals indexing cell size and granularity. These measurements are used to determine cell characteristics that define a cell's activation status, cell lineage, or subset identity. Flow cytometry offers the advantage of simultaneous multiparameter analysis of mixed cell populations without

Flow Cytometry in Drug Discovery and Development, Edited by Virginia Litwin and Philip Marder

the need for lengthy cell separation methods. However, use of the flow cytometer as a platform for developing immunogenicity assays does not always take advantage of these capabilities. Immunogenicity method development may employ cells expressing target proteins for biotherapeutic binding and the detection of the binding event or the resulting signal is recorded. Neutralizing antibodies would interfere with the drug target binding and result in lower recorded values. Applying bioassays in this manner for immunogenicity testing more closely mimics physiological conditions. The use of flow cytometry for immunogenicity testing also offers the advantage of cell viability assessment and cell gating options that are not typically available in nonflow cytometry-based bioassays. Changes in cell population growth characteristics are readily detected from the scatter profile and are used with the concurrent cell viability measurements. The acceptable ranges of these measurements can be incorporated into the overall assay run acceptance criteria when monitoring assay performance. The use of flow cytometry also offers the advantage of investigating both cell surface binding and intracellular signaling using permeabilized cells that are able to maintain their capacity to respond to cell surface receptor stimulation. Early signal events such as tyrosine and serine phosphorylation of cellular proteins [3], generation of cAMP, or inositol-phosphate hydrolysis can be measured in response to drug stimulation of permeabilized cells.

The advantages of flow cytometry assays are rapid and quantitative measurements. Depending on the mode of action of the drug and the selected readout, flow cytometry can offer an alternative assay format without using harsh denaturation steps and lengthy sample manipulation, as in the case of a fluorescent thymidine analogue that can be used to measure newly synthesized DNA *in situ* [4]. The use of flow cytometry as an assay platform for evaluating neutralizing activity requires a functional binding or signal transduction event involving a fluorescent reagent. The biological event to be measured must occur with sufficient signal in the presence of the intended biological matrix, so that a change in the signal due to neutralization can be detected [1]. The challenges of applying flow cytometry to immunogenicity testing lie in the understanding of the technical aspects of flow cytometry, readout selected, and internal variability of the assay and applying an assay cut point or threshold to define a sample as positive or negative. This chapter will describe the use of flow cytometry for the measurement of immunogenicity using as examples a biotherapeutic that binds a cell surface receptor and a biotherapeutic enzyme that is actively internalized.

10.2 SELECTION OF ASSAY REAGENTS

Numerous well-characterized fluorescent dyes are available for the purpose of preparing dye–reagent conjugates for use in assay development, validation, and sample testing [5]. Assay reagents can be either labeled directly with a fluorescent dye or used unlabeled with a secondary labeled reagent for detection. Common conjugation chemistries via functional groups on the protein as well as considerations for characterization of functional binding and specificity are necessary to be able to select optimum assay reagents. Conjugation of a protein may result in heterogeneity and/or altered reactivity to the target and therefore characterization of the reagent, particularly its binding activity, is important to select the most appropriate reagent and ensure the quality of long-term supply [6]. A common consideration when developing neutralizing antibody (NAb) assays that measure drug binding to cell surface proteins is the direct labeling of the drug. As loss of functional binding may occur as a result of label conjugation, site-directed labeling may be necessary to avoid conjugation of sites essential for target binding. If direct labeling strategies are unsuccessful, then use of secondary conjugated reagents to detect the bound drug may also be used.

10.2.1 Fluorescent Dye Selection

Multiple fluorescent dyes are commercially available, including organic dyes, phycolipoproteins, tandem dyes, and (more recently) nanocrystals, all with distinct excitation and emission spectra. Matching the fluorescent dye, brightness and photostability to the intended assay use, required sensitivity, and the optical characteristics of the detection system are all important parameters for successful assay development [7, 8]. Consideration of the target/signal event marker abundance may lead the user to a brighter fluorochrome for the detection of rare targets/signal event markers, while more abundant targets/signal event markers may be detected with fluorochromes that are less bright. In addition, when using several fluorochromes at the same time, it is critical to adjust the instrument and correct for spectral overlap to ensure accurate interpretation of the signal output [9, 10].

10.2.2 Assay Detection Reagent

10.2.2.1 Selection and Characterization The detection reagent in the assay may be one of the following types of reagent: antibody, soluble receptor, drug, ligand, or enzyme. An antibody reagent may be an intact whole molecule or a fragment of the antibody (Fab), which may offer the advantage of reduced Fc receptor binding on target cells during the staining procedure. During evaluation of candidate molecules for suitability as assay detection reagents, the specificity and selectivity of the reagent, the intended use, and the mode of action are all considered. Also, during the assay reagent selection process, it is highly recommended that long-term availability be considered to ensure availability throughout the assay life cycle.

10.2.2.2 Labeling the Detection Reagent with Fluorescent Dye Unless commercially available, labeling the assay detection reagent with fluorescent dye(s) at several different molar coupling ratios (MCRs) allows the assessment of dye selection incorporation level and with predefined assay performance criteria. Susceptibility of the reagent binding site to fluorochrome conjugation and interference with subsequent interactions with the target are important to assess during the optimization phase of assay development (Figure 10.1a). In addition, comparison of the binding affinities of the unlabeled and labeled assay reagents can be used to assess the degree to which the addition of the fluorochrome or the conjugation procedure influences the binding activity of the conjugate (Figure 10.1b). In some cases, increasing the fluorochrome

(a)

Molar coupling ratio	Reactivity in cell-based potency assay%
Unlabeled	100
0.33	100
0.66	100
1.4	74
2.3	42

(b)

Label	Molar coupling ratio	Molar incorporation ratio	Biacore relative binding	Affinity kD (pM)
Unlabeled	NA	NA	100%	579
Fluorochrome 1	10	2.3	93%	507
Fluorochrome 1	20	10.6	76%	747
Fluorochrome 2	4	1.1	85%	572
Fluorochrome 2	8	5.6	65%	1020

(c)

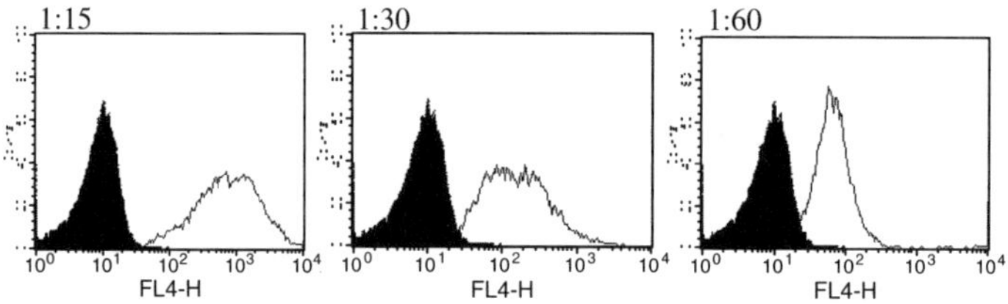

(d)

Lot	Molar coupling ratio	Molar incorporation ratio	GMFI
Lot 1	15	6.2	825
Lot 2	15	6.6	1052
Lot 3	15	9.0	1182
Lot 4	15	6.6	1090

FIGURE 10.1 The effect of increasing the molar challenge ratio (label:biotherapeutic) used during conjugation on biological activity (a), affinity to target (b), and cell binding profile (c). Reproducibility of conjugation procedure was determined and measured by cell binding intensity (d).

MCR may directly impact the binding ability of the reagent to the target (Figure 10.1c).

10.2.2.3 Characterization and Long-Term Usage Long-term use of any reagent requires characterization, stability assessment, lot-to-lot reproducibility studies, investigation of temperature and buffer storage conditions, effects of freeze–thaw cycles, and so on. Standardization of the preparation procedure and monitoring the characterization profile of new lots of assay reagent can minimize disruption of sample testing and maintain assay performance consistency (Figure 10.1d). Fluorescently conjugated reagents require the same level of characterization as other ligand binding reagents [6]. In addition, assessment of potential reagent photobleaching stability may be incorporated into the characterization profile, though in our collective experience this has not been shown to be an issue. As with all ligand binding assays, tracking and trending assay performance offers confidence in the quality and stability of the reagent in terms of both binding ability and fluorescent intensity over time and with different lots of reagents [11].

10.2.3 Positive Control

10.2.3.1 Selection and Characterization Antidrug and neutralizing antibody assays require positive controls (PCs) that both bind the drug and neutralize its activity, respectively. Assay positive controls may consist of antibodies (whole or fragments, monoclonal or polyclonal), receptors, ligands, or natural antagonists. Purposeful immunization of animals with a biotherapeutic to generate antibodies may also be successful at generating neutralizing antibodies, although maturation of the antibody response [12] to include neutralizing antibodies may take significantly longer time than the nonneutralizing antibodies.

Immunogenicity assays are designed to detect antibodies to the biotherapeutic molecule present in complex biological matrices that may be of nonclinical or human origin. Dilutions of the biological sample matrix can minimize matrix interferences from either the sample matrix or the positive control if using a polyclonal serum. Purification of the PC may be necessary when the neutralizing activity is low in order to enrich for neutralizing antibodies. Assessment of the effects of matrix interferences from the sample and the positive control biological matrix may be investigated as shown in Figure 10.2. Similar neutralizing activity was observed using either the hyperimmunized rabbit sera or the ligand affinity-purified PC. The data indicate that the rabbit sera did not interfere with the receptor-mediated uptake of the fluorescently labeled biotherapeutic.

10.2.4 Cells

10.2.4.1 Selection and Characterization Cell lines used in assays for immunogenicity testing can be obtained from different sources such as research labs, central repositories, or scientific collaborators. Investigation of cell binding or intracellular signal response measured in biological sample matrix determines

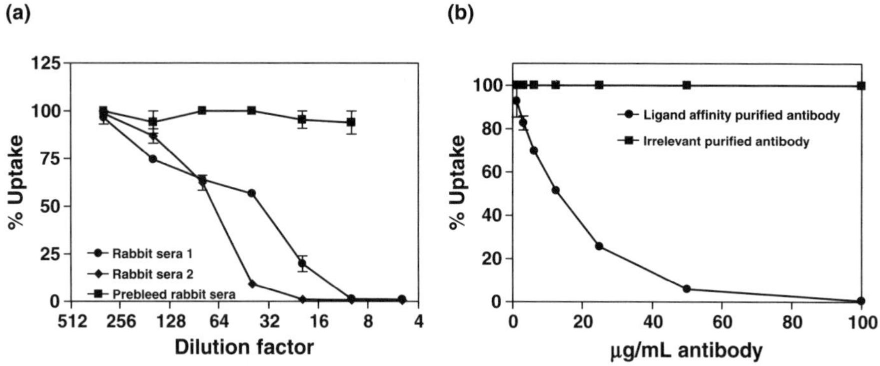

FIGURE 10.2 Evaluation of matrix interferences and specificity of positive control(s) used as sera (a) or ligand affinity purified (b).

whether the use of a particular cell line is feasible, although often times the most relevant cell line for immunogenicity testing is also the one used for the biotherapeutic drug potency assay. Neutralizing antibody assays typically measure a decrease in signal, but if the mode of action of the biotherapeutic results in a decrease in signal, neutralization of this activity would then be seen as an increase in signal [1].

The success or failure of each analytical run can be directly related to the characteristics of the cells in use. Therefore, the influence of critical cellular attributes needs to be fully examined during method optimization. This should include studying culture conditions to determine whether changes in ligand or receptor expression are related to culture conditions, cell passage, the degree of confluence, and changes in culture media formulation, in particular, the serum content and source, and whether these changes have any effect on the assay performance [1].

An example of the effect of passage on uptake of labeled drug is included in Figure 10.3. In this example, monitoring cell passaging (from passage 6 to 38) and cell confluency shows that the active uptake and accumulation of labeled drug is higher in postconfluent cells (Figure 10.3a). However, the percentage inhibition of drug uptake (Figure 10.3b) and inhibition by a neutralizing positive control (Figure 10.3c) were not significantly different across a range of cell confluencies or cell passage number as assessed using one-way ANOVA and Wilcoxon Rank Sum tests.

During assay feasibility studies, other cell line attributes may be associated with changes in assay performance, such as a decrease in proliferation rate or a change in cellular morphology. These changes could potentially have a direct influence on assay performance; at minimum, they should be documented and tracked along with assay performance. One example of the utility in tracking cellular growth attributes with assay performance is a suspension cell line that was grown and passaged during a NAb assay validation. Toward the end of the assay validation, a significant decrease in assay signal was observed and traced to a change in the cell growth characteristics – from loosely attached and clustered cells to suspended single cells (data not shown). These types of growth variations are subtle and sometimes hard to identify; therefore, it is highly recommended that growth characteristics, cell seeding density, and any

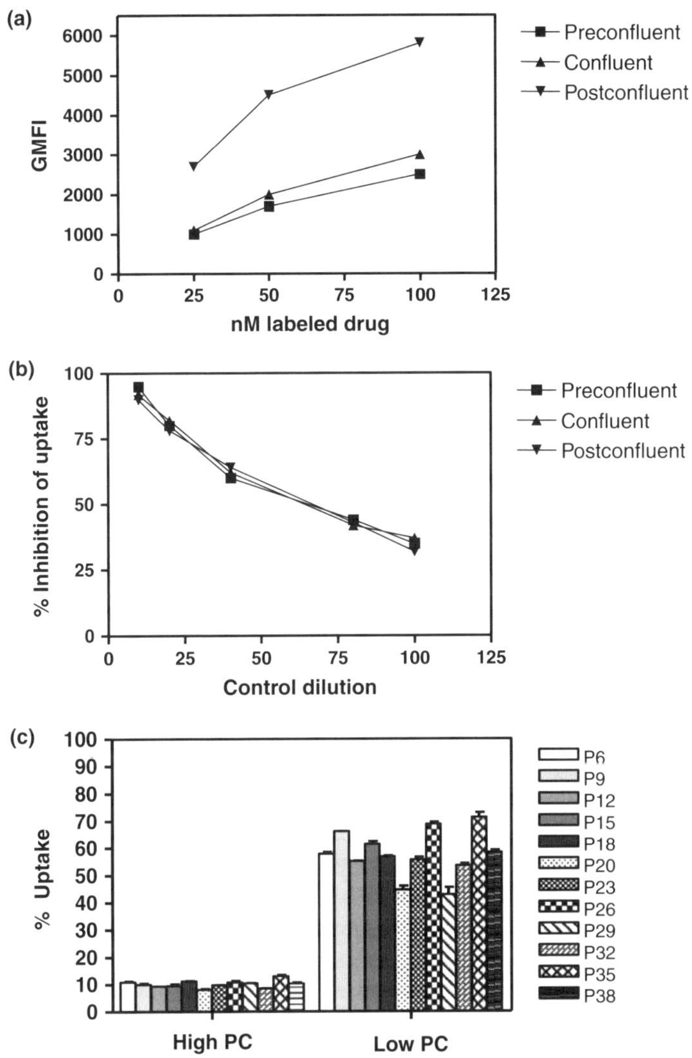

FIGURE 10.3 Effect of cell growth and culture conditions on assay readouts. The relationship is shown between cell confluency on signal output, GMFI (a) and percent inhibition of uptake (b). High and low positive controls (PC) are used with different cell culture passage numbers (passage 6 through 38), the association with percent uptake of fluorescently labeled biotherapeutic is shown (c).

other related parameters are recorded and tracked along with the assay readout. When passage-related effects are suspected or identified, larger cell banks can be made to minimize such effects. Creating master and working cell banks ensures the investigator that a well-defined and consistent population of cells is available for use in each assay.

10.2.5 Use of Factorial Design

During all phases of assay development, critical parameters may be assessed that have the potential to influence one another, and if tested together, can be individually identified [13]. Identification of these interactions can be exploited to make decisions allowing one to pursue more robust assay conditions. Design of experiment (DoE) is an approach that may be used to assess these factors and their influence on assay performance, such as assay variability and signal-to-noise (S/N) ratio. Such factors may include, but are not limited to, the following: cell growth media, serum (bovine or human) levels and exposure time, cell number, state of cell confluence, passage number, staining conditions, cell fixation, sample matrix interferences, or effect of different anticoagulants. Use of DoE during the optimization phase of assay development is advantageous since it can identify assay conditions in a relatively small-scale experiment and isolate critical factors from those having no effect, or those that may interact or influence one another in the performance of the assay. These interactions may not be easily identified by testing one factor at a time (OFAT). Appropriate factors and relevant levels for evaluation are selected at the start of DoE, followed by the identification of the appropriate responses to measure and how to set the numerical value [14–16]. Analysis of the data will highlight the factors that have the greatest effects as well as the relevant factor interactions that can be expanded upon with further experimentation. Once conditions have been identified and the effects on particular responses such as signal output have been assessed, further assay optimization can occur to ensure that the cell line, assay output, and other reagents in the system have been appropriately selected.

Since each drug candidate has a unique mode of action, the corresponding cell source and assay reagents may also be unique. Therefore, appropriate characterization of each component is required. Figure 10.4 shows the data output from a factorial experiment designed to investigate multiple factors, including two concentrations of cells (0.5×10^6 and 1×10^6), three concentrations of the fluorescent detector reagents (1.0 μg/mL, 2.5 μg/mL, and 5 μg/mL), and three different fluorochromes (Alexa 448, Alexa 647, and FITC). The assay responses measured were geometric mean fluorescent intensity (GMFI) and signal-to-noise ratio. Drug conjugated to Alexa 647 gave the highest GMFI in comparison to Alexa 448 or FITC (Figure 10.4a), resulting in a slight improvement of signal with a lower number of cells (Figure 10.4b). There was a noted interaction between drug conjugated to FITC and the number of cells, with the signal-to-noise ratio decreasing as the number of cells increased (Figure 10.4c). Identification of this type of interaction allows the assay developer to select against such combinations to develop a more robust assay. The condition selected for further optimization used Alexa 647 as the conjugated fluorochrome and 0.5×10^6 cells, and even if this did not give the highest GMFI or S/N ratio, it was the lowest drug–fluorochrome concentration that resulted in a reasonable signal. This lower concentration of drug–fluorochrome improved the sensitivity of the assay since less neutralizing antibody was required to inhibit a lower amount of drug than if a higher concentration was chosen.

Factor 1 A: Cell number	Factor 2 B: Type of Fluorochrome	Factor 3 C: [Drug-fluorochrome] (ug/mL)	Response 1 GMFI	Response 2 S/N ratio
0.5×10^6	FITC	1.0	95	10
1.0×10^6	FITC	1.0	62	6
0.5×10^6	Alexa 488	1.0	169	9
1.0×10^6	Alexa 488	1.0	100	10
0.5×10^6	Alexa 647	1.0	616	8
1.0×10^6	Alexa 647	1.0	339	7
0.5×10^6	FITC	2.5	199	20
1.0×10^6	FITC	2.5	107	10
0.5×10^6	Alexa 488	2.5	366	20
1.0×10^6	Alexa 488	2.5	159	16
0.5×10^6	**Alexa 647**	**2.5**	**1404**	**18**
1.0×10^6	Alexa 647	2.5	675	14
0.5×10^6	FITC	5.0	333	33
1.0×10^6	FITC	5.0	196	18
0.5×10^6	Alexa 488	5.0	559	31
1.0×10^6	Alexa 488	5.0	320	32
0.5×10^6	Alexa 647	5.0	2200	28
1.0×10^6	Alexa 647	5.0	1210	26

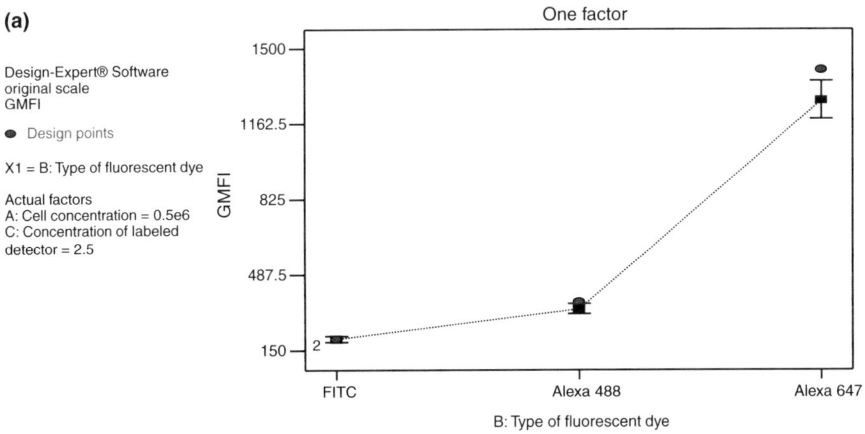

FIGURE 10.4 Solutions are tabulated from DoE with category factor and response levels. Plots show the relationship between maximum signal and cell number (a) or type of fluorescent dye (b) or cell number and signal/noise ratio (c). The experimental conditions selected from this experiment are shown in bold text in the table.

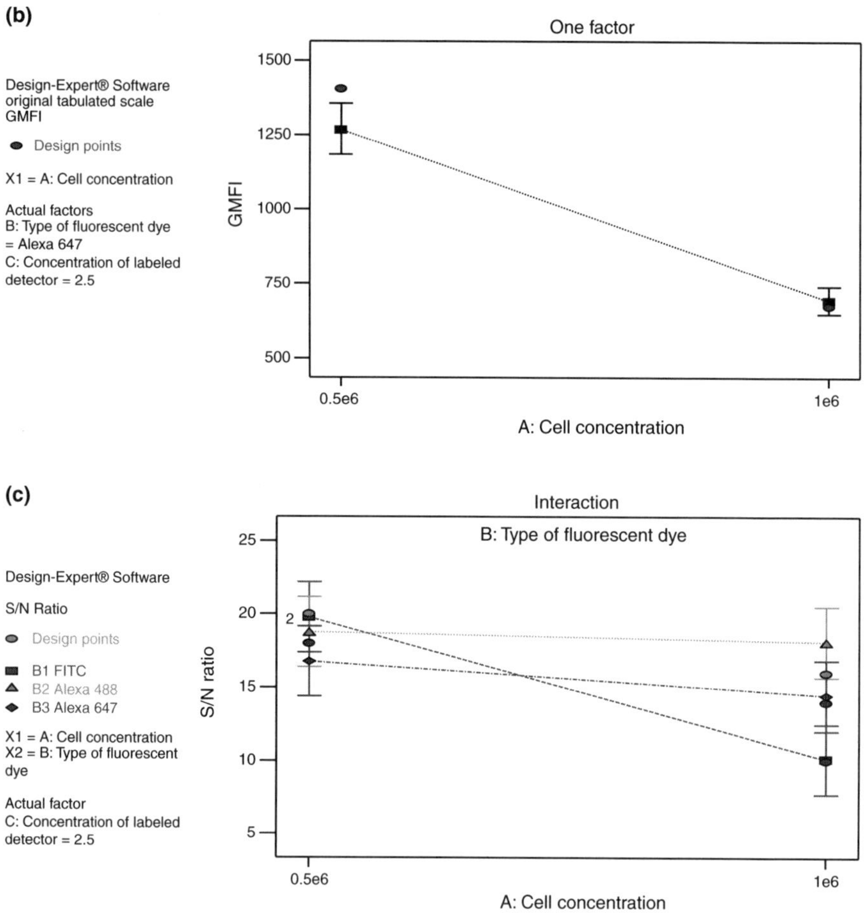

FIGURE 10.4 *(Continued)*

10.3 ASSAY OPTIMIZATION, DEVELOPMENT, AND VALIDATION

Guidelines and consensus publications already exist for immunogenicity method development, validation, and testing, including antibody [17–23] and neutralizing antibody assays [1]. These guidelines should be followed and can be applied to any assay platform used, including flow cytometry. Table 10.1 includes a listing of recommended studies that should be included in the optimization, development, and validation stages of cell-based flow cytometric neutralizing antibody assays. Studies overlap in the different assay phases and are repeated with larger n in the subsequent assay phase. Performance of a cell-based flow cytometric assay may be affected by cellular growth conditions, protein expression levels, biological sample matrices, and critical assay reagents. These parameters should be optimized and tested during the assay development phase.

TABLE 10.1 Recommended Studies, Listed by Assay Phase, to be Included During Feasibility, Development, and Validation of a Flow Cytometry-Based Neutralizing Antibody Assay

	Assay Phase		
Study Type	Optimization	Development	Validation
Select cell line	✓		
Identify appropriate controls	✓		
Prepare cell banks: master and working banks	✓		
Effect of FBS on assay performance	✓		
Human matrix reactivity/interference	✓		
Drug dose response	✓	✓	
NAb positive control dose response	✓	✓	
Detector reagent specificity, concentration, and incubation time	✓	✓	
Characterize cell line	✓	✓	
Demonstrate method specificity	✓	✓	✓
Effect of cell passage on assay performance	✓	✓	✓
Effect of cell confluence on assay performance	✓	✓	✓
Minimum required matrix dilution		✓	
Determine necessity of gating cells		✓	
Inter- and intra-assay variability		✓	✓
Stability of signal poststaining (time and temperature)		✓	✓
Test for prozone effect		✓	✓
Define assay response range		✓	✓
Drug interference		✓	✓
Evaluate different lots of critical reagents		✓	✓
Test for Instrument variation			✓
Determine assay cut point			✓
Analyst variation			✓
Sample and reagent stability			✓
Use of plate loader			✓

Studies that are specific to flow cytometry are shown in bold.

Changes that could affect the assay performance using flow cytometry are related to cell surface receptor endocytosis (capping) or shedding, anomalous changes to the membrane and/or receptor expression, or architecture caused by adherence to plastics or cells, cell confluency, or biological sample matrix causing stimulation or inhibition of protein expression or signaling. These effects on assay performance may be minimized by standardization of cell culturing, sample handling, and staining procedures. The effect of paraformaldehyde fixation on the stability of the signal poststaining can be evaluated during assay development and confirmed in validation by testing stained cells with and without the fixation step.

Drug responsiveness of a cell line is assessed initially in assay optimization and the drug concentration selected based on the signal and the need to detect neutralizing activity. An example of a drug–fluorochrome conjugate binding to cell surface receptor is shown in Figure 10.5a and b, together with a titration of the positive

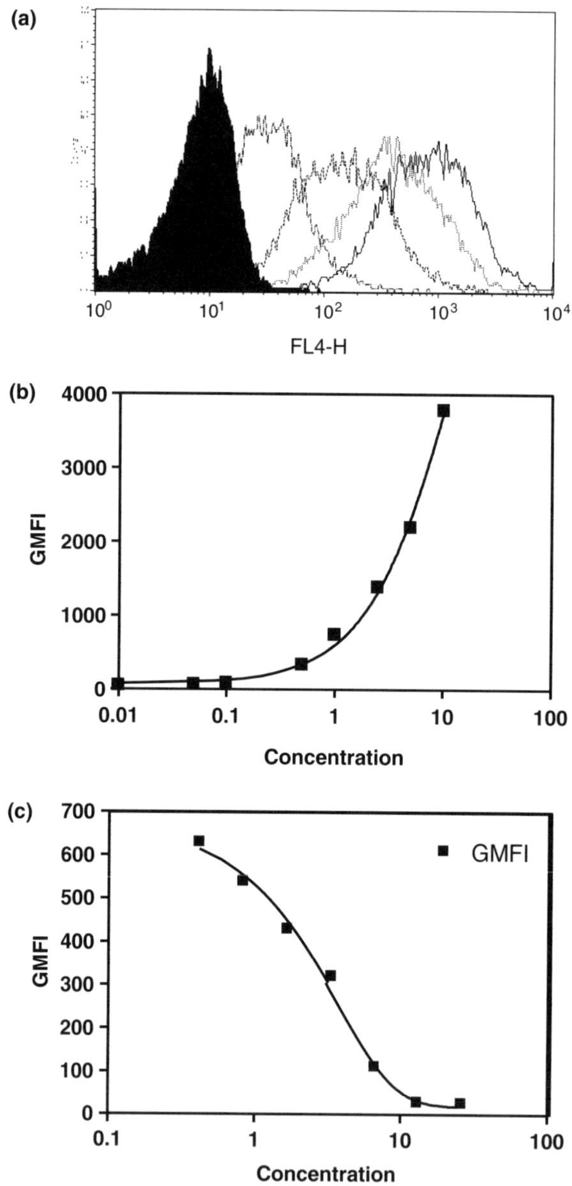

FIGURE 10.5 Labeled biotherapeutic binding to a relevant cell line. Labeled drug dose–response curve shown as overlayed histograms (a) or plotted as a dose curve (b). Titration of the positive control with a fixed amount of drug–fluorochrome (c).

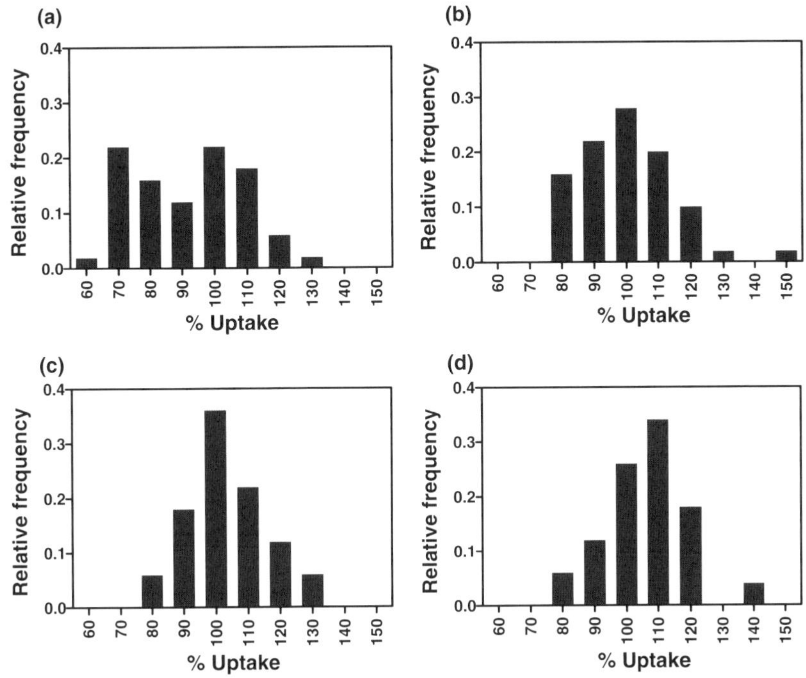

FIGURE 10.6 Effect of human serum matrix on normal frequency distribution at various minimum required dilutions: 1:10 (a), 1:20 (b), 1:40 (c), and 1:80 (d).

control antibody with a fixed amount of the drug–fluorochrome conjugate (Figure 10.5c). Biological matrix is added to the assay system and the effect observed. Figure 10.6 shows the effect of diluting human serum, 1/10, 1/20, 1/40, and 1/80, and the frequency distribution of percent uptake of fluorescently labeled drug observed at the various dilutions. This figure demonstrates the minimum required dilution (MRD) of the human serum matrix to achieve a normal distribution of response was 1/40. Throughout all phases of an assay's life cycle, reliable and sustainable sources of critical assay reagents are pivotal to the success of the nonclinical or clinical program that the assay supports.

10.3.1 Instrument Differences

Assays using GMFI signal output from flow cytometers are susceptible to complexities of the instrument, such as laser strength and alignment, instrument fluidics, and rigors of instrument maintenance. Due to the complexities of the instruments, the variation between instruments can be up to 50% and thus operational strategies need to be considered and outlined for longitudinal studies. During assay validation, signal output as a function of an instrument can be assessed using the same samples to compare different instruments, if available. For assays where changing instruments in the middle of the study influence the signal to the degree that comparisons cannot be made longitudinally, and the signal differences cannot be overcome by normalization, some investigators choose to dedicate specific instruments to individual studies.

10.3.2 Determining Neutralization Readout

Signal output from the flow cytometer can be reported as median fluorescent intensity (MFI) or GMFI, with either output being used to subsequently calculate percent inhibition of binding or percent of maximal signal.

Reported values from immunogenicity assays are qualitative and, at best, semi-quantitative for reasons that are well described in the literature [20]. The assay cut point value, which represents the neutralization capacity of 95% of the normal population, is used to score samples as either positive or negative.

As cut point values are critical to defining the neutralizing status of an antibody positive sample, it is essential that they are determined appropriately. Ideally, cut point values are determined using sera collected from the untreated patient population. If these samples are not available at the time of assay development and validation, normal donor sera may be used and the assay cut point confirmed or re-established at a later date using patient sample data.

To establish the cut point, 30–50 samples, inclusive of both male and female, are analyzed using the recommended assay dilution scheme. A total of three independent runs by at least two analysts are conducted. The data are tested for normal distribution and transformed if necessary. The cut point is then set at the 95% confidence interval or mean +/− 1.645 SD. Aggregate data from the different dilutions may be used as long as lack of a difference between them has been demonstrated statistically.

Various strategies can be taken to assess the presence of neutralizing antibody. For example, to obtain a qualitative (positive/negative) result, MFI or GMFI signal can be compared against an assay cut point that is calculated from the output signal of drug-naïve or normal individuals. Alternatively, to obtain quasi-quantitative results, percent inhibition can be calculated as [1 - (sample signal/max signal)] × 100. Background signal may be subtracted when appropriate. Another approach to obtaining a quasi-quantitative result is to compute percent of maximal signal or ratio of signal/maximal signal that is defined as (sample signal/max signal) × 100. Again background signal may be subtracted when appropriate. Regardless of the computation approach that is used, an assay cut point must be used for scoring patient samples.

10.3.3 Software

When testing samples from nonclinical and clinical studies, the ability to use a multifunctional data analysis software package that facilitates batch analysis, allows template-driven testing, and standardizes repetitive functions is advantageous. Two examples of such software are FlowJo and FCS Express. Other data analysis software packages are available for free downloading at the Purdue Cytometry Web site: *www.cyto.purdue.edu/flowcyt/software.htm*. Some of these applications have good report-generating functions as well, which may be important in some laboratory settings. Both FlowJo and FCS Express applications offer 21 CFR Part 11 compliant solutions that are necessary when testing samples in a GLP-regulated lab using validated flow cytometers.

10.3.4 Technical Considerations

10.3.4.1 Biological Matrices Matrix interference may be described as the effect of a substance present in the sample that causes a deviation of the measured value from the true value. The interferences may be endogenous or exogenous. Endogenous interferences include such substances as lipids, heterophilic antibodies, rheumatoid factors (RF), complement, binding proteins soluble biotherapeutic target, ligands, and receptors [24–27]. Exogenous interferences include anticoagulants, concomitant medications, and sometimes inappropriate sample storage conditions that change the biological components of the sample [27].

In neutralizing antibody assays, interferences present in samples, whether endogenous or exogenous, may mimic the binding of the drug or affect the detection of neutralizing activity in the sample. Depending on the nature of the interaction, a sample result may be falsely positive or falsely negative.

During immunogenicity testing using flow cytometry, assay interferences caused by matrix substances may present as a nonspecific signal when cells bind molecules that cross react with drug or when cells are responsive to these nonspecific components. This falsely elevated assay signal could potentially result in a reduced ability to detect neutralization activity by artificially increasing the apparent concentration of drug used in the assay. When matrix substances interfere with the binding of neutralizing antibodies present in the sample, again the potential exists to report erroneous sample results. Sample pretreatment may be necessary to remove or reduce the level of interfering substances, or a different assay format considered that is less susceptible to such interferences.

Furthermore, a specificity assay can be used to distinguish the pharmaceutical drug target binding event or signaling from that signal produced by other structurally related molecules. Similarly, studies can be designed to determine whether there is nonspecific interaction occurring between matrix components and neutralizing antibody. An assessment of specificity should be conducted using appropriate disease-state matrices so that the structurally related molecules are present in the amounts anticipated in the test sample. An example of a specificity assay for a cell surface NAb binding assay involves the use of an irrelevant cell surface marker. For enzyme uptake assays, an irrelevant enzyme conjugate whose cell-mediated uptake is unaffected by the presence of specific neutralizing antibodies may be used. In either assay, detector reagent and enzyme conjugate titrations are conducted to optimize conditions for the specificity assay and the validation parameters outlined in Ref. [1] are applied.

10.3.4.2 Drug Interferences In addition to the interfering substances mentioned previously, biotherapeutics with relatively long half-lives may still be in circulation at the time the sample is taken. This circulating biotherapeutic may interfere in the NAb assay by binding and sequestering drug-specific antibody or by binding to the target and inappropriately increasing the neutralizing signal. Where the biotherapeutic has a relatively short half-life, such as an enzyme, there is no apparent assay interference. In both the situations, timing of sample collection should be thoughtfully considered to reduce the impact of biotherapeutic on the detection of neutralizing activity in a study sample.

To minimize the potential for drug being present in the sample, it is recommended, when possible, that samples be collected prior to initiation of each treatment. For biotherapeutic molecules with short half-lives, this practice will ensure a drug-free sample. When the biotherapeutic has a long half-life, obtaining a drug-free sample is more challenging. It may be necessary to collect samples over the span of the clinical trial and evaluate them for circulating drug prior to testing for NAb. Alternatively, the study may be designed to continue collecting samples from the patient until drug is no longer detectable in circulation.

10.3.4.3 Interference by Circulating Target Similar to the challenges described with circulating drug, modulation of signal may occur when soluble target is present in circulation. The biological mechanisms of therapeutic molecules must be taken under consideration and, if possible, samples collected from treated and untreated patient populations should be evaluated for the presence of soluble target. If soluble target is confirmed to be in circulation, their impact can be minimized by removal or use of competitive assay formats.

10.3.4.4 Staining Patterns Attention should also be given to unexpected patterns of staining that may present as secondary peaks on either side of the main histogram or as a trailing, diminishing extension of the main histogram peak. Some causes of abnormal staining patterns to consider are uneven staining due to decreased receptor exposure resulting from cell clumping or overcrowding of treated cells in culture. Variability in staining patterns may also be explained by heterogeneity of the target cell population that can result from excessive cell passaging or from culture conditions favoring a given subpopulation. Since these artifacts can interfere with assay cut point assignment and determination of neutralizing activity in a sample, it is therefore important to determine the reasons for these types of anomalous staining patterns and incorporate assay strategies to minimize them.

10.4 CLINICAL APPLICATION

The safety and efficacy of administered biotherapeutics drive immunogenicity concerns and appropriate testing of immune responses. Figure 10.7 illustrates three patients with varying responses to antibody-mediated neutralizing activity over a

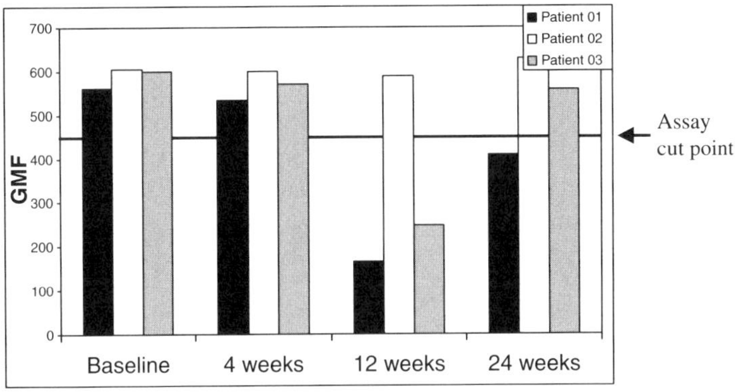

FIGURE 10.7 Neutralizing activity in three patients following administration of a biotherapeutic, measured by intensity of cell binding, and the assay cut point is indicated as a solid horizontal bar.

24 week period; patient 01 develops a persistent neutralizing response, patient 02 shows no response, and patient 03 exhibits a transient response. Assignment of sample score is a function of the quantitative readout from the instrument relative to the assigned cut point. As mentioned previously in immunogenicity testing, the final sample result is determined by a decision tree of ADA detection, specificity confirmation, and neutralizing antibody characterization. Assessing the correlation of a measured immune response following the administration of a biotherapeutic to any clinical sequelae is perhaps the most meaningful aspect of immunogenicity testing. Development of neutralizing antibodies to the biotherapeutic can be of severe significance to the patient. These antibodies can abrogate the efficiency of the therapeutic that may or may not be overcome by increasing the dosage. In cases where patients no longer benefit from life-saving treatment due to the presence of neutralizing antibodies, tolerance induction regimens are recommended [28].

10.5 SUMMARY

Flow cytometry is used successfully in immunogenicity testing to detect neutralizing activity in antibody positive samples. We have described examples of assay development using a biotherapeutic that binds a cell surface receptor and a biotherapeutic enzyme that is actively internalized. Although NAb assays do not typically take advantage of the multicolor capability of the flow cytometer, they do however utilize the simultaneous cell gating and cell viability functions that are easily incorporated into these types of assays. As with any assay development, application of factorial design can shorten assay time lines and offer early indications of conditions that can be used to develop robust flow cytometry assays. The technical considerations for NAb flow cytometry assays include the potential for assay interferences from biological matrices, circulating biotherapeutic and soluble target in the sample, in addition to unexpected staining patterns from variations in cell growth and protein expression profiles. Assay validation parameters follow those detailed in the literature and referenced here, using the flow cytometer with appropriate, applicable, and well-characterized reagents. Comprehending and mitigating the unique technical challenges, defining assay readout, and understanding internal variability of the assay combined with applying an assay cut point that defines a sample as positive or negative allow the bioanalytical lab to use flow cytometry as a powerful tool in immunogenicity testing.

ACKNOWLEDGMENTS

The authors would like to thank Terri Caiazzo, Alison Joyce, Keirsten Seiger, and Lisa Cherry for their technical assistance and John Nowak, Jim McNally, Corinna Fiorotti, Terri Caiazzo, Alison Joyce, and Yuanxin Xu for their review of this chapter. In particular, we appreciate the invaluable suggestions received from Bonita Rup and Susan Richards.

REFERENCES

1. Gupta S et al. Recommendations for the design, optimization, and qualification of cell-based assays used for the detection of neutralizing antibody responses elicited to biological therapeutics. *J. Immunol. Methods* 2007;321(1–2):1–18.
2. Ponce R et al. Immunogenicity of biologically-derived therapeutics: assessment and interpretation of nonclinical safety studies. *Regul. Toxicol. Pharmacol.* 2009;54 (2):164–182.
3. Krutzik PO, Nolan GP. Intracellular phospho-protein staining techniques for flow cytometry: monitoring single cell signaling events. *Cytometry A* 2003;55(2):61–70.
4. Rothaeusler K, Baumgarth N. Evaluation of intranuclear BrdU detection procedures for use in multicolor flow cytometry. *Cytometry A* 2006;69(4):249–259.
5. McCoy JP, Jr. Basic principles of flow cytometry. *Hematol. Oncol. Clin. North Am.* 2002;16(2):229–243.
6. Rup B, O'Hara D. Critical ligand binding reagent preparation/selection: when specificity depends on reagents. *AAPS J.* 2007;9(2):E148–55.
7. Holden TM, Tom F, Laurel EN, Trotter, J. Selecting fluorochrome conjugates for maximum sensitivity. *Cytometry A* 2004;62A:169–173.
8. Nomura L, V. C. Maino, and H. T. Maecker. Standardization and optimization of multiparameter intracellular cytokine staining. *Cytometry A* 2008.73(11):984–991.
9. Herzenberg L. A et al. Interpreting flow cytometry data: a guide for the perplexed. *Nat. Immunol.* 2006;7(7):681–685.
10. Tung JW et al. New approaches to fluorescence compensation and visualization of FACS data. *Clin. Immunol.* 2004.110(3):277–83.
11. Ritter N, Wiebe M. Validating critical reagents used in CGMP analytical testing: ensuring method integrity and reliable assay performance. *BioPharm.* 2001;14(5):12–21.

12. Harlow E, Lane D. *Antibodies: A Laboratory Manual*, Cold Spring Harbor Press, Cold Spring Harbor, NY, 1988.
13. Joelsson D et al. Optimizing ELISAs for precision and robustness using laboratory automation and statistical design of experiments. *J. Immunol. Methods* 2008.337 (1):35–41.
14. Anderson MJ, Whitcomb PJ. *DOE Simplified: Practical Tools for Effective Experimentation*, Productivity Press, 2000.
15. Anderson MJ, Kraber SL. Eight keys to successful DOE. *Quality Digest* 1999 (July).
16. Anderson M, Anderson H. Applying DOE to microwave popcorn. *PI Quality* 1993 (July/August);3.
17. EMEA. Guideline on Immunogenicity Assessment of Biotechnology-Derived Therapeutic Proteins, Committee for Medicinal Products for Human Use (CHMP), 2007.
18. FDA. *Bioanalytical Method Validation*, CDER, U.S. Department of Health and Human Services, 2001.
19. Koren E et al. Recommendations on risk-based strategies for detection and characterization of antibodies against biotechnology products. *J. Immunol. Methods* 2008.333(1–2):1–9.
20. Mire-Sluis, AR et al. Recommendations for the design and optimization of immunoassays used in the detection of host antibodies against biotechnology products. *J. Immunol. Methods* 2004.289(1–2):1–16.
21. Shankar, G et al. Recommendations for the validation of immunoassays used for detection of host antibodies against biotechnology products. *J. Pharm. Biomed. Anal.* 2008;48 (5):1267–1281.
22. ICH. Text on validation of analytical procedures, I.C.F. Harmonization, 1995.
23. ICH. Validation of analytical procedures: methodology, I.C.F. Harmonization, 1995.
24. Manshouri T et al. Circulating CD20 is detectable in the plasma of patients with chronic lymphocytic leukemia and is of prognostic significance. *Blood* 2003;101(7):2507–2513.
25. Albitar M et al. Free circulating soluble CD52 as a tumor marker in chronic lymphocytic leukemia and its implication in therapy with anti-CD52 antibodies. *Cancer* 2004;101 (5):999–1008.
26. Kaplan IV, Levinson SS. When is a heterophile antibody not a heterophile antibody? When it is an antibody against a specific immunogen clinical chemistry. *Clin. Chem.* 1999;45:3.
27. Weber TH, Kapyaho KI, Tanner P. Endogenous interference in immunoassays in clinical chemistry. A review. *Scand. J. Clin. Lab. Investig.* 1990;50 (201 Suppl.):77–82.
28. Wang J et al. Neutralizing antibodies to therapeutic enzymes: considerations for testing, prevention and treatment. *Nat. Biotechnol.* 2008;26(8):901–908.

11

PHARMACOKINETICS BY FLOW CYTOMETRY: RECOMMENDATIONS FOR DEVELOPMENT AND VALIDATION OF FLOW CYTOMETRIC METHOD FOR PHARMACOKINETIC STUDIES

YUANXIN XU AND SUSAN M. RICHARDS

11.1 INTRODUCTION

Flow cytometric methods can be used for quantitative analysis of analytes in biological matrices. This chapter will discuss development and validation of flow cytometric methods to quantify protein drugs for pharmacokinetic (PK) analysis for use in preclinical or clinical studies.

The assay principle is based on binding of a drug to target molecules expressed on the cell surface. Depending on the drug characteristics, this binding can be antibody (Ab)–antigen (Ag) or receptor–ligand mediated. Detection of the bound drug at different levels is usually achieved by a secondary antidrug antibody conjugated to florescent dye (FITC, PE, etc.). A dose–response curve (or standard curve), drug concentration versus signal, can be generated and drug concentration in unknown samples will then be quantified by interpolating the signal from the standard curve. Signal is typically MFI (median fluorescence intensity) or MESF (molecules of equivalent soluble fluorochrome) converted from MFI using calibration beads such as Quantum™ beads with QuickCal® (Bang's Laboratories, Inc., Fishers, IN, USA).

Flow Cytometry in Drug Discovery and Development, Edited by Virginia Litwin and Philip Marder

Similar principles could be applied to the use of target molecule-conjugated beads. The Ag, antidrug Ab, receptor, or ligand, could be conjugated to beads and the use of a fluorescent dye-conjugated secondary detection reagent allows drug quantification.

The focus of this chapter is to define key parameters for flow cytometric PK assay development and validation using cell-based analysis. Examples of flow cytometric PK analysis provided in this chapter are two antibody drugs that bind to cells expressing surface target antigen. Detection is accomplished by using a second Ab (FITC or PE conjugated) recognizing heavy chain of the Ab drug. Emphasis will be on assay development that is unique for flow cytometric PK assay. Validation of flow cytometric PK assays should follow the same guidance from the FDA and ICH on analytical method validation as described later in this chapter.

11.2 ASSAY DEVELOPMENT

11.2.1 Flow Cytometer

Most of the flow cytometers meet the instrument requirement for PK analysis. Typically, PK analysis can be done using a single color. Adding a cell viability dye, PI (propidium iodide) or 7-AAD (7-aminoactinomycin), in the analysis requires a second color. Cytometers with a 488 nm (blue) laser are sufficient for most analysis using fluorochrome such as FITC (fluorescein isothiocyanate), PE (phycoerythrin), and so on. A second 633 nm (red) laser will allow the use of additional fluorochrome such as APC (allophycocyanin). Selection of a stable fluorochrome that can retain signal over time is important to reduce assay variability and improve overall assay performance.

Cytometers that collect analogue data or digital data are appropriate for PK analysis. The use of a sensitive instrument will allow quantification of drug at low levels. Although some newer cytometers can generate a much higher florescence digital signal in comparison to the analogue signal, both types of instruments are adequate for PK analysis (shown in examples described later). The instruments can also provide a degree of automation. Depending on the blood volume of the sample, an instrument with a tube loader or 96-well high-throughput system has added advantages for sample acquisition in comparison to single-tube acquisition. Most of the analytical software for flow cytometric data analysis can be used. Software that allows batch analysis has added advantages.

11.2.2 Cell Characterization to Select Cells with Homogeneous, High, and Stable Surface Target Ag Expression

During assay development, cell characterization is important and the use of well-characterized cells is the most critical component in the development of a high-quality flow cytometric PK assay.

11.2.2.1 Homogeneous and High Target Antigen Expression Primary cells or cell lines that naturally express the target Ag can be used. Alternatively, appropriately engineered cells can provide a consistent reagent over time. Evaluation of cells (from multiple vendors, PBMC from apheresis of multiple subjects, etc.) is important to select ones with homogeneous and high surface Ag expression. The use of cells that express intracellular Ag is not recommended. Ab drug binding to intracellular Ag depends on the degree of cell membrane permeability; binding variability could be high. Antigen shedding and/or modulation (internalization) needs to be examined. Cells that shed Ag may be used if Ag turnover rate is minimized during staining. In general, fresh cells are more appropriate for PK analysis because they should best maintain the native target antigen conformation. If fixed or lyophilized cells are used, comparability of their binding to drugs in comparison to the fresh cells needs to be demonstrated. Some antigenic epitopes may change due to fixation or lyophilization.

A relatively homogeneous Ag expression profile is preferred. This is evident by a narrow symmetrical bell-shaped peak in the histogram (MFI versus cell number) demonstrating that the data are normally distributed where MFI mean and MFI median values are similar. A very broad peak indicates a wide range of Ag expression among cells, which could cause variability in MFI, resulting in poor assay reproducibility. Appearance of a tail or a shoulder and/or a double peak could also impact data analysis. Subcloning of cells may be necessary to select for a homogeneous and high-target Ag expressing cell population.

Unlike MFI mean, MFI median is less influenced by heterogeneous Ag expression and therefore is the preferred readout for PK assessment.

The presence of excess Ag for drug quantification is essential. This can be achieved by careful selection of high Ag-expressing cells and/or use of excess cell number. High level of surface antigen expression is preferred, which allows drug quantification across a broad dynamic range. It is recommended that signal/background (signal in the presence of drug/signal without drug) across the standard curve range should be at least 2.5-fold.

11.2.2.2 ***Stable Target Ag Expression*** It is essential to select cells with stable target Ag expression and with minimal shedding and/or modulation upon drug binding. This should be evaluated over time and through a number of cell passages; the use of cells with sustained consistent expression is recommended. For engineered cells, it is preferable to use cells with stable expression, instead of transient expression. Chinese Hamster Cells (CHO) or mouse EL4 cells are typical parental cells that can be used to generate target Ag expressing cells.

11.2.3 Cell Processing, Production, and Maintenance

Good aseptic tissue culture technique is critical. General practices of cell processing, production, and maintenance are described in detail in Ref. [1].

11.2.3.1 ***Primary cells*** Primary cells are generally available in limited quantity. Some cell types, such as fibroblasts, may be expandable up to certain passages while others such as PBMC undergo limited expansion or require cytokines/growth factors for their expansion, which may change cell phenotype such as surface antigen expression. Cell passage analysis should be performed to establish limits within which Ag expression remains high and stable. A master cell bank should be established and working cell bank generated to support both assay development and validation studies and in-study/clinical sample testing. Optimal cryopreservation and cell thawing procedure should be established with acceptable cell recovery and viability, and should maintain Ag expression. Freshly thawed cells (in a single-use aliquot) from the working cell bank may be used directly. Cells may also be kept in culture after thaw as long as target Ag expression and cell viability are maintained within the established time period or passage number. Fixed and/or lyophilized cells may be used once they are well characterized to demonstrate comparability with the characteristics of fresh cells such as Ag expression and binding to drug.

Some cell types, such peripheral blood mononuclear cells (PBMCs), are not easily expanded over time without changing cell phenotype. PBMCs can be isolated from venous blood following two commonly used procedures, Ficoll gradient centrifugation (commercially available for human PBMC) and CPT cell preparation tubes (BD Biosciences), following the vendor recommended procedures described in the product inserts. Apheresis products may also be used. Due to individual donor variability, PBMC from multiple healthy subjects should be evaluated and cells with relatively homogeneous high Ag expression selected. Cell banks with frozen PBMC in single-use aliquots should be established from the chosen PBMC donor. It is not recommended to refreeze thawed cells.

The example in Figure 11.1 illustrates binding of a polyclonal Ab drug (drug X) in a dose-dependent manner to PBMC that expresses the target Ag. The Ag expression profile is shown as histograms and MFI-median values are indicated (data collected on FACSCalibur™). This particular method is intended for PK analysis in humans.

11.2.3.2 Cell Lines and Engineered Cells For transformed cell lines, cell growth and maintenance procedures need to be established including cell feeding and culture scheme, Ag expression, and cell number, growth, and viability. Generation of a master cell bank and working cell banks is necessary to minimize lot-to-lot variability. Easy maintenance, good growth rate, expandability, and stability (both freeze/thaw and long-term storage) are additional criteria for selection of a suitable cell line. A master cell bank and working cell bank need to be established for method validation and in-study/clinical sample testing.

Figure 11.2 illustrates a monoclonal humanized Ab drug (drug Y) binding to engineered CHO cells expressing the Ag (CHO–Ag). Representative histograms are shown with MFI median indicated (data collected on FACSCanto™ II). This assay is also intended for PK analysis in humans. Similar to PBMC results, the use of CHO–Ag also demonstrated dose-dependent binding across the full drug concentration range. Ag expression on CHO cells is more homogeneous (a very narrow peak in the histogram) compared to the profile seen with PBMC containing different subpopulations of lymphocytes and monocytes, as expected.

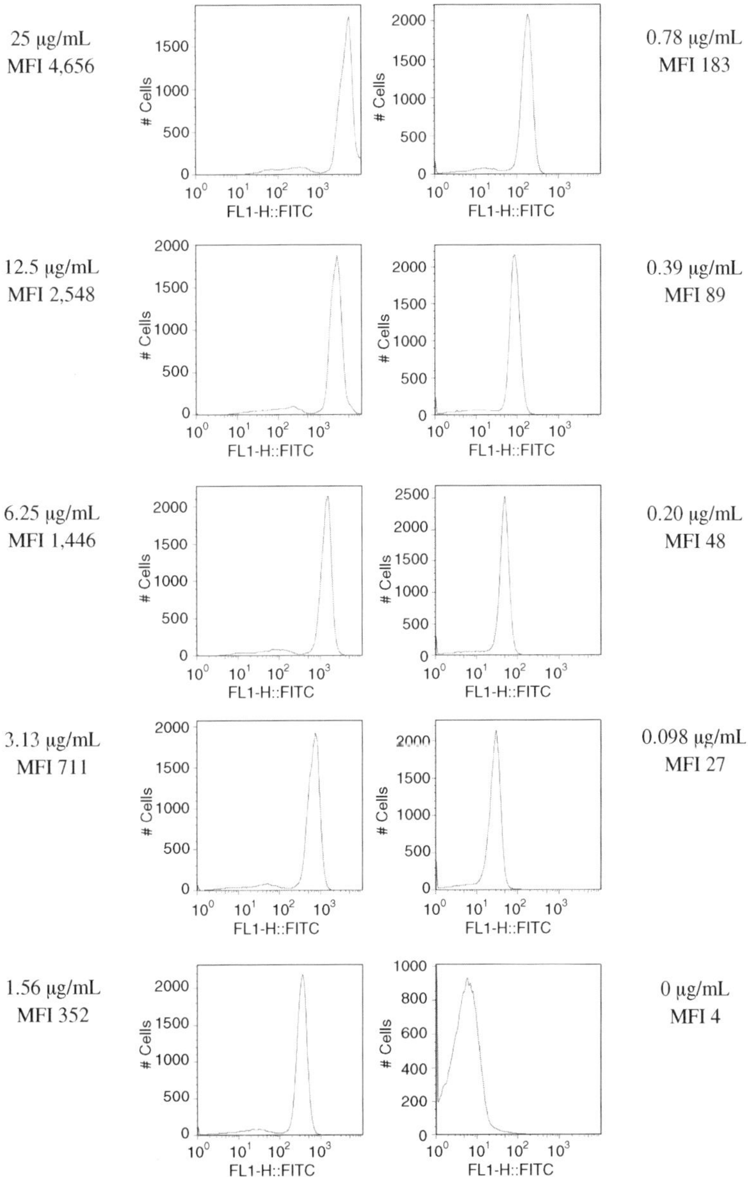

FIGURE 11.1 Dose-dependent binding of drug X to human PBMC. Dose-dependent binding of a polyclonal antibody drug X to human PBMC is shown in histograms (MFI median for FITC versus cell number). Drug X concentration (μg/mL) and the corresponding MFI median value are shown for each histogram.

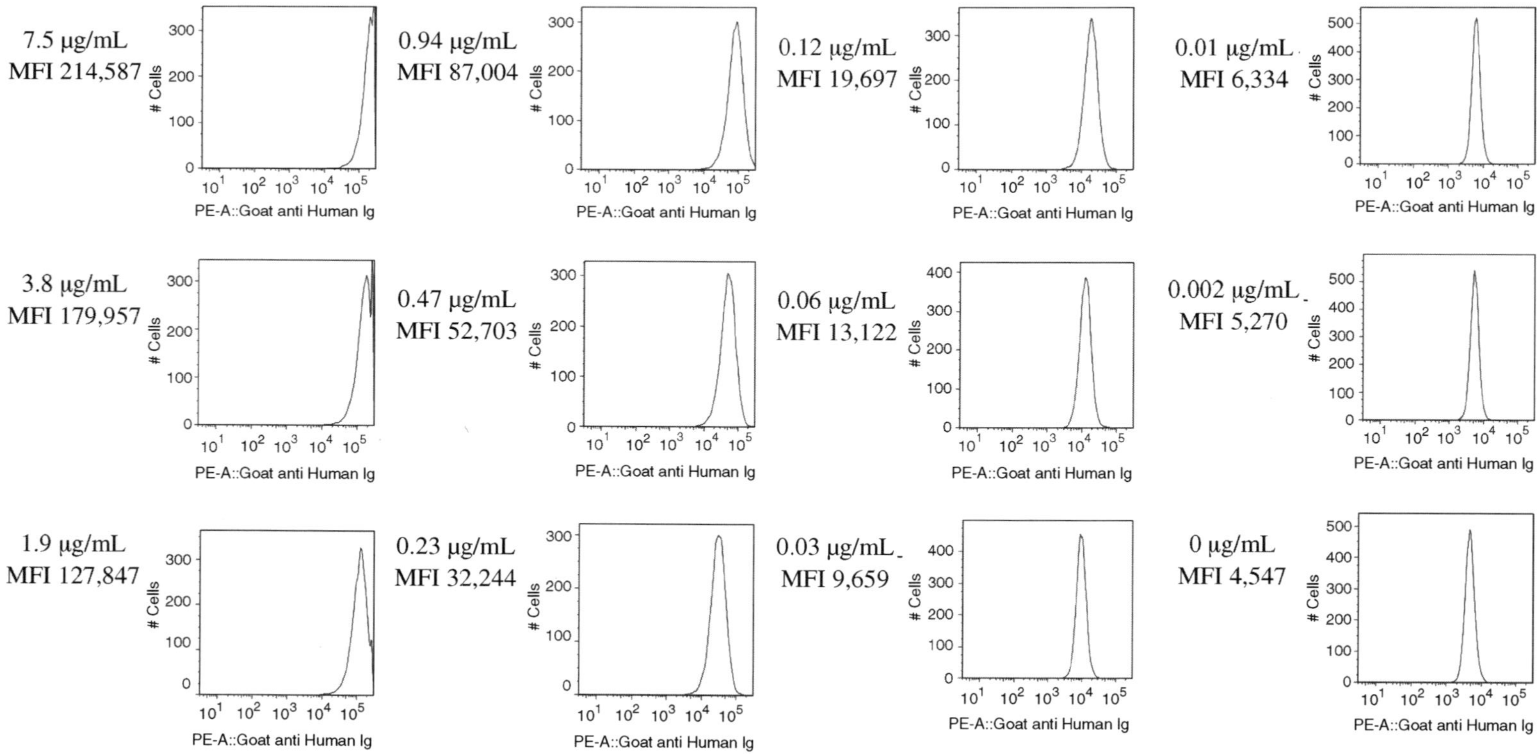

FIGURE 11.2 Dose-dependent binding of drug Y to CHO–Ag. Dose-dependent binding of a humanized monoclonal antibody drug Y to CHO–Ag is shown in histograms (MFI median for PE versus cell number). Drug Y concentration (μg/mL) and the corresponding MFI median value is shown for each histogram.

11.2.4 Selection of Secondary Ab Detection Reagent

11.2.4.1 Considerations for Selection of Secondary Ab as Detection Reagent Nonspecific binding to cells can be problematic. To reduce nonspecific binding of the secondary Ab to cells via Fc receptor, the use of antidrug IgG $F(ab)'_2$ is highly recommended. In addition, isotype-specific secondary Ab (e.g., antihuman IgG1 for a therapeutic human IgG1 MAb) may be useful to minimize background. These approaches would also be appropriate for animal studies. If the PK method is intended for preclinical studies, the use of antispecies-specific antibodies should be evaluated against the relevant immunoglobulin species (e.g., mouse IgG or monkey IgG). In general, either fluorochromes FITC or PE are typically used due to their stability.

The use of both commercial secondary Ab and custom conjugated Ab is suitable. For custom-made secondary Ab, conjugation methods should be optimized to achieve an appropriate conjugation ratio (FITC or PE/drug). If the ratio is too high, binding to drug may be affected, while having a ratio that is too low would result in poor assay sensitivity. It is recommended that the binding affinity of conjugated secondary Ab with different conjugation ratio should be evaluated and compared to unconjugated secondary Ab. In addition, binding capabilities of the secondary Ab (particularly if it is monoclonal) should also be evaluated and not compete with binding of Ab drug-target Ag.

Multiple secondary detection antibodies from several vendors should be evaluated to select those with high binding affinity, high specificity, and low background. Once the detection Ab is selected, additional lots, if available, should be evaluated to assess lot to lot variability.

11.2.4.2 Optimization of the Ab Binding Condition It is critical that all aspects of binding conditions be optimized including, but not limited to, incubation time, temperature, and so on. For high affinity binding, relatively shorter incubation time (30–60 min) would be sufficient. Low affinity binding will need longer time and lower temperature such as 4 °C. The use of DOE (design of experiment) or fractional factorial design would help optimize assay conditions and assess interaction of various factors.

11.2.4.3 Detection Ab Titration A broad titration of the detection Ab should be performed to ensure optimal signal and minimize background. Full dose–response curves (secondary detection Ab concentration or dilution versus MFI median) with drug at high concentrations would promote appropriate selection of secondary detection Ab concentration or dilution. In general, using an Ab concentration or dilution generating a maximal signal is recommended. The concentration or dilution should not be used close to the inflection point where small concentration changes could result in significant signal changes, thereby impacting assay precision.

11.2.4.4 Specificity This is an important aspect of reagent characterization. The secondary detection Ab specificity should be assessed in a number of ways. First, detection Ab should detect drug binding to cell surface target antigen (such as CHO–Ag). In addition, the detection antibody should not demonstrate irrelevant binding such as binding to cells without the target antigen (i.e., CHO alone) or binding to cells expressing the target antigen without the drug.

11.2.5 Preparation of Calibration (Reference) Standard and Quality Control

Well-characterized, sufficiently purified protein drugs spiked into appropriate biological matrix (such as serum or plasma) should be used for reference standard and quality control (QC) samples. The reference standard should be evaluated for any biological matrix effect (or interference) that could result in enhanced or reduced signal. Using drug spiked in buffer may also result in over- or underdetection of the drug in study samples and is not recommended.

Pooled normal human (or animal) serum (or plasma) is typically used as the biological matrix. In general, the pool is composed of individual samples from healthy subjects screened for lack of interference. A minimal sample dilution needs to be defined using criteria similar to that for immunoassays. The percentage of serum used in preparing reference standard and QC samples should be the same as study samples being tested.

Generation of a reference standard at different concentrations across the calibration standard curve concentration range is advised. The number of standard points needed for a good curve fit will depend on the curve-fit model and assay performance in general. An extended curve range should be evaluated, which can be narrowed to improve curve fit.

QC samples typically consist of high (H), medium (M), and low (L) levels of drug covering the linear portion of the standard curve. A negative control, pooled normal human or species-specific serum without the drug may also be used.

11.2.6 Curve Fit

Multiple curve-fit models should be evaluated. Typically, multiple standard curves from data collected by two–three analysts over multiple days are used for this analysis. Most ligand binding assays, including the drug-target Ag binding in the flow cytometric PK assay, have a sigmoidal dose–response, and a four-parameter nonlinear curve fit is a commonly used model.

An appropriate fitted curve is expected to show random deviation from the best-fitted curve (residual analysis) and acceptable percentage recovery of back-fit values compared to the theoretical concentrations. Interpolated values from unknown samples are ideally expected to have recovery within 80–120% from theoretical values. Assay sensitivity is demonstrated to support assay's intended use.

Figures 11.3 and 11.4 show the dose–response curve from drug X and drug Y, respectively. Corresponding histograms for drug X and representative histograms for drug Y are shown in Figures 11.1 and 11.2, respectively. Four-parameter curve fit is found to be suitable for both standard curves and criteria are met.

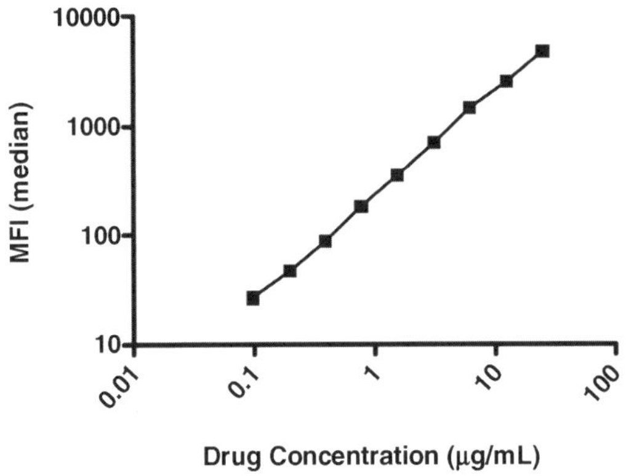

FIGURE 11.3 Drug X standard curve. A polyclonal antibody drug X dose–response is shown. Standard reference material was spiked at difference concentrations into pooled normal human serum. Standard curve is illustrated as concentration (μg/mL, *x*-axis) versus binding signal (MFI median, *y*-axis). Data were collected on FACSCCalibur™.

11.2.7 Assay Development Study

11.2.7.1 General Considerations Most considerations relevant to general assay development practices would apply to flow-based assays. During assay development, the following variables are typically examined to define the optimal conditions: cell number, drug concentration across the standard curve range, cell–drug incubation time/temperature/wash, cell-bound drug secondary detection Ab incubation time and

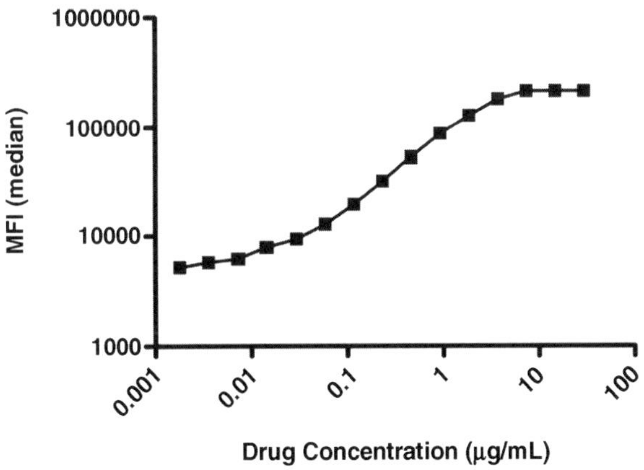

FIGURE 11.4 Drug Y standard curve. A humanized IgG1 monoclonal antibody drug Y dose–response is shown. Standard reference material was spiked at difference concentrations into pooled normal human serum. Standard curve is illustrated as concentration (μg/mL, *x*-axis) versus binding signal (MFI median, *y*-axis). Data were collected on FACSCanto™ II.

temperature, wash, sample acquisition cell number, and data analysis. Tolerance limits (assay robustness) should also be established for critical steps.

Selection of an appropriate cell number (per staining) would allow a drug standard curve with a broad dynamic range. Lack of prozone effect (or hook effect) where lower binding signal is seen at high drug concentrations is also important to investigate. The use of DOE is encouraged as it allows efficient assessment of multiple parameters, especially when these parameters interact with each other. Details on the use of DOE are described by Anderson and Whitcomb [2].

11.2.7.2 Assay Format Assays can be developed using either a tube or a microtiter plate format. The use of tubes results in a relatively low throughput, but allows a more thorough washing by using higher volume wash buffer. The use of 96-well plate for staining is more efficient, but due to the limited volume of wash buffer that can be added to each well, more than one wash may be needed. Evaluation of incubation time for each step and establishment of tolerance limits as part of assay robustness are important to define assay conditions. Each sample should be tested in duplicate or triplicate wells and results averaged. The average MFI (median) is considered a single data point with a $CV \leq 15\%$ being acceptable.

11.2.7.3 Staining Condition Staining can be done in PBS containing BSA and sodium azide at 4 °C. Room-temperature staining should be performed in a temperature-controlled incubator (22–26 °C). Incubation times should be determined to obtain maximal binding signal without increase in background.

11.2.7.4 Specificity and Selectivity Specificity is the ability of an analytical method to exclusively detect the target analyte and display complete discrimination from physiologically/chemically similar compounds that may be present.

Specificity considerations for the secondary detection antibody are previously described. It is good practice to evaluate binding to drug across the standard curve range with drug prepared in buffer versus in sample matrix (serum, EDTA plasma, heparin plasma, etc.). The use of the appropriate sample matrix (pooled serum or plasma that matches the study sample) to prepare reference standard drug and QC samples is recommended. If there is a sample matrix effect on the assay, interpolating data of study sample from a standard curve using drug in buffer may result in over- or underdetection.

Selectivity is examined to assess spike and recovery. Drugs at H, M, and L were spiked into sample matrix (serum or plasma) from multiple donors and results compared to samples spiked in buffer. Recovery of 80–120% is considered acceptable. Recoveries outside this range should be used with caution and warrant further interpretation of results. An interference study can be done to examine potential soluble interfering factors that could be present in patient's samples and influence results. For this type of study, drugs at H, M, and L levels were added to sample matrix in the presence of potential soluble interference factors (each at H, M, or L levels). This study can include sample composition interference, such as hemolysis or lipemia, or more specific interference such as soluble binding proteins or receptors.

DOE can be used to design an appropriate interference study. Recovery of 80–120% indicates lack of interference. Identification of potential interference factors that are present due to the disease and specifically in patient's samples is important.

11.2.7.5 Accuracy and Precision Accuracy is defined by the exactness (closeness) of the test value determined by the assay to the reference value (true value).

Accuracy is assessed using validation samples (drugs at three–five different levels across the standard curve linear range). Typically, each of the two analysts collects data from replicates for each sample on a single day (intra-assay) or over multiple days (interassay). An appropriate number of replicates selected should be relevant to clinical sample testing replicates. The use of a high number of replicates may result in acceptable accuracy in validation studies, which may not represent assay's intended use in clinical sample testing typically performed in duplicates or triplicates. Recovery (observed concentration/expected concentration) within 80–120% is considered acceptable, demonstrating assay is accurate.

Precision measures closeness of individual measures of an analyte when repeatedly assessed using multiple aliquots of the analyte in a single homogeneous volume of biological matrix.

Precision is assessed using validation samples (drugs at three–five different levels) as described in accuracy studies for repeatability (intra-assay precision) and intermediate precision (interassay precision). As described in accuracy assessment, an appropriate number of replicates should be relevant to the clinical sample testing replicates. For each sample examined by each analyst, CV (SD/mean $\times$ 100) $\leq$ 15% is considered acceptable for intra-assay and CV $\leq$ 20% for interassay precision.

Typically, CV $\leq$ 15% is considered acceptable for intra-assay and $\leq$20% for interassay and overall assay precision. Data from assays performed outside these generally accepted measures of assay performance should be evaluated cautiously. Precision is examined for both MFI median and interpolated mass values (concentration such as ng/mL). An example for drug Y is shown in Table 11.1.

The use of statistical approaches such as balanced design and various component analyses to design accuracy and precision studies is encouraged. Using such an approach, each multiple analyst performs the same number of intraday runs and repeats the intraday runs for the same number of days. Various component analyses allow the assessment of variability due to intraday, interday, and/or analyses that affect accuracy and/or precision.

11.2.7.6 Evaluation of Normal Sample and Treatment Naïve (Baseline) Disease Sample For human PK assays, it is recommended that 30–50 normal serum samples are examined. In general, normal donor samples are expected to be negative for the presence of target drug. Potential interference factors such as endogenous counterpart of the drug (e.g., cytokines or growth factors) are sometimes present in samples from healthy subjects and treatment naïve patients. These interference factors, especially those sharing similar characteristics with the target drug, may contribute to positive signals. Understanding any baseline sample positivity is important for development strategies to reduce interference in PK analysis.

TABLE 11.1 Assay Performance Characteristics Between the Two Validated PK Methods for Drug Y

Validation Parameter (Acceptance Criteria)	Flow Cytometric Pk Assay	ECL-Based PK Assay
Specificity: Method can detect a dose-dependent response	Pass	Pass
Accuracy: Assessed at 3–5 drug concentrations with data collected from 2–3 analysts (80–120% recovery of the theoretical value)	80–120% recovery Pass	80–120% recovery Pass
Precision: Assessed at 3–5 concentrations with data collected from 2–3 analysts (<15%CV for intra-assay; <20%CV for interassay)	Intra-assay (n = 5): <10%CV Interassay (n = 4): <16%CV Pass	Intra-assay (n = 6): <8%CV Interassay (n = 5): <13%CV Pass
LOD: Unspiked serum background MFI + 3SD	58.6 ng/mL	3.5 ng/mL
LOQ: The lowest drug concentration with acceptable accuracy (80–120% recovery) and precision (<20%CV)	300 ng/mL	60 ng/mL
Linearity of sample dilution:	Linear: $r^2 > 0.99$ and slope 0.95–1.09	Linear: $r^2 > 0.99$ and slope 1.31–1.33
Additional parameters examined:	Curve-fit, sample stability, normal sera distribution, and control ranges: Pass	Curve-fit, selectivity, spike/recovery, sample stability, normal sera distribution, and control ranges: Pass

11.2.7.7 ***Provisional Control Reference Ranges*** Data collected from H, M, and L QC samples in development studies may be used to establish provisional control ranges for assay pass/fail criteria in validation studies. Upon completion of the validation study, all QC data (from multiple interday runs by multiple analysts) should be analyzed for establishment of control reference ranges for in-study/clinical sample testing. A statistical analysis-based approach, such as mean $\pm$ 2SD, should be considered to establish ranges. If the data set for each drug level is not normally distributed, transformation may be required prior to establishing the appropriate range.

11.3 ASSAY VALIDATION

Technical resources, such as FDA, EMEA, and ICH [3–6] guidelines on analytical method validation, are available and should be referenced for flow cytometric method validation studies. Additional details can be found in a paper published by DeSilva et al. [7].

In general, the validation study design for flow cytometric PK assays is similar to other PK methods such as ELISA and mass spectrometry-based assays. It is recommended that the following assay performance characteristic should be evaluated in validation studies: specificity/selectivity, precision (intra- and interassay), accuracy, sensitivity [limit of detection (LOD) and limit of quantification (LOQ)], linearity of sample dilution, validation sample stability (ambient temperature: short-term 4 °C and long-term −20 to −80 °C), analysis of normal donor samples and baseline samples from treatment-naïve patients (or animal subjects), and establishment of control reference ranges.

Design of validation studies to assess assay performance characteristics described under assay development will not be discussed in this section. The emphasis of this section is to discuss challenges that are unique to flow cytometric methods.

11.3.1 Accuracy and Precision

It is not recommended to use cells from fresh blood as target cells for PK analysis. Instead, a cell bank of frozen PBMCs from apheresis is preferred. If cells from fresh blood have to be used as target cells, interday assays should be completed within 48 h after blood is drawn, due to cell stability concerns. Evaluation of different anticoagulants and preservatives may allow extension of fresh blood stability beyond the typical 48 h window. Another challenge is to fully examine normal donor-to-donor variability. The use of fresh blood from a single donor for PK analysis over time is not practical.

Accuracy and precision assessment for flow cytometric PK assay has unique challenges, especially if drug binding to cells is not homogeneous. Assay accuracy and precision may not meet the criteria established for other types of assays, recovery 80–120% and CV $\leq$ 20%, respectively. Under such circumstances, additional interpretation may be required for data generated.

Generally, a flow cytometric assay is typically performed in a single tube or at most in triplicates if 96-well plate format is chosen. Designing a study with a large number of sample replicates may result in acceptable accuracy and precision, but assay performance may not reflect actual in-study/clinical testing design. The appropriate number of repeats for intra- and interassay studies should be similar to repeats performed for clinical sample testing. Average of data collected from multiple runs with a large number of data points prior to calculation of recovery may not be appropriate since over- and under-recovery in each assay may cancel out, thereby making the assay appear accurate. The use of the absolute percentage bias is an alternative analysis to examine assay accuracy.

11.3.2 Limit of Detection

Limit of detection provides the minimal detection of signal that can be distinguished from background. Typically, MFI median plus 2 or 3 SD for the blank serum sample can be interpolated from the standard curve to generate a concentration value as LOD. At assay LOD, the method may not be able to quantify drug with acceptable precision and accuracy, especially when an endogenous counterpart of the drug is present. In these cases, the PK assay could detect positivity in normal donor and baseline patient's samples. In such cases, using normal or baseline samples for LOD assessment would artificially increase LOD and make the PK assay appear to be of poor sensitivity. An alternative approach would be to use samples depleted of endogenous target analyte for assessment of LOD. Other potential interference factors could also contribute to background positivity, which need to be examined. Defining minimal sample dilution is necessary.

11.3.3 Lower Limit of Quantification

A separate set of samples at low concentrations is prepared and assessed in multiple repeats for lower limit of quantification (LLOQ) assessment. The lowest interpolated concentration that meets precision ($CV \leq 20\%$) and accuracy (recovery 80–120%) criteria is considered assay LLOQ. If minimal sample dilution is needed, the dilution factor needs to be accounted for in calculating assay LLOQ. Limit of quantification at high concentrations also needs to be defined if a prozone effect (or hook effect) is seen. In these cases, drug at high concentrations shows decreased signal and the sample result would be misreported. For samples above the upper quantification range, repeated testing using diluted samples is recommended.

11.3.4 Linearity of Sample Dilution

If high concentrations (above the upper quantification range) of the analyte are expected, samples may need to be diluted prior to testing. Samples are prepared at high concentrations (at least 10-fold higher than the upper limit of the standard curve point) and tested at different dilutions. Recovery of samples tested at different dilutions is expected to be within 80–120% from the theoretical value. The dilution curves should be plotted with the standard curve to demonstrate continuity and linearity. The slope from the linearity study curve should be within 10% of the slope from standard curve. Lack of hook effect should be demonstrated. Serial and parallel dilutions will be both examined to demonstrate sample dilution linearity.

11.3.5 Stability

Sample stability is examined for benchtop (ambient temperature), short-term 4 °C and long-term −20 to −80 °C storage using samples at H, M, and L concentrations. Sample recovery between 80–120% from theoretical value (or baseline value) is considered acceptable. Cell-based assays generally are prone to inherent variability and this can also occur with flow cytometric assays. Consequently, it is not appropriate to use a single data value at time 0 as the reference value for evaluating sample stability. It is advisable to establish a reference baseline value using multiple runs within a short timeframe. Alternatively, the use of the "true" value (spiked concentrations at H, M, and L) could also be considered as reference baseline value for stability studies.

In addition to sample stability assessment, poststained cell stability, with or without fixatives, should also be examined. Data collected from immediate acquisition after staining and samples stored at 4 °C at various times prior to being acquired should be compared with reference baseline values (at H, M, and L); recovery within 80–120% is considered acceptable.

11.3.6 Control Reference Ranges

QC samples (at H, M, and L) tested in validation studies may be used to establish control reference ranges. Due to inherent variability seen in flow cytometric methods in general, at least 20 data values for each drug level from multiple analysts over a period of time are recommended. MFI values typically showed higher degree of variability in part due to instrument drifting as a result of decreased laser power output over time. Therefore, establishing control ranges on the basis of MFI would take this variability into account. Calculation of interpolated concentration (H, M, and L) inherently corrects for this variability. In general, control range setting for flow cytometric PK method is similar to practices for other analytical methods (see details in Section 11.2.7.7).

11.4 CONCLUSIONS AND DISCUSSION

Flow cytometry is an alternative technology platform that can be used for PK studies. The examples discussed demonstrate the utility of this approach relative to other PK methods. Both flow cytometric methods (for drug X and drug Y) were validated and met the proposed validation criteria. To further our understanding of its utility, a comparative study of the validated flow cytometric PK method for drug Y to an electrochemiluminescence (ECL)-based PK assay was completed. Results are shown in Table 11.1 for key performance parameters. Both assays are accurate and precise. The ECL PK method is more sensitive than the flow cytometric PK method primarily due to the use of an anti-idiotypic monoclonal antidrug antibody with high affinity as a capture reagent. Both methods were deemed acceptable for use in evaluating clinical samples.

In summary, flow cytometric PK methods can be successfully validated and hence assay performance meets validation criteria based on regulatory guidelines and published papers for analytical method validation [3–7].

REFERENCES

1. Current Protocols in Cell Biology, Juan S. Bonifacino, Mary Dasso, Joe B. Harford, Jennifer Lippincott-Schwartz, and Kenneth M. Yamada Editors, Wiley, Inc., 2010.
2. Anderson MJ, Whitcomb PJ. DOE Simplified: Practical Tools for Effective Experimentation, Productivity Inc., 2000.
3. FDA Guidance for Industry: Bioanalytical Method Validation. FDA CDER, US Department of Health and Human Services, 2001. Available at http://www.fda.gov/downloads/Drugs/GuidanceComplianceRegulatoryInformation/Guidances/UCM070107.pdf.

4. International Committee for Harmonization (ICH) of Technical Requirements for Registration of Pharmaceuticals for Human Use. ICH Harmonized Tripartite Guideline Q2A: Text on Validation of Analytical Procedures, 1995.
5. European Medicines Agency. Draft Guideline on Validation of Bioanalytical Methods, 2009. EMEA/CHMP/EWP/192217/2009.
6. International Committee for Harmonization (ICH) of Technical Requirements for Registration of Pharmaceuticals for Human Use. ICH Harmonized Tripartite Guideline Q2B: Validation of Analytical Procedures: Methodology, 1996.
7. DeSilva B, Smith W, Weiner R, Kelley M, Smolec J, Lee B, Khan M, Tacey R, Hill H, Celniker A. Recommendations for the bioanalytical method validation of ligand-binding assays to support pharmacokinetics assessments of macromolecules. *Pharm. Res.* 2003;20: 1885–1900.

PART III

VALIDATION AND REGULATORY COMPLIANCE

VIRGINIA LITWIN AND PHILIP MARDER

Among the most challenging aspects of incorporating flow cytometry into the drug development process is validating methods and instrumentation in accordance with regulatory requirements and within the aggressive timelines associated with present day drug development.

In drug discovery laboratories, flow cytometry screening methods are implemented in order to increase both throughput and selectivity (reviewed in Chapter 4). Given that this stage of the drug development lifecycle is not regulated, methods, in general, are not subjected to formal validation. Nevertheless, high quality, specific methods are essential in order to obtain successful outcomes.

Preclinical toxicology laboratories develop a wide variety of flow methods for the purpose of both evaluating the safety of a new compound and assessing pharmacodynamic effects in animal models (reviewed in Chapter 6). Safety assessment laboratories are required to follow GLP regulations. However, the appropriate regulatory requirements for nonsafety-related biomarker methods (e.g., pharmacodynamic and exploratory biomarkers) are less clear. The primary reason for including biomarkers in drug development and translational medicine programs is to support a more informed level of decision making, thereby lowering the risk of failure during clinical trials. To meet these objectives, methods must undergo rigorous validation.

Both chapters in Part III examine the current regulations and guidelines. Chapter 12 focuses on the important considerations for assay development and validation, while Chapter 13 discusses the benefits and objectives of instrument validation.

Flow Cytometry in Drug Discovery and Development, Edited by Virginia Litwin and Philip Marder

12

REGULATORY COMPLIANCE AND METHOD VALIDATION

CARLA G. HILL, DIANNA Y. WU, JOHN FERBAS, VIRGINIA LITWIN, AND MANJULA P. REDDY

12.1 INTRODUCTION

Philosophies surrounding method validation for flow cytometric assays conducted in the biopharmaceutical space are varied. The extent of method validation appropriate for a particular assay is a function of intended use and the stage of the drug development where the assay will be utilized. Likewise, determining which regulatory guidelines are applicable to a given flow cytometric method along the continuum of drug discovery and development adds to challenges to the validation process (because different regulatory guidelines apply in the lifecycle of potential clinical candidates). The portfolio of regulatory requirements that may be exerted in the discovery and development work stream is illustrated in Figure 12.1.

While considering the content for this chapter, the need to convey and discuss the gray spaces and adjustments in regulatory oversight as a function of the development and commercialization process was quite clear; the vast majority of cytometerists in industry do find themselves holding an overarching role that spans across development stages (which can be less frequent for other types of assays). While we all aspire to execute studies in full compliance of regulatory specifications, we also realize that guidance was originally written for toxicology and safety pharmacology methods, such as mass spectrometry to detect drug metabolites. So, although issues including the lack of true accuracy standards and sample stability of, for example, whole blood can be disconcerting – and perhaps even incongruous with method validation as applied to other assay types – effective approaches to minimize risk and generate data

Flow Cytometry in Drug Discovery and Development, Edited by Virginia Litwin and Philip Marder

with the confidence of its scientific integrity have been developed and are the topic of this chapter.

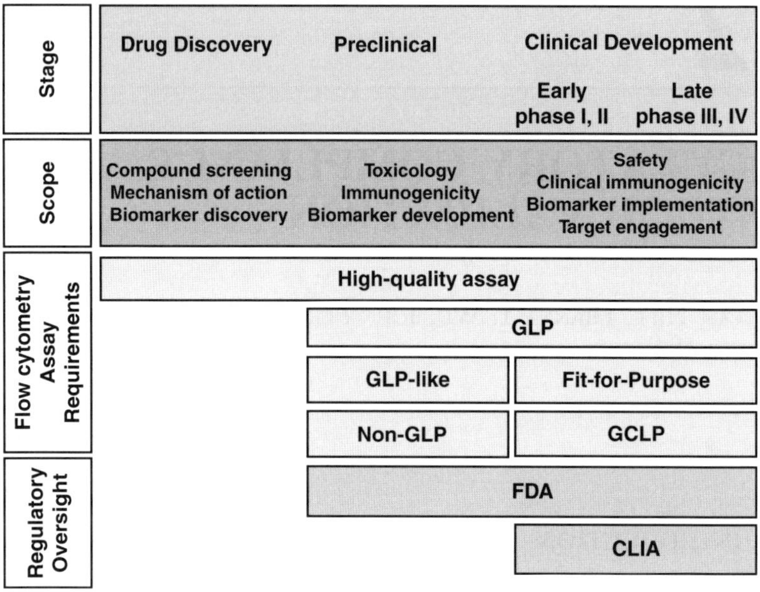

FIGURE 12.1 Drug development stages and regulatory oversight.

12.2 REGULATORY COMPLIANCE

The versatility of flow cytometry platform begs for its inclusion in campaigns to identify new drug targets or to discover potential biomarkers. There is an additional appeal, as this same platform can effectively support downstream efforts in the mechanism of action and biochemical coverage studies. While no specific regulatory guidelines are called out in this early stage, missteps cause a high impact because of their 1:1 influence on the downstream path. Although not absolute, the principles of validation are therefore regarded by many to represent an insurance policy against errant data sets derived from suboptimal specimen processing and data collection. Indeed, those with experience in the field frequently find themselves raising the warning flag that a flow cytometer will generate data regardless of the quality of the specimen. Application of a limited set of validation exercises – even in this early stage – will illuminate problems before they impact programs.

Regulatory practices for preclinical studies, on the other hand, are unambiguous. For toxicology studies designed to characterize lead compounds with data sets eligible for regulatory review, the need to implement Good Laboratory Practices (GLPs) is nonnegotiable. Although some perform flow cytometry as a "non-GLP component of a GLP study," most regard this as an inadvisable practice. In this latter setting, a better choice could be to conduct a separate non-GLP study to define the need to validate a flow cytometry assay for legitimate inclusion in a (subsequent) GLP study.

As lead compounds proceed to early clinical trials (phase I/II), "fit-for-purpose" validation is regarded by most to be sufficient provided the data are not diagnostic or representative of a safety measurement (in which case CLIA standards become enforceable). In later trials (phase III registrational studies and phase IV studies of commercialized products), Good Clinical Laboratory Practice (GCLP) and Clinical Laboratory Improvement Amendment (CLIA) guidelines are applied, where decisions reflect the intended utility of the data. So, a flow cytometric method may be used in more than one stage of the drug development process with the proviso that revalidation is exercised to meet regulatory requirements at the next stage [1]. Although validation and regulatory requirements may evolve as compounds progress to filing, the goal in all cases is to achieve reliable results that are calibrated and appropriate for the intended application of the data [2, 3]. Assay development, whether on a flow cytometer or other platform, is an iterative process where both the regulatory oversight and the need to formally document the reliability of the measurements are tethered to the context of the effort.

It cannot be understated that a key consideration in application of flow cytometry assays in the biopharmaceutical industry is the intended purpose of the assay. Exploratory data are sometimes not subject to the high precision and accuracy requirements of clinical data; clinical standards may not be rigorous enough to support identification and qualification of a new biomarker [4]. There is a very important distinction to be made, however, when assays are defined by the exploratory moniker in the clinical setting because in the true (clinical) sense an exploratory assay is one where the data are not a clinically validated end point or a surrogate marker of efficacy. Stated another way, "exploratory" sometimes means that the assay is the subject of scientific exploration, but in other cases an alternative definition that considers the measurement to represent a clinical end point of nonestablished meaning is intended. Although those in the field have this awareness and compensate for the potential confusion the terminology may bear, invention of a more precise terminology would be of benefit to all.

12.2.1 GLP

GLP guidelines were developed in the 1970s from best practices to improve reliability in toxicology data submitted to the U.S. Food and Drug Administration (FDA) [7]. The original documents were written for animal studies; however, revisions have been published to fit the unique characteristics of nonanimal-based *in vitro* methods, such as cell culture systems [5, 8]. The specific guidelines of GLP applicable to all aspects of nonclinical laboratory studies are outlined in Title 21 Part 58 of the Code of Federal Regulations (21CFR58). It provides an outline to justify a testing environment and its operations addressing facilities, personnel, equipment, standard operating procedures (SOPs), assay controls, records, and reporting requirements. In addition, GLP compliance requires an independent quality assurance component including formal internal audits.

The internationally accepted principles of GLP were developed by the Organization for Economic Co-operation and Development (OECD), outlining principles and quality assurance standards for nonclinical laboratory research performed by its member countries in the Dangerous Substances Directive 67/548/EEC [9]. OECD also issued a supplemental advisory document to address the application of GLP principles to the uniquely different characteristics of *in vitro* systems: positive and negative controls and benchmarks; care and maintenance of test systems; and quality assurance monitoring [10]. Following the extensive guidelines of GLP is often costly and resource intensive; however, research not conducted under these restrictions may render data inadmissible in support of a new drug filing.

Laboratories in compliance with GLP regulations are globally recognized as promoting trustworthy data supporting compounds for government regulation, although some international requirements differ in proof of compliance [11]. Some countries require laboratory accreditation for claiming GLP compliance; however, the United States does not have a GLP accreditation program in place [12]. In all cases, compliance with GLP requires extensive documentation and laboratory audits.

To supplement GLP regulations, the FDA released guidance for bioanalytical method validation in support of preclinical and clinical pharmacology studies [13]. The document provides general recommendations for performing precision, accuracy, selectivity, sensitivity, reproducibility, and stability assessments required in regulated method validation to ensure it is suitable and reliable for its intended application. In addition, this guidance describes the level of validation required for new and modified methods: full validation, partial validation, and cross validation. The provisions and acceptance criteria are sometimes difficult to directly apply to novel biomarker assays; nonetheless, the general recommendations can be modified as applicable to flow cytometric methods [14].

12.2.2 CLIA

In the United States, all testing performed on human specimens used for patient diagnosis, disease monitoring, health assessment, and subsequent decision making is subject to CLIA regulations. Although CLIA actually does not apply to clinical research testing, during clinical trials safety parameters and patient enrollment testing are likely to be conducted in CLIA-certified laboratories. The reason for this is two-fold: one is that data will be reported in real time to the clinical investigator, and thus might be used for patient care, and the other is that CLIA-certified laboratories abide by high-quality standards to positively impact data reporting [14]. The CLIA regulations define a quality system for laboratory testing to ensure the accuracy and reliability of results. They apply to all hospital, independent, and physician office laboratories performing analyses on human samples. Specific operational requirements are included for personnel credentials, continuing education, proficiency testing, method performance specifications, and reference range requirements. All aspects of laboratory operation and result reporting required under CLIA authority are outlined in 42CFR493. The Division of Laboratory Services, under the Center for

Medicare and Medicaid Services (CMS), is responsible for implementing the CLIA Program in the United States [6].

To facilitate implementation of CLIA standards in clinical laboratory operations, guidelines have been developed by the Clinical and Laboratory Standards Institute (CLSI), formerly known as the National Committee on Clinical Laboratory Standards (NCCLS). These guidelines are followed by hospital laboratories to maintain licensure and compliance. Internationally, some laboratories adopt CLIA recommendations to improve management of clinical and laboratory services in the interest of quality patient care [15, 16]. Several flow cytometric-specific documents are available through CLSI, providing guidance on performing immunophenotyping of lymphocyte subsets and neoplastic hematolymphoid cells, reticulocyte counting methods, and fetal red cell detection [17–20].

CLIA-regulated laboratories must be certified by an approved accrediting agency and follow appropriate guidelines based on the complexity of testing being performed [21]. On-site inspection every 2 years is a certification requirement for all moderate- and high-complexity testing laboratories. CMS-approved accrediting agencies include the College of American Pathologists (CAP) and Joint Commission on Accreditation of Healthcare Organizations (JCAHO) where guidelines are at least as stringent as CLIA. Examples of high-complexity flow cytometric analyses that may fall under CLIA guidelines include leukemia/lymphoma immunophenotyping, $CD34^+$ stem cell counts, and $CD4^+$ lymphocyte counts. CAP provides detailed guidance for performing these flow cytometry analyses in accordance with CLIA regulations [22, 23]. However, if analyses are considered for research testing in clinical trials and not used for patient care management, compliance is not required.

12.2.3 GCP

Good Clinical Practice (GCP) guidelines are internationally recognized quality standards for conducting, recording, and reporting human clinical trials, but the guidelines do not pertain to method validation and laboratory testing. The FDA is responsible for GCP guidance for U.S. trials, outlined in 21CFR312 [24]. Globally, the International Conference on Harmonization (ICH) defines GCP [25, 26]. Several countries have gone so far as to adopt GCP guidelines as law for conducting clinical trials. Adherence to GCP in clinical trials will facilitate the acceptance of clinical data by the regulatory authorities in ICH participating countries.

12.2.4 GCLP and GLP-like

Although GLP and CLIA standards do not directly apply to assays conducted in support of clinical trials, many nonregulated laboratories operate under “GLP-like” conditions to add additional support for production of high-quality data [14]. Documented protocols and standard operating procedures, validation of assay performance, instrument standardization, and assay quality controls are implemented in clinical trials in preparation for FDA filing and patent application requirements, if necessary [27].

A new approach to bridge the regulatory gap for human trials considered clinical research is the proposal of GCLP recommendations. The British Association of Research Quality Assurance (BARQA) wrote initial GCLP guidelines in 2003, blending GCP trial conduct standards and GLP facility and procedural requirements to create a framework for laboratory analysis in support of clinical trials [28]. Although drafted in a generic format, the goal of GCLP was to facilitate comparison of laboratory data generated in global clinical research trials [29].

More specific GCLP recommendations were later put forth by the National Institute of Allergy and Infectious Disease (NIAID) and PPD, Inc. to address data comparability concerns in immunogenicity assays between phase I and II vaccine studies conducted in global HIV trials [30, 31]. Guidelines specifically address flow cytometry analyses including diagnosis of HIV-1 infection, blood processing immunogenicity, and cellular immunogenicity assays for trials supported by the Division of Acquired Immune Deficiency Syndrome (DAIDS) [32]. These GCLP guidelines embraced both research and clinical aspects of FDA compliance: 21CFR58 and 21CFR493. Recommendations from other accrediting agencies, such as CAP and the International Organization for Standardization (ISO), were also incorporated: CAP CLIA-based Laboratory Accreditation Program, CAP Accreditation to ISO 15189 Standard Program; ISO 15189 Quality Management and Competence of Medical Laboratories; ISO 17015 Quality of Laboratory Data; and ISO 17025 Requirements for Testing and Calibration [33–35].

Consistent GCLP application in research and development laboratories will provide reliable and comparable data, paramount to a successful clinical trial. The key elements of GCLP in laboratory testing include organization and personnel, laboratory equipment, testing facility operations, quality control program, verification of performance specifications, records and reports, physical facilities, specimen transport and management, personnel safety, laboratory information systems, and quality management. Conduct of clinical trials within the GCLP guidelines will harmonize clinical-trial flow cytometry operations among multiple laboratories producing high-quality data and ensure smooth transition of a compound in drug development [29, 31, 34].

12.3 METHOD DEVELOPMENT AND VALIDATION FOR DRUG DEVELOPMENT

Flow cytometric assays in drug discovery and clinical development not only provide valuable information pertaining to the population of cells targeted by the drug, binding of drug to its receptors or ligands, and modulation of the antigen by the drug but also contribute to the understanding of drug effects on cellular functions. Various assays, such as immunophenotyping, apoptosis, activation, proliferation, cell signaling, intracellular cytokines, and receptor occupancy are being extensively performed in the drug development process utilizing different types of flow cytometers. Flow cytometry is not only becoming more widely applied in a clinical diagnostic setting for disease diagnosis and monitoring but also being used in biomarker research and pharmacodynamic marker evaluation during drug treatment [36–39].

High-quality flow cytometry assays are important to all stages of drug development and are essential for appropriate decision making. A high-quality assay is well controlled, reproducible, well documented, and characterized to meet the intended use. The same assay can be utilized as a drug advances from the exploratory discovery stage to early clinical studies and late-stage trials; however, assay validation considerations may differ to meet intended objectives. Development of a versatile flow cytometric assay for clinical utility requires identification of (i) the appropriate method; (ii) intended use of the results from the assay; and (iii) an optimization, validation, and implementation strategy. An assay development plan is highly recommended for all stages of drug development to ensure appropriate execution of flow cytometry experiments in a timely manner. The development plan is a useful tool to obtain relevant well-characterized reagents, assay controls, and samples (cell lines or primary cells or tissue); to establish time lines; and to identify critical steps in the assay optimization process to further plan validation experiments.

12.3.1 Method Development and Optimization (Method Feasibility/Prevalidation)

During assay development and optimization, conditions for all aspects in performing the assay should be established (specimen collection/processing, reagent selection, gating/data reporting, instrument setup and monitoring). The following sections discuss various parameters of a flow cytometry assay that are important to evaluate in generating a good-quality flow-based assay that is reproducible and reliable.

12.3.1.1 Specificity and Reagent Selection The strict definition of accuracy is defined as the closeness of agreement between the averages of the test results when compared to a well-defined official reference standard. Accuracy assessment is not possible in flow cytometry assays in this regard, as reference standards are not readily available. In practice, many of the biomarker-type assays supported by the flow cytometry laboratory report results relative to a study subject's baseline sample. In this vein, assay precision, and the reliability of a measurement under test-retest conditions, steps forward as a very important parameter.

Specificity is another key assay parameter that can be addressed during assay development. Most flow cytometric methods use well-characterized monoclonal antibodies or nucleic acid binding dyes. It is important to examine appropriate antibody reagents that identify markers of interest, whether used in single-color analysis or as part of a multicolor panel. Reagents that are determined to be specific and that show optimal (good signal-to-noise) staining intensity should be selected. Prior to selecting the reagent source, consideration should be given to whether it will be able to support validation and subsequent clinical testing of samples. Although the use of antibodies that have not received a cluster designation (CD) assignment is often necessary, such reagents are a liability to the flow cytometry laboratory because the weight of responsibility to understand specificity rests squarely upon the investigator's shoulders.

Another parameter that affects antibody specificity arises from their ability to bind many cell types by their nonspecific (Fc) ends. This is especially true of monocytes. In this setting, the use of Fab or F(ab′)2 fragments (antibodies without their Fc ends) is recommended, whenever possible, to help alleviate this problem. Addition of an Fc-receptor blocking agent may also be included to minimize nonspecific binding. Finally, dead cells, with compromised membrane integrity, tend to be sticky and bind reagents nonspecifically. Careful cell preparation and "gating out" dead cells in data analysis – either by propidium iodide staining to positively identify dead cells by their membrane permeability or by avoiding cells with low FSC intensity – will exclude the dead cell population from analysis.

12.3.1.2 Compensation In a multicolor flow cytometry assay, the spectral overlap of fluorochromes contributes to unwanted fluorescence spillover into adjacent channels. The process of compensation shall be followed to subtract the spillover of fluorescent signal to the detectors that are not assigned to the particular fluorochrome. Compensation has historically been manually performed by visual comparison of positive and negative stained populations, when assay panels were limited to three or four fluorescent-conjugated mAb. With the advance of digital cytometers providing multiple lasers/detectors and the availability of additional fluorochromes, it is now possible to design an assay panel to assess detailed cell phenotypes with up to 18 colors. Meanwhile, application of correct compensation becomes a much more complex process, even with the implementation of computer software to set up compensation automatically [40, 41]. To minimize undesired spectral spillover and resolve the compensation issues, it is important to choose the best fluorochrome and mAb combination to ensure assay sensitivity and specificity [42]. Usually, FITC, PE, and APC are reserved for conjugating mAb to detect protein with low expression level (i.e., intracellular cytokine, phosphorylated protein, etc.). Tandem dyes (i.e., PE-Cy5, APC-Cy7, etc.) are chosen for the detection of well-characterized markers (i.e., CD3, CD4, etc.). Although the intricacies of compensation are not fully addressed here, the end game is to use fluors with strong emission profiles to detect antigens with low expression level (and vice versa). The alternative possibility to apply strong fluors to highly expressed antigens can create instances where excessive compensation is necessary to rectify fluorescence spillover; in some cases, even a high degree of compensation cannot remove the artifactual signal.

12.3.1.3 Staining Conditions Appropriate staining conditions are important for optimal sample acquisition and for clearly distinguishing positive from negative staining. In general, cells should be at the highest concentration to use the lowest fluidics pressure possible to pass through the cytometer while minimizing doublets/cell aggregates to achieve a suitable flow rate (typically about 1000 cells per second). Using high pressure to achieve high flow rates will affect resolution of staining intensity. Also, cell staining suspension volume should be as low as possible (50–100 μL is ideal). Smaller staining volumes require less antibody reagent to achieve saturating concentration that is most important. Staining volumes can be adjusted at acquisition time to get optimal flow rate. Multiple samples of controls will help in getting the instrument settings and compensation for a new assay especially for multifluorochrome analysis.

Staining of samples can be performed in test tubes that fit onto the cytometer platform or in multiwell plates. The use of multiwell microtiter plates is ideal for staining a large number of samples; however, wash volumes are limited by well volume. Nonetheless, if plates are used and the flow cytometer is equipped with a multiwell plate sampler, the samples can be directly acquired on the flow cytometer. Alternatively, if tubes are preferred, additional automated solutions are available. Samples manually stained in test tubes can be lysed, washed, and fixed using automated instruments such as Lyse Wash Assistant (BD Biosciences), then directly loaded onto a FACSLoader for automated acquisition.

Optimization of temperature, time, and antibody concentration is important for establishing a good quality flow cytometry assay. Staining should be performed at colder temperatures, ideally 4 °C, and with buffers that contain sodium azide to prevent antibody capping. Antibody titrations using these time(s)/temperature(s) are appropriate for determining the saturating concentration of antibody during assay development. When a laboratory initiates a work stream to study a new antigen (or one they are unfamiliar with), it is strongly suggested that the impact of receptor shedding/soluble receptors on antibody titrations are considered. Soluble receptors impact the saturating concentration of an antibody reagent, and the concentration of a soluble receptor can vary in a population of normal versus diseased individuals. Measures that ensure that saturating concentrations of antibodies are used will ensure good signal-to-noise ratio and overall assay performance. In cases where whole blood is the desired specimen and the level of saturating receptor is high and/or variable, a simple wash with PBS immediately prior to staining safeguards against erroneous titrations and subsequent staining conditions when the assay moves to the production stage.

12.3.1.4 Sample Fixation It is not always possible to analyze samples on a flow cytometer immediately after staining, therefore cells may have to withstand delays in storage or shipment (hours to days). It is critical to stabilize labeled cells by fixation in a low concentration of formaldehyde. Most laboratories store fixed samples at 4 °C, and all should store the samples protected from light. Fixation also alleviates potential biohazard concerns regarding the samples and is always recommended for clinical sample preparation. Protein markers are affected differently by fixation techniques; therefore, it is important to evaluate stability of all markers over a range of fixation time. Because of these effects, some laboratories opt to fix all samples even if they are analyzed soon after processing. Once stability in fixative is established, these assay conditions can be used for subsequent validation.

12.3.1.5 Sample Collection, Processing, and Stability It is critical that all relevant cellular and antigenic properties are maintained on cell lines or primary cells to be used for method development. For cell lines, particularly adherent cell lines, it is important to generate suspension cultures using reagents that preserve the expression of cell surface markers. For peripheral blood samples, anticoagulants must be evaluated and selected on the basis of antigenic integrity and stability.

Common anticoagulants for immunophenotyping and functional assessment are ethylenediaminetetraacetic acid (EDTA), heparin, and acid citrate dextrose (ACD) anticoagulants. Cyto-Chex ®BCT tubes (Streck, Inc.) are a newer product that can allow extended stability; however, this product is not suitable for all antigens or for functional studies as the product contains a mild fixative [43, 44]. Determination of marker stability in appropriate anticoagulants during the assay development phase will be beneficial for methods intended for use in clinical protocols. Sample stability should be assessed in the collection container during shipment and storage, during sample processing, and postfixation (if applicable). Understanding the limitations for maintaining stable marker expression when collected in a particular additive will help define appropriate sample collection instructions for clinical protocols.

Whole blood is usually considered the first choice of sample matrix for flow cytometry assays, as it closely mimics the functional status of cells *in vivo*. Moreover, if a biomarker assay involves cell stimulation (e.g., phospho flow), the readout is viewed as more appropriate in whole blood because the circulating concentration of the drug is carried throughout the assay. The conundrum is that whole blood is a perishable specimen; therefore, consideration should be given to isolated cellular samples in lieu of a whole blood matrix. When cell counts are the assay end point and the cell population under study is of low frequency, some laboratories opt for a concentrated cellular sample to facilitate analysis of the desired cell population.

If whole blood is not the matrix of choice, cell isolation procedures must be optimized to safeguard against cell loss and to identify the most stable processing conditions for the markers of interest. It is uncommon to hold isolated cells in buffer or culture medium prior to phenotyping them because of the possibility that cells will up- or downregulate their antigen profile. In the case that isolated cells cannot be immediately stained, storage in liquid nitrogen is common. If the assay is to be implemented in clinical studies using frozen PBMC, stability of markers after cryopreservation is then an additional parameter to be investigated during assay development before proceeding with validation of the assay.

12.3.1.6 Impact of PBMC Isolation Process on Cell Phenotype and Function

When using frozen PBMC as the sample matrix, optimization of immunophenotyping assays should include similar assessments to those made for peripheral blood (i.e., appropriate anticoagulant, selection of reagents with high specificity and staining intensity, etc.) with a few additional sample condition considerations: (i) postcollection marker stability prior to PBMC isolation; (ii) impact of PBMC isolation techniques, freezing conditions, storage temperature, and thawing method on cell phenotype and function; and (iii) marker stability for the entire duration of PBMC storage. Continuation of a stable environment is recommended for frozen PBMC samples to preserve the integrity of markers, especially in functional assays [45].

The process of PBMC isolation should be evaluated to optimize conditions for analysis. Blood samples can either be directly collected in a cell preparation tube containing anticoagulant and Ficoll gradient for PBMC separation through centrifugation (i.e., BD Vacutainer® CPT tube containing heparin or citrate) or be collected in an alternative anticoagulant for manual Ficoll gradient application and cell separation. In both processes, marker stability should be evaluated prior to and including PBMC isolation processing as any fluctuation in handling and storage can affect marker instability and greatly impact result data [46]. Cell yield and viability results will also vary depending on PBMC isolation conditions. Additional benefit may be gained in obtaining PBMC from diseased donors (if available) to provide insight into the expectation of marker expression in clinical samples and the effects of isolation techniques on sensitive markers.

Reagents used during the PBMC isolation process may affect some markers of interest and should be evaluated. Expression of some markers, such as CCR5, may be affected by regents used for PBMC isolation (i.e., Ficoll) and may not be suitable for sample collection in a particular flow cytometric method. In addition, the cell preservation reagents, thawing media, and final cell suspension should be evaluated for optimizing marker expression. One should not underemphasize the need to assess stability of markers before and after PBMC isolation if this route is chosen.

A process of controlled freezing of isolated PBMC samples (e.g., using 1 °C cryogenic freezing containers at −70 °C) followed by a prompt transfer to liquid nitrogen storage is beneficial in maximizing PBMC recovery and marker stability [47]. Mononuclear cells stored in liquid nitrogen are known to preserve the majority of known surface markers and certain cell functionality aspects; however, extended cryopreservation may be detrimental to certain phenotypes or functions affecting intracellular immunophenotyping, functional, and tetramer assays. Therefore, stability of marker expression should be determined over a period of extended cryopreservation, if applicable [48]. If markers are stable after cryopreservation during the testing period in the assay development stage, the method validation can be initiated with greater confidence using batch frozen PBMC. In addition, *ex vivo* stimulation methods can be implemented in batch processing using frozen PBMC when long-term marker stability has been established.

12.3.1.7 Assay controls Inclusion of appropriate controls for reagents, samples, instrument setup, and gating is critical for a flow cytometric assay. Requirement for these controls is based on the assay type, the purpose of the experiment, and the intended utility of the data generated. Assessment of relevant controls should occur during the assay development phase and selection should be made before validation begins. A well-designed flow assay at a minimum must include controls for reagents and method. Main categories of controls include (i) negative or autofluorescence control, (ii) compensation or positive controls, (iii) specificity or gating controls, and (iv) sample controls.

(i) Negative control is set up using cells (typically untreated, but can include treated cells also) not stained with any antibodies to detect background/autofluorescence of the cells. Autofluorescence tends to lower discrimination between stained and unstained cells interfering with measurement of low-level expression of bound fluorescent antibody. Autofluorescence is largely evident when monocytes/macrophages or activated cells are used. Comparison of negative control autofluorescence (unstained cells) with a positive control signal (stained cell) in each fluorescence detector will confirm that the stained sample cells will be on scale for each parameter. In addition, the unstained cell controls are used for light scatter gating and defining the threshold signal, above which true positives are determined.

(ii) Compensation controls are positive controls to remove spectral overlap in multicolor flow cytometry analysis. Beads (i.e., Comp Beads) or cellular samples can serve as compensation controls when labeled with single-color monoclonal antibody reagents used in the method [42]. Fluorochrome emission cross talk into other detectors must be compensated for to achieve an accurate fluorescence signal. A contemporary practice is to exercise the fluorescence-minus-one (FMO) paradigm, where each antibody–fluorochrome conjugate is dropped in succession. This allows one to identify the culprit of fluorescence spillover by deduction; this is also a very useful gating control as discussed below.

(iii) Specificity or gating controls are reagent controls used to verify antibody reactivity and to verify method staining is specific for the target marker and cell type. Labeling with a species relevant immunoglobulin of the same isotype is used to distinguish nonspecific binding (due to Fc receptors on cells of interest) that is gated out to measure specific binding. Isotype controls are usually used for experiments using "sticky" cells such as monocytes and dendritic cells. The isotype control antibody should be of the same concentration and have the same fluorescence-to-protein ratio as the antibody of interest. Another gating control to determine positive staining is fluorescence minus control (FMO), which is a control labeled with all but one antibody reagent used in the actual sample panel. FMO controls are applicable for experiments using four or more colors and are usually used to set discriminate gating for positive/negative expression of the missing antibody [49]. Events that display fluorescence above these controls are considered positive for the marker of interest. For indirect staining techniques, specificity controls include samples with just primary antibody or secondary antibody. An unstimulated sample is an ideal control for setting the gate for positive events in a stimulated sample.

(iv) Sample controls are used as an internal control of the assay method. A stabilized positive sample can be commercially obtained (CD-Chex from Streck, Cyto-trol from Beckman Coulter) and labeled with the same reagents as the test samples. These method controls should be considered for use both in assay development and in validation [49]. Unstimulated controls may also be used as negative method controls.

12.3.1.8 Data Reporting The method of reporting signal readout is established during assay development. Besides reporting cell percentages, the use of mean or median fluorescence intensity (MFI) can be used to estimate expression level on a per cell basis. There is a liability with the MFI readout, however, as the degree of reported fluorescence intensity could change at different PMT settings on the same cytometer and vary from instrument to instrument. For this reason, standardization methods have been developed. Three useful approaches are to convert signal to molecules of equivalent soluble fluorochrome (MESF), to report the antibody bound per cell (ABC) using a standardized curve, and to calibrate the instrument using beads to standardize signal output before sample analysis [50].

12.3.2 Method Validation (Prestudy Validation)

Optimized clinical trial methods and preclinical toxicology experiments require validation. The key elements of an assay validation are an approved validation plan, assay validation experiments, predetermined acceptance criteria of results, and documentation of the validation. The stringency of the validation assessment depends upon at which stage of drug development the assay is used and the intended applicability of the assay output (Figure 12.1).

Because flow cytometry applications are varied (and include immunophenotyping, pharmacokinetic, pharmacodynamic, biomarker, immunogenicity, and functional assays), validation parameters will be somewhat different and depend on assay type. Thankfully, general guidelines have been described in earlier reports [27, 51–57], with additional advances along the way as these efforts mature. Although flow cytometry assays to be used in discovery and early research phase need not be validated as per GLP, they should be well characterized and reproducible using a calibrated, properly functioning flow cytometer. If this assay is to be used for testing clinical samples in early-phase clinical studies for exploratory biomarker purpose, the assay needs to be validated using a fit-for-purpose approach at a minimum and performed in a GLP-like environment [1]. Furthermore, if the assay is to be used in later stage registrational studies, the need for validation is tangibly heightened. For a diagnostic application of the assay, the validation of the assay is performed as per CLIA guidelines. The validation of a flow cytometry assay for application in a clinical setting is described in the following section, whereas validation of an assay for CLIA certification is beyond the scope of this chapter (refer to Section 12.2).

12.3.2.1 Validation Plan The validation protocol is an essential tool used to guide the validation process. It should describe all the elements of the validation process including sample type and source, list of equipment, identification of scientists performing validation experiments, approvers of the results of validation, validation parameters, statistical methods, assay and result acceptability criteria set prospectively, and assay magnitude during validation (number of samples or plates mimicking intended use of the assay). Validation study parameters to be considered in the validation assessment include precision, sensitivity (limits of quantitation), specificity, accuracy, stability, and reference ranges. Flow cytometric assay validation plan considerations and suggested acceptance criteria are summarized in Table 12.1. Prior to the start of a validation assessment, sufficient reagents should be obtained to perform all experiments intended for validation, and the analyst should be

TABLE 12.1 Method Validation

Validation Parameters	Focus	Value Added
Stability	Biomarker stability in assay matrix	Determine how to ship/store clinical samples after sample collection Determine whether stained samples could be fixed and/or whether fixed samples could be stored short term prior to sample acquisition
Intra-assay precision	Within run variability	Determine sample variability and need for replicate analysis
Inter-assay precision	Between run variability	Determine the need to analyze all samples in same analytical run Establishes expected variation for comparison to drug effect
Intra-subject variability	Physiological variability	Identify diurnal variations to avoid or target in analysis
Instrument-to-instrument, or lab-to-lab comparison	Quality assurance for alignment in analysis	Ensure data generated from different cytometers/ laboratories are comparable
Analyst-to-analyst comparison	Training analyst(s) on assay performance	Ensure data generated by different analysts are comparable

well trained to consistently perform the assay to ensure consistency and high quality in validation results. In some cases, tests for lot-to-lot variability are warranted and accommodation would therefore be required for increased efforts that include trending analysis.

12.3.2.2 Validation Assessment

Precision The variability in an assay can be statistically measured and can be assessed using standard deviation and percentage coefficient of variation (%CV). Precision of an assay is usually determined by assessing both intra- and inter-assay variability. Intra-assay precision is the variability within an assay (i.e., variability between replicates). Inter-assay variability is the variability between multiple assays (i.e., variability between experiments). Apart from intra-assay and inter-assay variability, inter-operator, inter-instrument, and inter-laboratory precision should also be determined during assay validation. Evaluation of these variability factors is described in the following sections.

INTRA-ASSAY PRECISION The intra-assay precision is an assessment of within-run variability. Consideration must be given to the frequency of the cell population being measured to determine acceptance criteria for intra-assay precision. When cell subsets are at least 10% of the parental population, precision of < 20%CV may be attainable. However, when measuring rare cell populations (less than 10% of parental population), 30%CV may be more likely. When assay imprecision is greater than expected, and the lack of reliability does not exclude the analysis from the clinical plan, analysis of samples in replicates is advised. Even when samples are assayed in replicates, clinical teams must be made aware of the liability imposed by intra-assay variability when interpreting assay results in their trials. Perhaps, the only acceptable place where assay imprecision can be discounted is in the case of cell-depleting agents because intra-assay variability approaches zero in samples devoid of the population of interest.

INTER-ASSAY PRECISION Inter-assay precision is an assessment of between-run variability. Multiple experimental runs are performed using the same source samples; therefore, the measured parameter must be stable enough to allow analysis at different times while generating the same results. Consideration of cell population frequency and sample stability is important in determining acceptance criteria for inter-assay precision. Whole blood samples with limited postcollection stability present challenges for conducting inter-assay precision assessment where all analytical runs must be assayed within the limited sample stability window, usually 1–2 days. Multiple runs must be performed on the same day and can lead to long hours in the laboratory. Frozen PBMC, however, usually have longer storage stability and, therefore, PBMC assays offer more flexibility in sample management performing multiple analytical runs for inter-assay precision assessment. When two or more analysts participate in this stage of assay validation, care should be taken to ensure all analysts perform analysis comparably. To reduce the analyst-to-analyst variability, all analysts should be formally trained in assay performance, and training results be documented with acceptance criteria. Whenever more than one analyst is performing an assay for precision assessment or training, although aliquots of the same sample will be used, special consideration should be made to ensure they are truly "independent" analytical runs.

INTER-ANALYST COMPARISON Operator training is a key component in obtaining high-quality data. Visual observation followed by practice is recommended to obtain reproducible assay results. A training plan with documented performance and acceptance criteria upon assay transfer can be valuable to successful analysis no matter where it is applied in the drug development process. Precision validation assessment performed by one analyst is the best-case scenario for reproducibility in an assay when transition is made to the clinic. Although one analyst may primarily be responsible for sample analysis, it is important to have more than one analyst trained on properly conducting the assay as a backup. Frozen PBMC-based assays, in many cases, can be scheduled and batched, facilitating the one-analyst scenario. For whole blood assays that have limited sample stability, it is more likely that a second analyst will be required to perform sample analysis. In addition, a primary analyst who is responsible for assay validation may not be the primary analyst for sample analysis; therefore, training and documentation of meeting acceptance criteria are important. The more practice and assay experience an analyst has performing the assay, the less variability between analysts is expected. The data generated between analysts should be within the inter-assay variability or predetermined acceptance range (Table 12.1). Finally, the sample materials for precision testing should mimic the expected samples in clinical trials, including both samples from subjects (donors) and QC samples.

INTER-INSTRUMENT COMPARISON Differences in instrument performance can add variability to an assay increasing the imprecision assessment. While a primary cytometer may be used for the majority of validation tests and clinical sample analysis for a given assay, a flow cytometry lab may have more than one instrument to handle large numbers of samples from the clinical trials. The additional cytometer(s) also serve as backup instruments when the primary cytometer is subject to maintenance services. To ensure the cytometers are comparable within the same lab or between different labs, the instrument-to-instrument comparison should be performed at least every 6 months and after maintenance service. Clear documentation of instrument maintenance and service is a good practice to obtain high-quality data.

Accuracy The strict definition of accuracy is as the closeness of agreement between the averages of the test results when compared to a well-defined official reference standard. Accuracy assessment is not possible in flow cytometry assays in this regard, as reference standards are not readily available. For those laboratories engaged in proficiency testing services, comparability of results from one site against the group can imply accuracy. In instances where an antigen is not included in the typical proficiency testing panel, a biopharmaceutical company may elect to prepare samples in-house and distribute them for analysis by as many flow cytometers and by as many analysts as are available. Although this may be better described as an exercise in robustness, it is quite important to to understand whether a given data set would agree with another generated in a different time and place. The lack of accuracy standards do not excuse those in the field from ignoring this important parameter during validation exercises, and every effort should be made to speak to the accuracy of an assay despite limitations of our field.

Specificity Cellular assays are often challenged by a complex matrix (i.e., whole blood) presenting specificity and sensitivity issues. Reagents must be specific enough to recognize the intended target in the presence of interfering substances (i.e., compounds, proteins, etc.) and sensitive enough to measure low levels of expression over the background signals. Antibody specificity and cross reactivity within a complex biological matrix greatly influence assay sensitivity in flow cytometric assays. Since both are related to proper reagent selection, specificity and sensitivity are customarily discussed together.

Proper reagent selection is critical to the development of a flow cytometry assay. The antibody clone must be specific for the intended target [5]. It is important to understand assay specificity to determine the reliability of measurements in the presence of the biologic matrix or interfering substances. Where possible, assays intended for use in clinical trials should evaluate possible drug and metabolite interferences prior to implementation. Generally, laboratories rely on cluster of differentiation (CD) designations or the antibody vendor to characterize reagent specificity. It is not difficult to find those in our field that are uncomfortable with this approach, but there is no solution to this issue other than performing specificity experiments in-house. In some cases, testing a variety of antibody clones (when available) can lend credence to the reproducibility of the result; in other cases, Western blots on cell lysates have been commonly used.

Sensitivity Sensitivity is the measure of assay performance with known negative samples used in determining the lowest limit of measurement in known positive samples. GLP guidelines define the lowest limit of quantitation as the lowest level at which a method can accurately measure, whereas CLIA defines sensitivity as the lowest value able to be reported [6, 8]. Reagent sensitivity is important in developing a flow cytometric assay to find the minimum staining intensity above nonspecific staining able to be detected in an assay. Sensitivity assessment should determine both the maximum fluorescence intensity of a positive population and the maximum separation between positive and negative populations. These results can be used to determine the appropriate amount of reagent in a particular assay to be most sensitive [58]. Even though commercial reagents are usually provided with recommended amounts for use in staining protocols, the reagents should be titered to confirm saturating conditions in a particular protocol.

Stability Sample handling and storage conditions can greatly impact the assay-specific phenotype and/or function being monitored. Therefore, stability for a biological marker must be validated from the time of sample collection to the time of analysis. Validation may include assessment of physiological expression, effect of matrix, storage and shipping effects, sample handling, processing conditions, and effects of reagents. Early determination of stability limitations will provide valuable information for assay transition to clinical trials.

It is common practice to begin assay validation with intra-assay and inter-assay precision experiments; however, evaluation of sample stability limitations first may provide valuable insight into how to manage clinical protocol sample collection, handling, and shipping conditions. Consider the alternative scenario where post-collection sample stability is validated first, and 48 h postcollection is determined to be the maximum limit for marker stability. Subsequent precision validation experiments can then be performed with samples that mimic the 48 h postcollection conditions of acceptable clinical samples. The use of samples that are less than optimal (not freshly collected) in validation will provide a more realistic estimation of assay performance in clinical trials.

Storage, Handling, and Shipping Conditions Sample stability should be validated in the collection container to determine marker stability during shipment and storage. Many cell surface markers are stable when samples are stored at ambient temperature;

however, a preliminary evaluation including additional temperature conditions (i.e., chilled or frozen) and expected shipment duration can be illuminating. Stability of markers stored at ambient temperature should be compared to chilled conditions where samples are shipped with ice packs, if appropriate. Samples can be shipped in the original collection container or may be processed prior to shipment to extend marker stability before arriving at the analysis laboratory. Validation of postcollection stability becomes increasingly important in global clinical trials, where samples are shipped to a central laboratory and may be received next day, or second or third day, after blood draw. In some instances, cellular samples may be fixed prior to shipping to extend stability up to 7 days (Streck cell preservation media) [59]. Additional value is gained when the assay is transitioned to clinical trials and markers are stable more than 24 h postcollection. Extended stability facilitates batch sample analysis and more efficient sample management in clinical trials.

To evaluate PBMC marker stability in the collection container, samples should be processed immediately after collection (as baseline), as well as at different post-collection time points that would mimic clinical conditions. PBMC samples collected in Vacutainer® CPT tubes offer several options for shipping in clinical studies; therefore, marker stability in all sample storage conditions should be evaluated. CPT tubes can be shipped as anticoagulated whole blood or PBMC can be separated via centrifugation and resuspended in the plasma layer above the gel barrier and remain in the collection container prior to shipping. PBMC can also be isolated from the plasma and transferred to a secondary container for processing, storage, or shipping. Finally, PBMC can be isolated and frozen for shipping, especially in global studies where it is not feasible to transport samples in the original CPT collection tube to the laboratory for processing within the marker stability time frame. In this case, stability of markers after cryopreservation must also be investigated. Finally, if processed and stained samples cannot be acquired immediately after fixation, the fixed sample stability should be evaluated to permit sample storage for later acquisition.

Documentation A validation summary report becomes the reference document describing assay performance and limitations as defined by validation. It should describe the assay method, include raw data from the validation, and summarize results and assay limitations. Assay acceptance criteria are a very important element of this overall exercise; the validation fails if assay acceptance criteria are not met. In addition, the report should cite the laboratory notebooks where validation is documented and identify binders or electronic records of raw data. Accurate validation experiment documentation can be valuable for performing revalidation experiments and assay adaptation for future drug development studies, as required. It is this document that will be used within a clinical trial to determine if results are an effect of treatment or expected assay variability. Validation reports are considered permanent records and are reviewed and approved by senior management.

Documentation and implementation of a quality assurance plan focused on assay and instrument quality during validation are useful tools to ensure reliable flow cytometry data are generated at every stage in the drug development process. A validation plan should be considered to outline experimental procedures and acceptance criteria, identify how the assay will be reported, determine what regulatory guidelines are applicable, and describe controls required to monitor and standardize results between experiments. Written standard operating procedures are recommended to ensure methods are performed as expected. This validation plan will become the guideline for successful validation of a high-quality assay.

12.4 SUMMARY

With the rapid development of instrumentation, reagents, and analysis software, flow cytometry has been widely used in different stages of drug discovery and development programs, including target identification, compound screening, biomarker development for proof of principle/mechanism of action, preclinical and clinical toxicology, immunogenicity, and clinical diagnosis. It is challenging, however, to implement the flow cytometry assays to various stages of drug programs. On the basis of the intended use of data, the stringency of assay validation varies and the lab shall be compliant with different regulatory guidelines (Table 12.2). In addition, due to lack of specific regulatory guidelines for biomarker development, it is commonly accepted to implement fit-for-purpose approaches for biomarker assay development and validation. Therefore, it is critical to develop an implementation plan for assay qualification, validation, and clinical sample analysis. Prior to the initiation of assay

TABLE 12.2 Regulatory Agencies

Guidelines	Agency	Regulatory Jurisdiction	Objectives	Reference
GLP	FDA	Preclinical animals, toxicology, and *in vitro* testing (United States)	Consistency and reliability of results	21CFR58 www.fda.gov
	OECD	Nonclinical research studies (International)		Directive 67/548/EEC www.oecd.org/ department
GCP	FDA	Clinical trials on humans	Safety and protection of trial subjects while obtaining credible data	21CFR312 www.fda.gov
	ICH			CPMP/ICH/135/95 www.ich.org
CLIA	Center for Medicare and Medicaid Services	All human laboratory testing except research (United States)	Patient safety management and disease monitoring	42CFR493 www.cms.hhs.gov/ clia

validation, which could be labor intensive and time consuming, the assay should be evaluated and qualified. This includes the assessment of assay robustness, the impact of flow cytometer setup and standardization on data quality; sample stability in the context of sample matrix, sample collection, and shipment logistics; and assay quality controls.

On the basis of the intended use of data and regulatory compliance, it is recommended to develop an assay validation plan identifying the required resources (time, budget, and full-time employees), to develop validation parameters (stability, precision, sensitivity, specificity, accuracy, limits of quantity, reference range, and acceptability criteria, if required and applicable), and to appropriately document the validation approaches and results. To ensure high data quality and minimize instrument-to-instrument and analyst-to-analyst variability, instrument performance and analyst performance should be compared and qualified prior to the stage of sample analysis. During the stage of sample analysis, both the assay performance and the instrument performance require close monitoring using predetermined acceptability criteria. It is only after orchestration of all parameters of assay development, validation, and performance that the flow cytometry laboratory can present a legitimate immunological story to support the development and commercialization process as applied to experimental therapeutic agents.

REFERENCES

1. Lee JW, Devanarayan V, Barrett YC, Weiner R, Allinson J, Fountain S, Keller S, Weinryb I, Green M, Duan L, Rogers JA, Millham R, O'Brien PJ, Sailstad J, Khan M, Ray C, Wagner JA. Fit-for-purpose method development and validation for successful biomarker measurement. *Pharm. Res.* 2006;23:312.
2. Cummings J, Ward TH, Greystoke A. Ranson M, Dive C. Biomarker method validation in anti-cancer drug development. *Br. J. Pharm.* 2008;153:646–656.
3. Owens MA, Vall HG, Hurley AA, Wormsley SB. Validation and quality control of immunophenotyping in clinical flow cytometry. *J. Immunol. Methods* 2000;243:33–50.
4. Biomarkers Definitions Working, Group. Biomarkers and surrogate end points: preferred definitions and conceptual framework. *Clin. Pharmacol. Ther.* 2001;69(3):89–95.
5. Code of Federal Regulations Title 21 Part, 58. Good laboratory practices regulations: final rule. *Fed. Regist.* 1987;52(172):33768. Available at http://www.fda.gov/downloads/ICECI/EnforcementActions/BioresearchMonitoring/ucm133730.pdf.
6. Code of Federal Regulations Title 42 Part, 493. Medicare, medicaid and clinical laboratory improvement amendments of 1988 (CLIA) program laboratory requirements. *Fed. Regist.* 2004;57:1017. Available at http://wwwn.cdc.gov/clia/pdf/42cfr493_2004.pdf.
7. Urley AK, Curren RD, Harbell JW, Mun G, Raabe HA.Applying good laboratory practices (GLPs) to *in vitro* studies: one laboratory's perspective. Institute for In-Vitro Sciences Inc. Gaithersburg, MD. Presented at 5th World Congress on Alternative Animal Uses in the Life Sciences, Berlin, Germany, 2005.
8. Code of Federal Regulations Title 21 Part 58. Good laboratory practice for non-clinical laboratory, studies. *Fed. Regist.* 2001;58(1):303.

9. Organization for Economic Co-operation and, Development. The Application of the GLP Principles to Short-term Studies. Series on Principles of GLP and Compliance Monitoring. Number 7 (Revised). ENV/JM/MONO(99)23, 1999. Available at http://www.olis.oecd.org/olis/1999doc.nsf/LinkTo/NT000010C2/$FILE/09E98856.PDF.
10. Organization for Economic Co-operation and, Development. The Application of the Principles of GLP to *In Vitro* Studies. Series on Principles of Good Laboratory Practice and Compliance Monitoring. ENV/JM/MONO(2004)26.1, 2004.
11. Department of Health The United Kingdom Good Laboratory Practice Monitoring, Authority. Guide to UK GLP Regulations 1999. Statutory Instrument No. 1999/3106, 2000.
12. Fox A. GLP regulations vs. ISO 17025 requirements: how do they differ? *Accred. Qual. Assur.* 2003;8:303.
13. FDA. Guidance for Industry: Bioanalytical Method Validation, 2001. Available at http://www.fda.gov/downloads/Drugs/GuidanceComplianceRegulatoryInformation/Guidances/UCM070107.pdf.
14. Swanson BN. Delivery of high quality biomarker assays. *Dis. Markers* 2002;18:47–56.
15. Clinical and Laboratory Standards Institute. *Continuous Quality Improvement: Integrating Five Key Quality System Components. Approved Guideline*, 2nd ed., Clinical and Laboratory Standards Institute, Wayne, PA, CLSI document GP22-A2, 2004.
16. Clinical and Laboratory Standards Institute. *Laboratory Documents: Development and Control. Approved Guideline*, 5th ed., Clinical and Laboratory Standards Institute, Wayne, PA, CLSI document GP02-A5, 2006.
17. Clinical and Laboratory Standards Institute. *Enumeration of Immunologically Defined Cell Populations by Flow Cytometry: Approved Guideline*, 2nd ed., Clinical and Laboratory Standards Institute, Wayne, PA, CLSI document H42-A2.27(16), 2007.
18. Clinical and Laboratory Standards Institute. *Clinical Flow Cytometric Analysis of Neoplastic Hematolymphoid Cells: Approved Guideline*, 2nd ed., Clinical and Laboratory Standards Institute, Wayne, PA, CLSI document H43-A2.27(11), 2007.
19. Clinical and Laboratory Standards, Institute. *Methods for Reticulocyte Counting (Automated Blood Cell Counters, Flow Cytometry and Supravital Dyes): Approved Guideline*, 2nd ed., Clinical and Laboratory Standards Institute, Wayne, PA, CLSI document H44-A2. 24(8) 2004.
20. Clinical and Laboratory Standards, Institute. *Fetal Red Cell Detection: Approved Guideline*, Clinical and Laboratory Standards Institute, Wayne, PA. CLSI document H52-A. 21(26) 2001.
21. Centers for Disease Control and, Prevention. Good laboratory practices for waived testing sites: survey findings from testing sites holding a certificate of waiver under the Clinical Laboratory Improvement Amendments of 1988 and Recommendations for Promoting Quality Testing. *MMWR* 2005;54(RR-13):1–20.
22. College of American Pathologists Commission on Laboratory, Accreditation. *College of American Pathologists: Laboratory General Checklist*. College of American Pathologists, Northfield, IL, 2009. Available at http://www.cap.org.
23. College of American Pathologists Commission on Laboratory, Accreditation. *College of American Pathologists: Flow Cytometry Checklist*. College of American Pathologists, Northfield, IL, 2009. Available at http://www.cap.org.

24. Code of Federal Regulations Title 21 Part, 312. Investigational new drug application. *Fed. Regist.* 2009;74:68750. Available at http://www.accessdata.fda.gov/scripts/cdrh/cfdocs/cfcfr/CFRSearch.cfm?CFRPart=312&showFR=1.

25. ICH. International Committee for Harmonization (ICH) of Technical Requirements for Registration of Pharmaceuticals for Human Use. ICH Topic E6 (R1) Guideline for Good Clinical Practice (GCP), 2002. Available at http://www.emea.europa.eu/pdfs/human/ich/013595en.pdf.

26. ICH. International Conference on Harmonization of Technical Requirements for Registration of Pharmaceuticals for Human Use. ICH Harmonised Tripartite, 2009. Available at http://www.ich.org.

27. Lee JW, Weiner RS, Sailstad JM, Bowsher RR, Knuth DW, O'Brien PJ, Fourcroy JL, Dixit R, Pandite L, Pietrusko RG, Soares HD, Quarmby V, Vesterqvist OL, Potter DM, Witliff JL, Fritche HA, O'Leary T, Perlee L, Kadam S, Wagner JA. Method validation and measurement of biomarkers in nonclinical and clinical samples in drug development: a conference report. *Pharm. Res.* 2005;22(4):499–511.

28. Stiles T, Grant V, Mawbey N. *Good Clinical Laboratory Practice (GCLP): A Quality System for Laboratories which Undertake the Analyses of Samples from Clinical Trials.* British Association of Research Quality Assurance, Ipswich, UK, 2003, pp. 1–17. Available at http://www.barqa.com/.

29. Sarzotti-Kelsoe M, Cox J, Cleland N, Denny T, Hural T, Needham L, Ozaki D, Rodrigues-Chavez IR, Stevens G, Stiles T, Tarragona-Fiol T, Simkins A.Evaluation and recommendations on good clinical laboratory practice guidelines for phase I–III clinical trials: PLOS medicine guidelines and guidance 6(5) e1000067, 2009. Available at http://www.plosmedicine.org.

30. Department of Health and Human Services, NIH. DAIDS Guidelines for Good Clinical Laboratory Practice Standards. N01-AI-50022, 2008. Available at http://www3.niaid.nih.gov/LabsAndResources/resources/DAIDSClinRsrch/PDF/GCLP.pdf.

31. Ezzelle J, Rodriguez-Chavez IR, Darden JM, Stirewalt M, Kunwar N, Hitchcock R, Walter T, D'Souza MP. Guidelines for good clinical laboratory practice: bridging operations between research and clinical research laboratories. *J. Pharm. Biomed. Anal.* 2008;46:18–29.

32. Horton H, Thomas EP, Stucky JA, Frank I, Moodie Z, Huang Y, Chiu YL, McElrath MJ, De Rosa SC. Optimization and validation of an 8-color intracellular cytokine staining (ICS) assay to quantify antigen-specific T cells induced by vaccination. *J. Immunol. Methods* 2007;323:39–54.

33. College of American, Pathologists. *CAP 15189: Accreditation to the ISO 15189:2007 Standard.* College of American Pathologists, Northfield, IL, 2008. Available at http://www.cap.org.

34. International Organization for, Standardization. Medical Laboratories: Particular Requirements for Quality and Competence. ISO-15189, 2007.

35. International Organization for, Standardization. General Requirements for the Competence of Testing and Calibration Laboratories. ISO-17025, 2005.

36. Lemster B, Huang LL, Irish W, Woo J, Carroll PB, Abu-Elmagd K, Rilo HR, Johnson N, Russell-Hall R, and Fung JJ. Influence of FK 506 (tacrolimus) on circulating $CD4^+$ T cells expressing CD25 and CD45RA antigens in 19 patients with chronic progressive multiple sclerosis participating in an open label drug safety trial. *Autoimmunity* 1994;19(2):89: 98.

37. Takashima A, Morita A. Genomic, phenotypic, and functional analyses of T cells in patients with psoriasis undergoing systemic cyclosporin A treatment. *J. Invest. Dermatol.* 1991;96:376–382.

38. Khoury JS, Guttmann RGC, Orav EJ, Kikinis R, Jolesz AF, Weiner LH. Changes in activated T cells in the blood correlate with disease activity in multiple sclerosis. *Arch. Neurol.* 2000;57:1183–1189.

39. Strauss K, Hulstaert F, Deneys V, Mazzon AM, Hannet I, De Bruyere M, Reichert T, Sindic CJ. The immune profile of multiple sclerosis: T-lymphocyte effects predominate over all other factors in cyclophosphamide-treated patients. *J. Neuro. Immunol.* 1995;63 (2):133–142.

40. McLaughlin BE, Baumgarth N, Bigos M, Roederer M, Rosa CS, Altman JD, Nixon DF, Ottinger J, Oxford C, Evans TG, Asmuth DM. Nine-color flow cytometry for accurate measurement of T cell subsets and cytokine responses. Part I: Panel design by an empiric approach. *Cytometry A* 2008;73:400.

41. McLaughlin BE, Baumgarth N, Bigos M, Roederer M, Rosa CS, Altman JD, Nixon DF, Ottinger J, Li J, Beckett L, Shacklett BL, Evans TG, Asmuth DM. Nine-color flow cytometry for accurate measurement of T cell subsets and cytokine responses. Part II: Panel performance across different instrument platforms. *Cytometry A* 2008;73:411.

42. Mahnke YD, Roederer M. Optimizing a multi-colour immunophenotping assay. *Clin. Lab. Med.* 2007;27(3);469 -v.

43. Streck News, Release. *Streck Blood Collection Tube Cleared by FDA*. Streck Inc., Omaha, NE, 2008.

44. Warrino DE, DeGennaro LJ, Hanson M, Swindells S, Pirruccello SJ, Ryan W. Stabilization of WBCs and immunologic markers for extended analysis using flow cytometry. *J. Immunol. Methods* 2005;305:107–119.

45. Weinberg A, Song LY, Wilkening C, Sevin A, Blais B, Louzao R, Stein D, Defechereuz P, Durand D, Reidel E, Raftery N, Jesser R, Brown B, Keller MF, Dickover R, McFarland E, Fenton T. Optimization and limitations of use of cryopreserved peripheral blood mononuclear cells for functional and phenotypic T-cell characterization. *Clin. Vaccine Immunol.* 2009;16(8):1176–1186.

46. Bull M, Lee D, Stucky J, Chiu Y-L, Rubin A, Horton H, McElrath MJ. Defining blood processing parameters for optimal detection of cryopreserved antigen-specific responses for HIV vaccine trials. *J. Immunol. Methods* 2007;322:57–69.

47. Reimann KA, Chernoff M, Wilkening CL, Nickerson CE, Landay AL, and ACTG Immunology Advanced Technology Laboratories. Preservation of lymphocyte immunophenotype and proliferative responses in cryopreserved peripheral blood mononuclear cells from human immunodeficiency virus type 1-infected donors: implications for multicenter clinical trials. *Clin. Diagn. Lab. Immunol.* 2000;7(3):352–359.

48. Owen RE, Sinclair E, Emu B, Heitman JW, Hirschkorn DF, Eplng CL, Tan QX, Custer B, Harris JM, Jacobson MA, McCune JM, Martin JN, Hecht FM, Deeks SG, Norris PJ. Loss of T cell responses following long-term cryopreservation. *J. Immunol. Methods* 2007;326:93–115.

49. Maecker HT, Trotter J. Flow cytometry controls, instrument setup and the determinations of positivity. *Cytometry A* 2006;69A:1037–1042.

50. Wang L, Gaigalas AK, Marti G, Abbasi F, Hoffman RA. Toward quantitative fluorescence measurements with multicolor flow cytometry. *Cytometry A* 2008;73:279.

51. DeSilva B, Smith W, Weiner R, Kelley M, Smolec J, Lee BM, Tacey R, Hill H, Celniker A. Recommendations for the bioanalytical method validation of ligand-binding assays to support pharmacokinetic assessments of macromolecules. *Pharm. Res.* 2003;20:1885.
52. Mires-Sluis et al. Recommendations for the design and optimization of immunoassays used in the detection of host antibodies against biotechnology products. *J. Immunol. Methods* 2004;289:1–16.
53. Gupta S, Indelicato SR, Jethwa V, Kawabata T, Kelley M, Mire-Sluis AR, Richards SM, Rup B, Shores E, Swanson SJ, Wakshull EJ. Recommendations for the design, optimization, and qualification of cell-based assays used for the detection of neutralizing antibody responses elicited to biological therapeutics. *J. Immunol. Methods* 2007;321(1–2):1–18.
54. Cunliffe J, Derbyshire N, Keeler S, Coldwell R. An approach to the validation of flow cytometry methods. *Pharm. Res.* 2009;22(12):2551–2557.
55. Clinical and Laboratory Standards Institute. *Preliminary Evaluation of Quantitative Clinical Laboratory Measurement Procedures: Approved Guideline*, 3rd ed., Clinical and Laboratory Standards Institute, Wayne, PA, CLSI document EP10-A3, 2006.
56. Shankar G, Devanarayan V, Amaravadi L, Barrett YC, Bowsher R, Finco-Kent D, Fiscella M, Gorovits B, Kirschner S, Moxness M, Parish T, Quarmby V, Smith H, Smith W, Zuckerman LA, Koren E. Recommendations for the validation of immunoassays used for detection of host antibodies against biotechnology products. *J. Pharm. Biomed. Anal.* 2008;48:1267.
57. Reddy M, Davis C, Davis H, Pendley C, Prabhakar U. T-cell surface markers in human peripheral whole blood using flow cytometry. In *Validation of Cell-Based Assays in the GLP Setting: A Practical Guide*, John Wiley & Sons Ltd., 2008, pp.85–105.
58. Clinical and Laboratory Standards, Institute. *Enumeration of Immunologically Defined Populations by Flow Cytometry: Approved Guideline*, 2nd ed., Clinical and Laboratory Standards Institute, Wayne, PA, CLSI document H42-A2, 2007.
59. Ryan W, Warrino D. Stabilizing WBCs and immunological markers for flow cytometry. *Clin. Lab Products* 2005. Available at http://www.clpmag.com/issues/articles/2005-01_06.asp.

13

INSTRUMENT VALIDATION FOR REGULATED STUDIES

John Ferbas and Michelle J. Schroeder

13.1 INTRODUCTION

When flow cytometers are used in preclinical and clinical studies, a variety of assurance levels can be exercised to prove that the data are reliable, reproducible, and with defined precision and relative accuracy. The variety of levels of regulatory oversight can be categorized as Good Laboratory Practices (GLPs) or Good Clinical Practices (GCPs) when investigational agents are administered in the setting of a preclinical or clinical study, and as CLIA (Clinical Laboratory Improvement Amendments) derived processes when the data have diagnostic value. Although overlap in principle exists between CLIA, GLP, and GCP efforts, laboratories that operate without a fundamental understanding of the similarities and differences of the oversight processes could inadvertently place studies at risk. It is often the case that a CLIA-certified flow cytometry laboratory prepares and analyzes specimens from persons who received an investigational agent, but it is inappropriate to assume that CLIA regulatory practices provide appropriate coverage for the GLPs or GCPs (and visa versa). Moreover, the need to understand and comply with 21CFR Part 11 requirements [1] is mandated for GLP studies and may be applied with discretion to GCP investigations (see below), but CLIA processes are not governed by these requirements.

Practices common to one approach are not interchangeable with another. For example, specific documentation of processes defined by IQ (installation qualification), OQ (operation qualification), PQ (performance qualification), and UAT (user acceptance testing) do not reside in the CLIA nomenclature, but are very important for GLP efforts. For those involved in biomarker investigations, where "intended use"

Flow Cytometry in Drug Discovery and Development, Edited by Virginia Litwin and Philip Marder

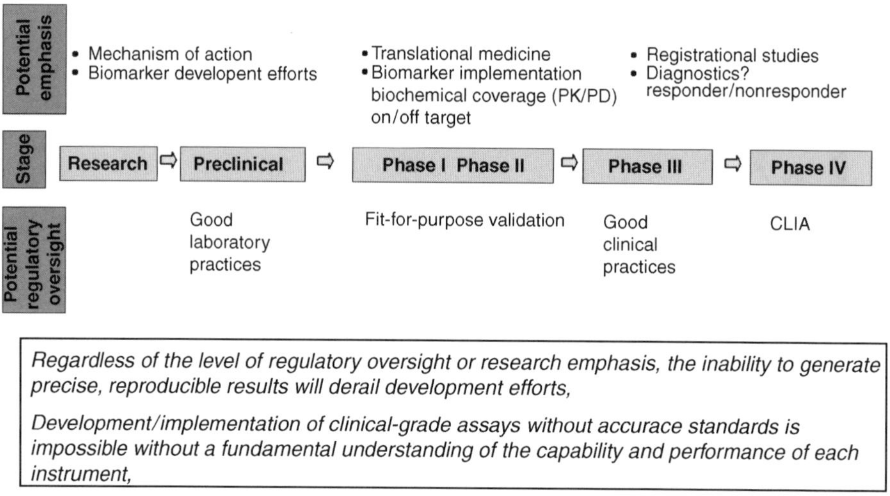

FIGURE 13.1 The versatility of flow cytometers makes them an ideal platform for integration into the biopharmaceutical drug development process. It is important to calibrate administrative oversight to each stage of drug development. This is key with respect to instrument validation where GLP studies place the highest demand on validation activities. Given the intensive nature of GLP validation exercises, it would be inadvisable to apply this guidance to other stages of the development process; however, the core principles of GLP-driven instrument validation are of value throughout the development and commercialization scheme (see text for discussion).

validations are often cited, a question of relevance is whether any common denominator can be gleaned from efforts with the CLIA, the GLPs, or the GCPs or whether these concepts introduce unnecessary cost. Thus, the roadmap in deciding how a flow cytometer should be used in a validated setting starts with the basic knowledge of knowing where you are in the regulated world (Figure 13.1).

13.2 ATTRIBUTES OF CLIA, GLP, AND GCP INSTRUMENT VALIDATIONS

In industry, it is not unusual for a team to be charged with a service function for specimens derived from a variety of sources/studies. Activities need not be segregated, and work stream amalgamation represents a value-oriented strategy because infrastructural needs are minimized through efficiency. In other words, if a laboratory has only a 20% need for CLIA-governed analyses, the excess capacity could host GLP or GCP studies provided that efforts are appropriately defined.

13.2.1 CLIA

When human specimens are analyzed to provide diagnostic information, to guide treatment decisions during disease, or to perform health assessments, the laboratory is legally obligated to comply with the set of standards as defined by CLIA, and to obtain licensure from an agency such as the College of American Pathologists (CAP) through demonstration of their proficiency [2]. Stated another way, CLIA is the law and CAP is an accrediting agency. The roots of this regulatory oversight can be traced to the late 1960s when it became apparent that malignancies were sometimes overlooked due to high error rates in evaluation of Pap smears [3], with the Clinical Laboratory Improvement Amendments of 1988 (CLIA'88) as an important historical milestone.

Today's flow cytometry laboratory may rely heavily on demonstration of proficiency through CAP processes, including test method validation and instrument calibration. A significant number of clinical scientists that utilize flow cytometers reside in or around the CLIA environment and it is therefore not surprising that many of the principles and practices of CLIA are leveraged in an effort to introduce quality into biomarker campaigns. These processes, however, are exclusive of instrument and system validation as defined under the GLPs and GCPs. Laboratories that perform immunophenotyping on specimens from persons (or animals) treated with investigational agents may need to comply with the GLPs and GCPs to support their regulatory filings and therefore are not subjected to CLIA mandate.

13.2.2 GLP

The Drug Amendments of 1962 were drawn in part from the European drug market approval of thalidomide, a sedative prescribed to pregnant women to treat morning sickness that caused severe birth defects [4]. These measures provided more time to review new drug applications, more safety and efficacy testing, and authorized records inspections of the sponsoring company. The FDA later added requirements for the manufacturer to prove drug effectiveness (in addition to safety) and to present postapproval reporting to the FDA. This forms the bedrock of GLPs.

The basic contemporary document dealing with international GLPs is the OECD Principles of Good Laboratory Practice, revised in 1998 [5]. With respect to these principles, there is no question that GLP studies are the native soils that nourish the practices and principles of instrument validation. Perhaps, the most important philosophy that can be provided to those new to instrument validation as governed under OECD/GLP is that validation exercises must follow predicate rule (Figure 13.2), with execution governed by a variety of system experts (Table 13.1). That is, validation is a carefully orchestrated process where results will either pass or fail predefined acceptance criteria. A second important point is that those who provide guidance for instrument validation use the term "computerized systems," meaning that the flow cytometer is one component of a process whereby electronic data are generated, analyzed, and stored. Finally, it is important to know that validated systems must be "closed systems," meaning that access is granted to only those who are qualified via their experience and have received appropriate training.

Many flow cytometry vendors fail to provide technical controls for even the most basic elements of a validated system. For example, granting entry into a flow cytometer via a password often provides a user access to the entire set of electronic

Document	Description	Notes
Validation plan	Defines overall validation strategy, system, scope, and required deliverables	Includes purpose, roles, and responsibilities; can reference validation SOP(s) if applicable
Requirement/design specification	Defines functional and performance requirements for the system based on intended use. Documents, at a high level, how each requirement will be met by the system	Cytometric functionality, data managment. security, training, and regulatory requirements are defined in this document
Traceability matrix	Traces each requirement from the requirements/design specification to a corresponding test in one or more of the test protocols	For auditing puropses. Requirements may also be met by SOPs (which are considered part of the "system")
Test protocols (IQ/OQ/PQ)	Traces each of the requirement defined in the requirements/design specification document to provide documented eidence that the system meets each requirement	Consists of vendor testing and IQ/OQ/PQ protocols
Execute test protocol: **Installation qualification (IQ)**	Components and configuration of system	Hardware: components, configurations, manufacturer, model, and serial number
		Software: versions and configurations
		SOPs: Effective dates and document numbers
Execute test protocol: **Operational qualification (OQ)**	Tests of applicble system functions	Cytometer functions: Light scatter, fluorescence, and statistics
		Computer functions: User access control and password management
		Data integrity and security
Execute test protocol: **Performance qualification (PQ)**	Tests to demonstrate that system performance is suitable for its intended use	FACSComp calibration
		Chicken red blood cell calibration within specifications
Validation summary report	Summarizes the test that was performed and deliverables that were generated during the course of the validation	Describes test faiulres and/or deviations from the validation plan or test protoco is that may have occurred during test execution, how they were addressed, and their impact on the validation

FIGURE 13.2 Breakdown of an instrument validation work stream. A critical aspect of instrument validation is that it follows the predicate rule, meaning that the process is governed by a defined set of procedures and exercises. This table represents an example by which a predicated validation could be executed. The traceability matrix (third row in the table) is a nonobvious tool that facilitates all downstream activities through an auditing function.

TABLE 13.1 Responsibilities of the System Experts Involved in an Instrument Validation

	Responsible Team Member				
Components/Deliverables	System expert	Quality assurance	Project manager	Scientific management	Consultants
Validation plan			x	x	
Requirements/design specification	x	x	x	x	x
Test protocols	x		x		x
Traceability matrix	x		x		x
Validation summary report			x		
Document review		x			
Time line creation and monitoring			x		

The components and deliverables listed in column 1 of the table are called out in Figure 13.2. Delineation of roles and responsibilities ensures an efficient validation process without duplication of effort.

records in that instrument/server, and this can provide an opportunity to those not authorized to access specific data sets to modify files or temper with them. In this case, the instrument would be considered to have breached the requirement of a closed system.

The practical solution to a technical breach of validation requirements need not be the exclusion of flow cytometry from GLP studies. Although some have chosen to identify flow cytometry efforts as "non-GLP components of GLP studies" due to gaps in technical controls, this may be inadvisable. Rather, a judicious use of procedural controls demonstrating that the management (1) is aware of the gaps that exist, (2) the management has developed predicated in-house processes and correlated the staff training curricula, and (3) the management has allotted time to review and enforce such controls as to meet the regulatory requirements is the advisable path forward. Examples of procedural controls include limited access to the system to authorized personnel and revocation of privileges set forth by standard operating procedures (SOPs). Another example is to require (by SOP) that users change their passwords at least once every 90 days. The third example is to require that raw data files shall not be altered, deleted, or overwritten. It is, therefore, a combination of technical and procedural controls that ensure appropriate system control and adherence to GLP requirements.

13.2.2.1 Instrument Validation 101: The Process Once the compliance risk has been assessed and a validation team has been assembled, meaningful consultations (i.e., the system compliance assessment) about the validation plan can begin. In this setting, definition of validation requirements, review (or authorship) of instrument-specific SOPs, and any procedural controls that are envisioned are detailed. Instrument validations can be focused by restricting validation to a particular instrument functionality that will be used within the regulated space. For instance, one

TABLE 13.2 Accommodation for the Computer's Operating System in a Flow Cytometer Validation

	Operating Systems	
Attributes	Windows™	Macintosh®
User accounts	Global NT user groups	
Password management	Global (corporate) NT password policies apply	Local password policies configured through OSx terminal
Data storage and handling	Data saved directly to file server	Third party program (e.g., DAVE®) interfaces with Windows server
Disaster recovery	Norton Ghost™	Carbon Copy Cloner (freeware)
Procedural controls	Automatic time-out set by administrator for all users	Automatic time-out set per user

such discussion in our group led us to exclude efforts to evaluate the mechanical sorter that was a component of one of the FACSCalibur instruments in our validation plan. In another meeting, we identified certain software configurations and related issues that had to be set independently for each user or specific to each computer system to establish compliance (Table 13.2).

The suite of activities triggered by the validation plan (VP), requirements and design (RD), requirements traceability matrix (RTM), and test protocols (IQ/OQ/PQ) are the true heart and soul of the instrument validation process.

Compliance is achieved through the verification and acceptance of intended use and documented upon approval of the validation summary report (VSR); compliance is maintained through change control practices, forward-thinking training oversight, and documented maintenance and monitoring of the instrument.

13.2.2.2 Access Controls, Integrity, Authentication, and Transmission Security The validation process consists of verification and control over the entire system, including both user and data controls. The basic principle of user access is defining, documenting, and managing who should have access to what. Defining the minimum business requirements for managing user access includes defining local (instrument-specific) user groups, for example, defining role-based access privileges and responsibilities such as Administrator, and defining instrument configured parameters (e.g., password requirements, allowable logon attempts, etc.). Once user groups are defined and configured, a documented procedure defining the role-based training requirements, method of access approval, and administration of users, including removal and periodic verification of the process, ensures access control over time.

Assurance of data integrity starts with a careful review of the data flow, from raw data generation to archive. At any point where data may be manually manipulated or lost, a technical or procedural control is necessary to ensure accuracy and retrievability. Whenever possible, automated data generation and transfer should be aggressively pursued to enable validation of the entire process, reduce risk of error, and increase efficiency. When a validated pathway is not an option, as is the case with most flow cytometers, areas of risk must be identified and manual verification steps added to the process. In addition, electronic data supporting a GLP study are required to be archived in a secure location and indexed to ensure expedient retrieval. To ensure this requirement is met, a procedure defining electronic file nomenclature (how files shall be named) and transfer to a secure file location (how, when, and where files shall be transferred) for storage/archiving is essential. For example, in the case of nondigital flow cytometers, we chose to save raw data files and exported text files to the local hard drive as recommended by the vendor. However, procedure dictated a manual transfer of files to a secure file location until a technical (automated) solution could be identified.

Ultimately, appropriate access control and data integrity are best accomplished by carefully analyzing the data generation process from system to archive, identifying technical gaps, and applying procedural controls as necessary to assure GLP compliance.

13.2.2.3 21CFR Part 11: What is Part 11 Compliance? Title 21CFR Part 11 of the Code of Federal Regulations provides important guidance for those who are responsible for the equivalency of electronic records to paper documents [1]. The guidance provides clear requirements for those charged with establishing and upholding the reliability of electronic records and signatures; however, the FDA also recognizes the need for flexibility in execution. For those required to ensure compliance for activities governed by the FDA predicate rule (i.e., GLP), the need to develop unambiguous philosophies regarding audits and audit trails, system validations, electronic signatures, and software systems that warehouse and process electronic data is paramount. There are no turnkey systems that can adequately cover the case-by-case requirements of any individual organization or work stream. All drug makers, including medical device manufacturers and biotech and pharmaceutical companies, are charged with meeting the requirements of these regulations by assembling a coherent program after selecting from of a mosaic of commercially available resources in combination with system components that may be organically derived from in-house expertise. Although the tangible degree of uncertainty generated from this approach may be initially unsettling, the growing expertise in this field is noteworthy and can be reassuring. A wide variety of experts are emerging as a function of maturation, and presentations/resources addressing practical solutions to Part 11 compliance [1, 6] that do not unnecessarily overengineer solutions have become common in public domain. Thus, development of a risk-based approach anchored by sensible but robust philosophies regarding Part 11 compliance is the kernel by which an individual can build an appropriate and effective Part 11 compliant system.

One should be wary of any vendor that offers a turnkey "Part 11 compliant system." To practice Part 11 compliance, both procedural controls (i.e., notification, ongoing training, governance via SOPs, and administration) and administrative controls are required in addition to technical controls provided by the instrument and associated software. The role of a vendor in Part 11 compliance resides in their duty to summarize the technical attributes of their system in a way that the end user can leverage them to build their own system. In a flow cytometer validation we conducted at Amgen, we defined our system as everything used from the generation of the electronic data to its final resting place in the archives. This included validated file servers where the data were saved, a third party file backup and archiving engine, an electronic file security software service, third party components (e.g., plate loaders), and the SOPs we wrote that govern the use of the instruments. For this reason, we were not able to simply purchase the vendor's validation package and consider our system validated.

13.2.2.4 Are There Exemptions from Part 11 Compliance? It may be surprising to some that Title 21CFR Part 11 does not require organizations to use electronic records or signatures. In the case that an organization exempts itself from the need to demonstrate the accuracy and trustworthiness of electronic records that are governed by the predicate rule, Part 11 guidance defines the requirements that must be met. In this case, the organization cites paper documents as the sole authoritative source for regulatory purposes. The problem is that this creates the obligation to demonstrate that hard copies produced from an electronic source are complete and accurate, and that computer use was incidental. The burden of proof here is high and underlain by resource intensive activities with high potential for error. In this day and age, it is difficult to envision an organization opting to exempt itself from Part 11 compliance; we rely on electronic records and cannot retract from our responsibility to ensure their compliance.

An important distinction should nonetheless be made between "enforcement discretion" and exemption from Part 11 compliance. This topic has roots tracing back to February 2003 when FDA re-examined the breadth of 21CFR Part 11 enforcement [7]. The key branch point in this argument is whether the work stream was conducted under the predicate rule; if not, a documented risk assessment is advisable and is discussed in the guidance document. In the case of flow cytometry experiments, the heavy reliance on a variety of software packages to recover data through gating processes and fluorescence compensation algorithms markedly tip the balance toward enforcement of many of the elements of Part 11 compliance. Even if not guided by formal mandate, these principles simply make a sound practice for those leveraging flow cytometers in their drug development campaigns.

13.2.3 GCP

The idea that GLPs are required for the conduct of clinical studies is a misconception. The introduction to the OECD principles of GLP [5] makes it clear that they apply only to preclinical studies, excluding clinical studies such as pharmacokinetic and efficacy studies. The relevant documents for clinical studies are the various codes such as the International Conference on Harmonisation (ICH) [8]. Because data supporting a clinical trial have the same assurances as what is more typically cited at GLP – such as accuracy, traceability, reproducibility, and retrievability – many investigators find it convenient to exert the practices and principles of GLP to GCP efforts (and there is nothing inherently wrong in this approach).

If one wishes to take full advantage of the nuances of GLP and GCP requirements and the roles and responsibilities of the staff concerned, a productive partnership between the laboratory and an experienced quality assurance unit cannot be underemphasized because guidance is sometimes less clear for GCPs. For example, an important core feature of both GLPs and GCPs is the establishment of predicate processes that identify and define requirements for instrument validation. Although it is quite clear that management holds the approval role with the quality assurance unit exerting an independent oversight function for GLP studies, it is not, in fact, specified who approves release for GCP efforts. Moreover, a simple memo can suffice in a GCP study in places where one would defer to more regimented analytical procedures and standard operating practices for GLP studies. So, while the work breakdown presented in Figure 13.2 is quite useful for those engaged in GCP studies, variance in the level of detail, method of documentation, and assignment of responsibilities can be introduced as a function of the biopharmaceutical concern hosting the studies. For instances where the same staff might be involved in both GLP and GCP efforts, it is often more efficient and safer to extend the GLP work stream to the GCP efforts to simplify processes for all involved.

13.3 HYBRID SOLUTIONS

From one perspective, it could be stated that every instrument/system validation problem is met with a hybrid solution because of the need to stitch a variety of elements together to ultimately represent a coherent and infallible process to guarantee data authenticity. In practice, the terminology of a hybrid solution is used primarily to identify areas where one supplements the validation with information or assurances provided by the vendor and/or the mosaic of technical and procedural controls that might be used to reach the end goal of instrument/system validation. One must take note, however, of the liabilities associated with hybrid approaches inasmuch as, for example, reliance on a vendor for IQ/OQ of a flow cytometer cascades to the correlated need to qualify the vendor. That is, as the vendor holds no responsibility toward your system, outsourcing the IQ/OQ in this way incurs the additional activity of vendor qualification. The phrase of caution sometimes bantered about in the field that "validation means validation" seems to hold special importance for those who approach their validation exercises with an excessive reliance on hybrid solutions, so the watchword is to proceed with caution in this space.

13.4 FEEDING AND CARE OF YOUR VALIDATED SYSTEM: CHANGE CONTROL

The milestone of instrument/system validation is a significant accomplishment that enables flow cytometry laboratories to fully engage in regulated studies. Lack of appropriate change control processes can void these efforts almost instantly, so the care taken for a planned installation, including appropriately documented IQ/OQ with any required configuration details, needs to be combined with a documented evaluation of any future change to hardware and software from the very start.

Thus, as part of the validation process it is important to identify what future changes may be made to the system that modify the validated state and therefore require additional testing and/or calibration. In this regard, software upgrades and major hardware replacement such as lasers, filter sets, or photomultiplier tubes should be obvious. On the other hand, sheath/waste tanks and other consumables are not likely to void the validated state, and time should be taken to identify these as nonconsequential changes to the system. This ensures an appropriate level of control without spending resources to overdocument changes.

Another essential procedure generated prior to or during the validation defines the operation and maintenance of the system, including scheduled and unscheduled maintenance. To ensure appropriate documentation, procedure-driven forms may be utilized to capture the details of scheduled and unscheduled maintenance including methods to assess performance after such interventions. To complete the exercise, completed forms can be maintained simply in a logbook associated with the instrument and reviewed and archived periodically.

Ultimately then, it is the documented evidence of changes to and maintenance of the system that not only serves as a heath assessment of the instrument but also represents the assurance that a system validation performed today is applicable to the data that are generated from that instrument tomorrow.

13.5 IN CLOSING: VALIDATION MEANS VALIDATION

A comprehensive understanding of instrument validation is very important to those conducting flow cytometry investigations for biopharmaceutical companies. Although the complexities of validation efforts may be initially daunting, the value of taking time to understand the regulatory landscape with respect to a given project cannot be underemphasized. It is easy to incorrectly assign regulatory oversight to an analysis work stream, but ample resources are available to those who need to calibrate their activities in the regulated space. Perhaps, the best advice to those entering the regulated environment is to live by the maxim that validation means validation: those who study the requirements, parse out the responsibilities, partner with those best qualified to meet the set of validation requirements, and make adjustments to maintain the validated state will succeed and add value to the work group that depends on this expert function.

REFERENCES

1. Guidance for Industry. Part 11, Electronic Records: Electronic Signatures – Scope and Application. U.S. Department of Health and Human Services, Food and Drug Administration, Center for Drug Evaluation and Research (CDER), Center for Biologics Evaluation and Research (CBER), Center for Devices and Radiological Health (CDRH), Center for

Food Safety and Applied Nutrition (CFSAN), Center for Veterinary Medicine (CVM), Office of Regulatory Affairs (ORA), 2003, pp. 1–9. Available at http://www.21cfrpart11.com/files/fda_docs/guide_doc_scope_appFeb2003.pdf. Accessed Jan. 13, 2009.

2. Title 42 Code of Federal Regulations, Pt 493. Centers for Medicare and Medicaid Services, Department of Health and Human Services. Available at http://wwwn.cdc.gov/clia/regs/toc.aspx. Accessed Jan. 13, 2010.
3. Dorsey D. B. Evolving concepts of quality in laboratory practice: a historical overview of quality assurance in clinical laboratories. *Arch. Pathol. Lab. Med.* 1989;113(12): 1329–1334.
4. Kaplan AH. Fifty years of drug amendments revisited: in easy-to-swallow capsule form. *Food Drug Law J.* 1995;50(Special Issue):179–196.
5. OECD Principles of Good Laboratory Practice, Organisation for Economic Co-operation and Development, 1998, pp. 1–41. Available at http://www.olis.oecd.org/olis/1998doc.nsf/LinkTo/NT00000C5A/$FILE/01E88455.PDF. Accessed Jan. 13, 2010.
6. General Principles of Software Validation: Final Guidance for Industry and FDA Staff, U.S. Department of Health and Human Services, Food and Drug Administration, Center for Devices and Radiological Health, Center for Biologics Evaluation and Research, 2002, pp. 1–43. Available at http://www.fda.gov/downloads/MedicalDevices/DeviceRegulationandGuidance/GuidanceDocuments/ucm085371.pdf. Accessed Jan. 13, 2010.
7. Guidance for Industry. Part 11, Electronic Records: Electronic Signatures – Scope and Application, U.S. Department of Health and Human Services, Food and Drug Administration, Center for Drug Evaluation and Research (CDER), Center for Biologics Evaluation and Research (CBER), Center for Devices and Radiological Health (CDRH), Center for Food Safety and Applied Nutrition (CFSAN), Center for Veterinary Medicine (CVM), Office of Regulatory Affairs (ORA), 2003, pp. 1–9. Available at http://www.21cfrpart11.com/files/fda_docs/part11_final_guidanceSep2003.pdf. Accessed Jan. 13, 2009.
8. Guideline for Good Clinical Practice, ICH Harmonised Tripartite Guideline, ICH Expert Working Group, 1996 pp. 1–53. Available at http://www.ich.org/LOB/media/MEDIA482.pdf. Accessed Jan. 13, 2010.

PART IV

FUTURE DIRECTIONS

Virginia Litwin and Philip Marder

These are exciting times in the field of flow cytometry. The ongoing development of fluorescent probes and digital instrumentation now allows 8-color analyses on benchtop models and around 18-color analyses on research models (reviewed in Chapter 2). At the same time, however, these advances present more challenges. Among them one would be in developing new ways to analyze data since the traditional method (Boolean gating) does not allow effective mining of large and diverse data sets. The second challenge would be to how best apply the new capabilities in order to answer basic research questions and optimize support during drug development.

In Part IV, possible resolutions to these challenges are considered. For example, in Chapter 14, a new paradigm for data analysis is presented. In Chapter 15, the importance of multiparametric analysis in evaluating intracellular signaling is explained as it relates to drug development and, ultimately, patient care. Finally, the chapter contemplates the evolution of flow cytometry into mass cytometry.

Flow Cytometry in Drug Discovery and Development, Edited by Virginia M. Litwin and Philip Marder

14

PROBABILITY STATE MODELING: A NEW PARADIGM FOR CYTOMETRIC ANALYSIS

C. Bruce Bagwell

14.1 INTRODUCTION

Cytometry is likely to be the technology of choice for observing and understanding cellular processes in the foreseeable future because of its ability to relate complex sets of measurements to a single cell. With the advent of new cytometers capable of measuring numerous fluorescence wavelengths and the ever-increasing number of antibodies and fluorochromes, it is becoming increasingly difficult to deal with the dimensionality of cytometry data. Traditional analysis techniques that are based on examining one measured parameter versus another run headlong into what is commonly referred to as the dimensionality barrier. In some cases, hundreds of histograms are necessary to assess the status of a single specimen.

This chapter describes a new paradigm called probability state modeling (PSM) that is capable of visualizing and analyzing the rich multiparameter information embedded in cytometric listmode files. Model-derived parametric plots merge intensity, percentage, and variance information into easy-to-understand graphical formats that represent virtually every bit of information contained in these files. These models incorporate expert metainformation that helps elucidate the hidden meanings in the data even to those not trained in cytometry. Model-based analyses are likely to be the method of choice in bringing automation to cytometry.

Flow Cytometry in Drug Discovery and Development, Edited by Virginia Litwin and Philip Marder

14.2 PARAMETRIC EQUATIONS

In mathematics, you can easily write equations that define a line in n dimensions. The trick is to define a set of equations that all depend on the common variable, t.

$$p1 = p1_0 + p1_1 \times t$$
$$p2 = p2_0 + p2_1 \times t$$
$$\ldots$$
$$pn = pn_0 + pn_1 \times t$$

These equations are called parametric equations [1]. They allow us to easily describe objects such as lines, surfaces, and volumes in any number of dimensions. A general characteristic of these equations is that rather simple parametric equations can describe highly complex objects (see Figure 14.1).

$$p1 = \sin(7t)$$
$$p2 = \sin(8t)$$

What do parametric equations have to do with the field of cytometry? In this chapter, I hope to convince you that future of cytometric analysis is about to change in a radical way and these parametric equations are at the very heart of this change.

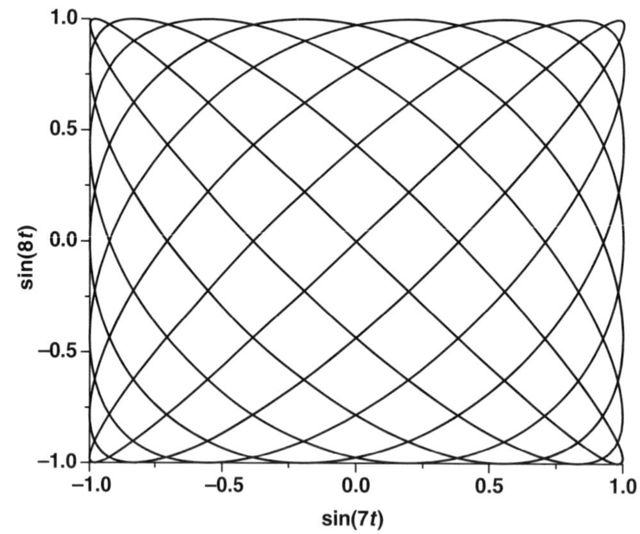

FIGURE 14.1 Complex graphical object constructed from simple parametric equations. The relatively simple parametric equations, $\sin(7t)$ and $\sin(8t)$, can create a complex graphical object when one is plotted against the other.

14.3 SHOWING COORDINATED EXPRESSIONS

What is interesting about this paradigm shift is that we have always known the proper way of analyzing and displaying our data. In the 1980s, Loken and Wells [2, 3] described B lymphoid maturation with charts such as the one shown in Figure 14.2. The changes in antigenic expression of markers such as CD19, CD34, CD10, CD20, CD22, and CD45 were shown against a common axis. There were never any units for this axis. It was simply used as a means of showing coordinated changes in expression. We always knew that these charts provided valuable information about the coordinated expression of a large number of parameters, but we never realized that it was possible to compute this axis or one like it directly from the data.

14.3.1 Time As a Parametric Parameter

What units should we use for this axis? Since it normally shows some kind of progression, our first guess would be some sort of time unit. This makes sense since cellular progressions are really just a set of kinetic reactions. There are a host of papers that attempt to describe systems with time-dependent differential equations [4–6]. The problem, of course, is that when we take our snapshot set of measurements on cells, we lose time. This means that all the rate constants for these differential equations must be measured from other experiments and it must be assumed that the rate constants are applicable to the cells being measured. Valuable information can be gleaned from this approach [7], but it does put limits on the scope and complexity of the system being investigated.

14.3.2 Relative Time or Cell Age

In order to explain how to create this common axis and parameter, we will first visit a figure that I first used in my PhD thesis [8] back in 1979 to clarify the implicit cell cycle relations in a DNA histogram (see Figure 14.3, top panel). On the x-axis is relative cell age, ranging from 0 to 1. When a cell has a cell age of 0, it has just been created after its parent cell's division and when it has a cell age of 1, it is just about to divide into two daughter cells. The cell depicted in this figure begins to synthesize DNA at a cell age of 0.4 and stops at 0.8 when the DNA is duplicated. The phases of the cell cycle are defined by these two cell ages and are normally represented as G1, S, and G2M.

14.3.3 Number of Cells Versus Cell Age

We can use this same cell age parameter to look at the number of cells that are at each relative cell age unit (see Figure 14.3, bottom panel). We used to state that for exponentially growing population with no cell loss, the number of cells is a decreasing exponential. Since that ideal relation is rarely found in practice, we will make no

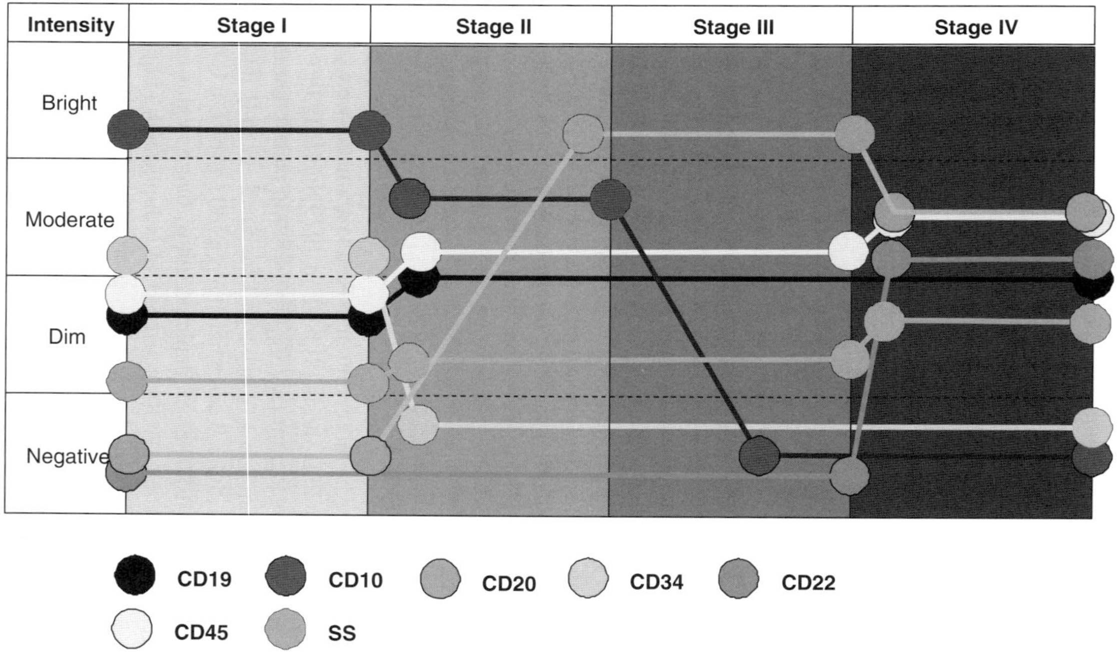

FIGURE 14.2 Chart of changes in marker expression with normal B-cell lineage. This figure shows the changes in CD19 (black), CD10 (blue), CD20 (green), CD34 (cyan), CD22 (green), CD45 (yellow), and SS (SSC) (orange) with normal B-cell development in the marrow. Adapted from Ref. [2], Figure 3. (See the color version of the figure in the Color Plates section.)

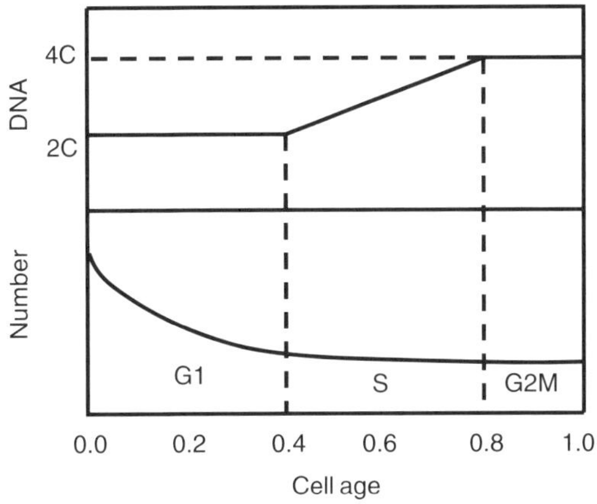

FIGURE 14.3 DNA and number of events versus cell age. This figure shows two parametric relations: DNA versus cell age and number of cells or events versus cell age. An ideal DNA histogram is the solution of these two relations where the G1 and G2M peaks are located at 2C and 4C, respectively. The areas of G1, S, and G2M are equal to the areas under the number–cell age curve partitioned by the cell age = 0.4 and 0.8 boundaries.

assumptions about the relation between cell age and number of cells. It could be a complicated or simple relationship. It really does not matter.

14.3.4 Eliminating Cell Age

I remember giving lectures back in the late 1970s in which I would show that DNA versus cell age and number of cells versus cell age were two parametric equations, where you could eliminate the common parameter of cell age and end up with a DNA histogram with DNA content on the x-axis and number of cells on the y-axis. It never occurred to me or any one else that we should not have been so hasty to eliminate this important common parameter.

14.3.5 Cell Age As a Parametric Parameter

Let us not make that mistake this time. Let us keep cell age around and see where it leads us. You can easily imagine that if we could measure cell age, the best way to look at coordinated expressions of cell cycle proteins would be to use cell age as a common axis much like our parametric equation variable, t. If we produced a graphical overlay with parameter intensities on the y-axis and cell age on the x-axis, we would be able to visualize how cyclins and other cell cycle proteins modulate as cells progress through the cycle. We could examine all the correlations of a very large number of parameters. We really would not be limited by number of parameters. The more the better. . . If we had cell age as a parameter, we would have never gone down the road of displaying one measured parameter versus another.

14.4 EQUAL PROBABILITY STATES

We cannot measure cell age, but we can get close. Imagine we constructed another axis where each bin or state is defined by cell age boundaries. These boundaries are chosen such that the number of events is equal in each state. Associated with each of these states are the parameter averages evaluated over the same boundaries. Another way of stating this situation is that we can rebin the cell age axis such that each bin or state has an equal probability of being associated with a measured event [9].

14.4.1 Probability State Axis and Percentages

From a practical point of view, this probabilistically determined axis is just as effective in showing coordinated expression of numerous parameters as cell age. In fact, it has a few advantages over cell age. Since the number of events is uniform along this axis, any segment of the axis is proportional to the measured population's cumulative fraction. For example, if we divide this axis into five equal segments, the first segment defines the first 20% of events, the second, the next 20%, and so on until the last 20% of events in the progression.

14.4.2 The Fusion of Intensity Patterns and Percentages

In the early 1980s, cytometerists began to realize that there was a lot more to cytometric analysis than reporting percentages. Reporting $CD4^+$ percentages was valuable, but understanding the disregulation associated with lymphomas and leukemias required an appreciation of changes in intensity patterns rather than just percentages. There has always been an uneasy balance between these two ways of quantifying data, but with this new probability state axis, percentages and intensity patterns are merged together into one coherent picture of the status of a specimen.

14.4.3 Synthesizing Data Sets

Another advantage of the probability state axis is that it allows us to easily synthesize complex data sets from the model. The way this is done is quite easy to understand. A computer algorithm picks one of the states at random. Once it has selected the state it can then look up the parameter values associated with that state. It can then add some uncertainty to these values by using Gaussian distributed random numbers. These synthesized values can be put into a listmode file and stored on disk. These data sets are valuable because the underlying "truth" associated with their creation is known.

14.4.4 Calculating the Probability State Axis

The major advantage of the probability state axis, however, is that we can calculate it from the observed data. This notion at first appears to be untenable, since we initially

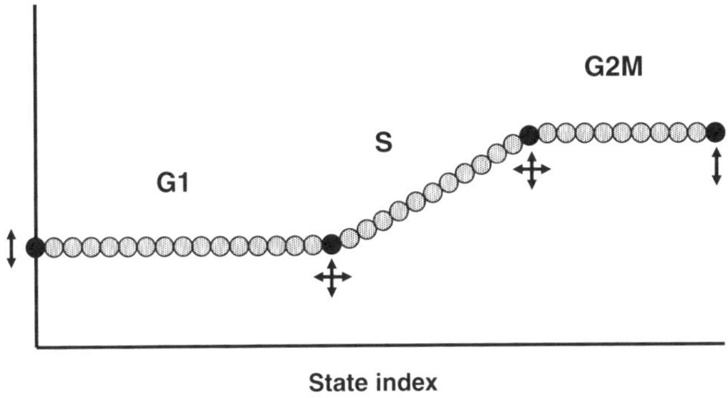

FIGURE 14.4 Control definition points in a DNA histogram. The set of control definition points that represent important transitions in a DNA histogram are shown as black circles.

defined the state or bin boundaries using cell age. How can we possibly measure these bins from the data if we cannot measure cell age?

14.4.4.1 Control Definition Points Let us keep our simple DNA example to see how this is done. Suppose we add to our state model a set of key inflection points called control definition points (see Figure 14.4, y-axis is relative intensity) that define the important transitions in DNA as the cell progresses along the cell cycle. For the moment, assume that we have some magical way of associating an event with one of the states on the x-axis. Initially, these control definition points are positioned exactly in the right locations such that all the events are distributed equally for all the states (Figure 14.5a, bottom plot). Even though the probability is equal for all states, there still will be counting error that accounts for the chop in the frequency data.

14.4.4.2 Importance of the State Frequency Distribution What happens if we move the second control definition point to the left and then let the model reclassify the data (see Figure 14.5b). Since there are now too few states to distribute the G1 events, the average number of events in this region is higher than it should be. In contrast, there now are too many states to distribute the S events so their average is less than it should be. The more we move the control definition point from its ideal location, the greater the disturbance in the state frequencies. If we move the second control definition point up or down (not shown), it disturbs the classification system so it no longer matches the data. Again, the result is a disturbance in the state frequencies.

14.4.4.3 Searching for Uniformity Suppose we had an accurate method of quantifying how uniform the state frequencies were. We could then find the control definition point locations that minimize this value. There are a number of minimization methods that will do this for us [10, 11]. When these algorithms find this minimum value, we have a solution to the problem. We have calculated our probability state axis from the data.

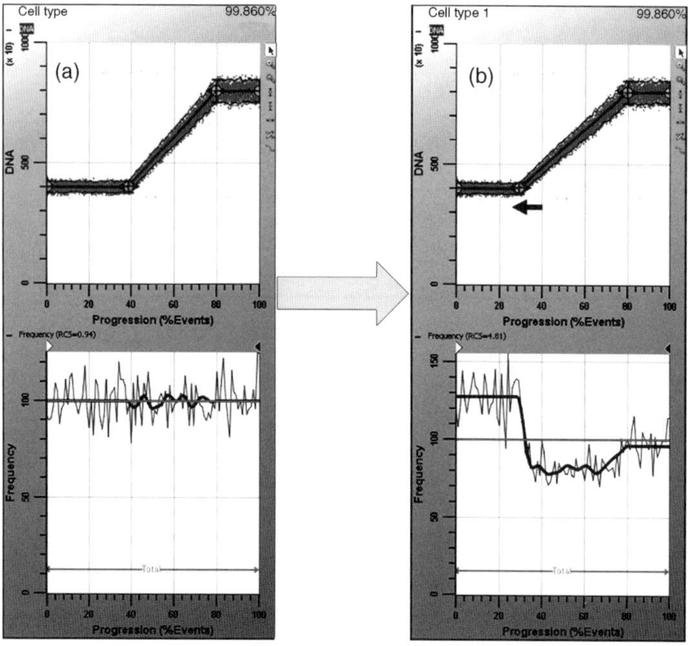

FIGURE 14.5 Effect of control definition point positions in a DNA histogram model. (a) Probability state model for a DNA histogram where the control definition points are at their optimal location. Notice the uniform frequency distribution (bottom, b). In (b), the second control definition point (arrow) is moved to the left. Notice the nonuniform frequency distribution (bottom, b).

14.4.4.4 Classifying Events into a State Of course, we still have the issue of how to associate a particular event with a probability state to figure out. After all, we cannot just use magic. The first step in the process is quite simple. We design a computer algorithm that takes one event at a time and compares it with each state's parameter value. It asks the question, "how close is this event to the model parameter intensity value?" Since each of these parameter values is surrounded by an uncertainty probability distribution, it uses this distribution to calculate the probability weight. The higher the weight, the more likely the event is associated with the state. If there are a number of parameter values associated with the state, it simply multiplies all of these weights together to account for how close it is to all of them.

14.4.4.5 Method of Maximum Likelihood Fails If there are a hundred states for this axis, then we will have a hundred weights calculated for each event. This process sounds more complicated than it really is. The next step, however, is a difficult one to grasp immediately. How do we use these weights to figure out what state our event belongs to? At first, you might be tempted to say, find the highest weight and associate the event with that state. In all honesty, that was my first choice when developing the event selection algorithm. When you use this maximum likelihood approach, the selection system fails miserably. It turns out that there are often states that tend to collect too many events and states that collect too few. The other problem is that it does not define what should happen when there are multiple highest weights. Which state do you pick?

14.4.4.6 Stochastic Selection The answer to this dilemma was to use a stochastic process [12] to make the selection. Again, this sounds much more difficult than it really is. An example helps make this clear. Suppose we only had five states and an event generated the following five probability weights (0, 0.3, 0.4, 0.3, and 0). Notice that the sum of these weights is unity. A stochastic selection process would randomly choose the first state none of the time, the second state, 30% of the time, the third state, 40% of the time, the fourth state, 30% of the time, and the last state, 0% of the time.

Usually, when we are analyzing data, we try to eliminate uncertainty. Cytometerists typically like nice, smooth data and contours, as they should. But in this case, we let uncertainty guide our selection choice. As it turns out, stochastic selection was the key that led to the ultimate success of this algorithm.

If we use this selection process to classify events into states and we let the system find the optimal control definition point locations that minimize the nonuniformity of the state frequencies, something quite unexpected happens.

14.4.4.7 Accounting for Population Overlap First, let me give some background. For over 20 years, we have been perfecting cell cycle analysis methods that yield accurate G1, S, and G2M estimates for complex DNA histograms. We have done everything possible to make these estimates as accurate as possible. We have hundreds of generated histograms that we use to test our analysis procedures. Early in 2006, we decided to test the probability state modeling system against our cell cycle analysis product [13] to see how the S-phase estimates compared (see Figure 14.6). We were astounded to discover that its accuracy was very comparable to the cell cycle modeling package. This figure shows the correlation plot of S-phase estimates from 150 synthesized DNA histograms. You can see an almost perfect correlation between the two methods, indicating that the probability state modeling system compensates for population overlap, at least as well as a DNA modeling package.

How is this possible? If you look carefully at the analyses of a single DNA listmode file (Figure 14.7), you can begin to appreciate why the probability state model accounts for most of the population overlap between G1 and S, and S and G2M. Since the probability state model represents parameter values and their uncertainty on the y-axis (see vertical arrows) and we find our percentages on the x-axis (see horizontal arrows), we are automatically accounting for overlap due to measurement error. You will need to examine this figure for a while to convince yourself of this fact. There is uncertainty in the control definition point's position along the x-axis, but most of that uncertainty is due to counting error, not measurement error.

What is exciting about this attribute of probability state models is that the subregions or zones along the probability state axis can be defined by any number of correlated parameters. Thus, this method solves the population overlap problem in

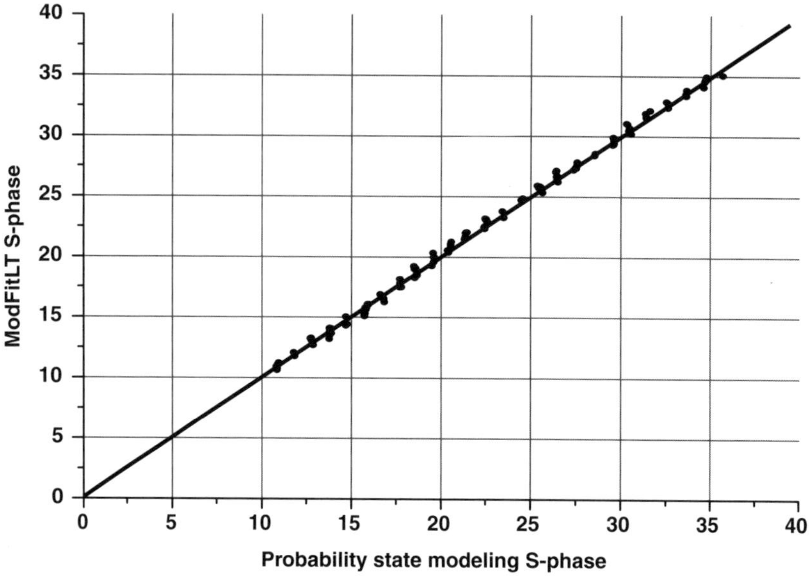

FIGURE 14.6 Comparison with conventional least squares modeling. One hundred and fifty DNA listmode files with 10,000 events and 3%CV were generated with %S-values ranging between 10% and 35% and were analyzed by the probability state modeling system (x-axis). The same data were read and analyzed by a cell cycle modeling package (Zz13) and the %S-phases were compared. The results indicate similar accuracy between these approaches.

a general way. It largely eliminates the error and subjectivity associated with creating analysis regions or quadstats. These types of gate-based constructs have hard boundaries that immediately create false positives and negatives. Hierarchical gating compounds these errors since the output of one gate is the input to another. Probability state modeling boundaries can be considered soft and malleable and can change as more parameters are added to the model. We will see plenty of examples of this a little later in the chapter.

14.5 TIME-DEPENDENT PROGRESSION

So far, I have been using cell age and cell cycle as an example to help you understand how probability state models work. It is time to take the training wheels off and consider the real complexity of many of our samples. I used cell age as an example of a time-dependent axis, but the reality is that this concept is much more general than cell age. The proper way of abstracting cell age is to consider it as an example of some time-dependent progression. This abstraction allows us to look at other progressions, such as cell differentiation and maturation, in the same way as we have with cell cycle progression. It really does not matter to us what progression we are interested in since we can always map these time-dependent progressions to our probability state axes.

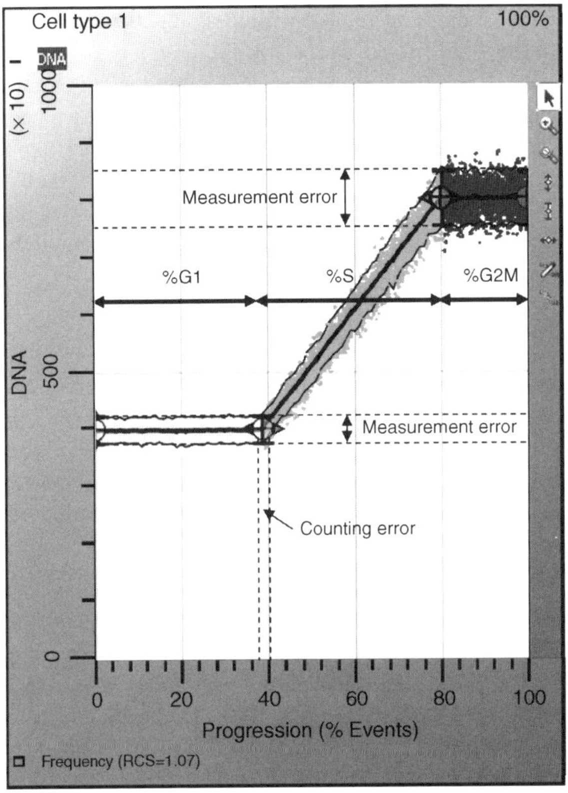

FIGURE 14.7 How probability state modeling accounts for population overlap. Because measurement error is represented on the y-axis (vertical arrows) and percentages are estimated from the x-axis, probability state models account for population overlap due to measurement error.

14.5.1 Two or More Progressions or Cell Types

What happens when we wish to evaluate two or more progressions in a sample? How can the probability state model include this additional level of classification in its algorithms? Again, the answer is easier than you might think. When there is more than one progression, the modeling system evaluates each progression or cell type separately. It finds all the state probability weights for each cell type and then summarizes these weights as a single cumulative weight. Another stochastic selection process uses these cumulative weights to choose a cell type. Once the cell type or progression is chosen, it then uses the same method as described earlier to find the state within that cell type.

14.5.2 It Is Too Good to be True

We will investigate this powerful capability more in a few moments. Before we look at how this technique works for some complex applications, it is important to clear up one last issue. Those of you who have been reading along should have a nagging suspicion that there is something that is not quite right with this story. At the very least, it seems to be too good to be true. In a way, you are right. It is too good to be completely true. Let me now try to pull the rug out from under the stack of cards that I have built.

Let us go all the way back to our original example where we had cell age versus DNA and cell age versus cell number. I showed how we can remap the cell age to a probability state axis. The boundaries of these states were originally defined by cell age, but later on, I showed that we can compute these boundaries without using cell age. That was the point where I left something very important out of the story. It is now time for a full disclosure.

14.5.3 Loss of State Sequence Information

At the moment we make our measurements, we lose time or, for our example, cell age. What this means is that there is no information in the data that allows us to sequence our states. When we lose time, it is like pulling the string out from the necklace of states that define our model. Equivalent data can be synthesized by any random sequence of these states. Figure 14.8 shows three examples of different state

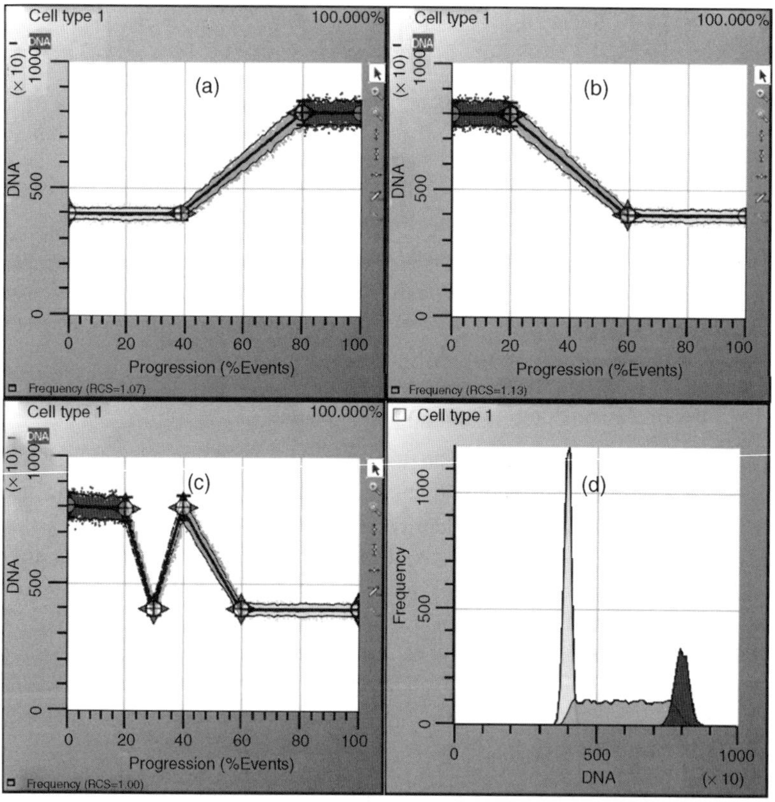

FIGURE 14.8 Different models can produce equivalent data. The three models (a, b, and c) can produce equivalent data (d) because listmode data no longer contain state sequence information. (See the color version of the figure in the Color Plates section.)

models (a, b, and c) that generate equivalent data for the DNA histogram (d). This means that there is no way for a computer algorithm to determine which states are first and which are last.

And we were so close . . .

14.6 METAINFORMATION

Of course, we are not computer algorithms. Most of you did not realize that I left that little fact out of the story because the model seemed totally reasonable. G1 comes first, then S, and then G2M. We all know that.

But, how do we know? That information is not in the data, so where is it? Most of us have seen the original pulse labeling experiments where we watched the pulse move from G1 to S and then to G2M. Most of us know that cells have a resting 2C complement of DNA first and then it is duplicated. Most of us have seen synchronized DNA histograms showing the sequence of G1, then S, and then G2M. This metainformation is in our heads. We learned it. Our brains impose these relations on our measured data without us being aware of it.

When we analyze a DNA sample, we choose a step-up type of parameter profile for DNA. As soon as we make that choice, we have used this metainformation in our minds to help us sequence our probability states. Do we need to know what these parameter profiles look like for every single parameter we measure? The answer to this question may surprise you. The best way of explaining how metainformation is used is for us to examine a real sample. We will come back to the "how much" metainformation question on the other side of this example.

14.6.1 B-Cell Lineage in Bone Marrow

Let us examine a 12-parameter listmode file from bone marrow. The markers are CD19, TdT, CD10, CD20, CD45, CD43, CD23, CD34, Kappa, Lambda, FSC, and SS. Our objective is to identify and visualize the B-cell lineage progression contained in this complex data set.

14.6.1.1 Getting Started The most common question I get when I talk about probability state modeling is "how do I get started?" How do I choose the correct marker to start with? A good answer to this question comes from a particularly apt analogy. If you are solving a crossword puzzle, how do you begin? If you think about it, you scan all the clues and pick the clue that you are most sure of, which is normally something simple and self-evident. From that simple beginning, you have a reference to help "fill in the blanks."

14.6.1.2 Constant Selection Parameters When we scan our B-cell markers, the simplest are CD19 and SSC. B-cells have CD19 on their surface, so this is a good place to start. We first assume that CD19 expression is constant throughout our B-cell progression. We refer to these constant parameter profiles as selection parameters. As soon as we add CD19 to the model (Figure 14.9a), we can then look

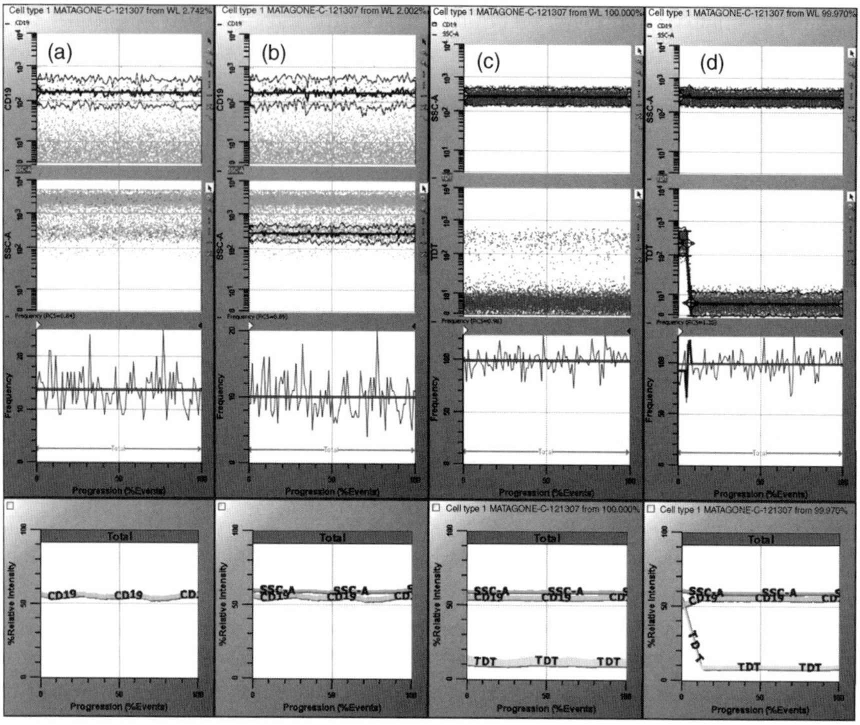

FIGURE 14.9 Sequence of analysis steps for finding the normal bone marrow B-cell lineage in a 12-parameter listmode file. This figure shows the first four analysis steps (data kindly provided by Fred Preffer, Harvard University). See text for details.

at SSC. We see some CD19 positive events in the granulocytes, so we include a SSC constant parameter profile in the lymphocyte region to eliminate these events from our B-cell selection (Figure 14.9b). Keep in mind that we could have started with SSC and then done CD19 and the final solution would have been the same. Notice that our frequency distribution (Figure 14.9, bottom panels) remains uniform as we go through this process.

A parameter overlay plot can be used to show our progress thus far. CD19 and SSC are essentially constant. The plots are bumpy because B-cells comprise only about 2.3% of the complete sample (Figure 14.9a and b bottom overlay plots) and counting statistics is adding a lot of variability to our model at this point.

We can eliminate much of this statistical variability by enriching our sample for just B cells. Notice that after enrichment our % classification becomes almost 100% and the SSC and CD19 plots become much smoother (Figure 14.9c).

14.6.1.3 TdT Step-Down Parameter When we examine TdT, we initially see two distributions in the data (Figure 14.9c). Note, only two parameter profiles are shown at a time. The initial CD19 distribution has been collapsed in order to show the new parameters at full scale. There are relatively few TdT + events and a lot of TdT− events (second panel from the top). We now have a choice. Are the B cells initially TdT + and then negative or is it the other way around? Perhaps, it is even more complicated. Perhaps, the progression begins with TdT +, becomes negative, and then becomes TdT + again. Since our data sets no longer contain the string of time, we must resort to using other information outside this data set to infer this sequence. The best that we can do is to assume that the data from the literature is relevant to our data set. Most of this work was done in the late 1980s by Loken and Terstappen [2, 3]. Using a number of techniques, Loken and Terstappen showed that TdT and CD34 were lost early in B-cell differentiation. We now need to use some of that information to create our B-cell progression model.

We choose a step-down type of parameter profile and then let the model optimize so that the frequency is as uniform as possible (Figure 14.9d). As soon as the system shows the final solution, we can see something very interesting. Notice that CD19 is no longer constant. Very early B cells have less CD19. Also notice that there is some SSC structure during this phase as well. The fact that CD19 is lower for early B cells also happens to be a well-published observation and thus serves as confirmation that we are on the right track.

14.7 TRADITIONAL GATING STRATEGIES AND METAINFORMATION

Before continuing on, it is important to discuss the relevance of this metainformation that we are using in our model definition. You might be tempted to say that traditional gating techniques do not make this assumption and therefore are not as dependent on these initial assumptions. The truth is, however, that when an experienced cytometrists looks at a histogram that has TdT as a parameter, they will impose this relation on the data to understand what is going on with the sample. If they did not, all that they would see would be dots. Those of you who are new to cytometry know exactly what I mean. Experienced cytometerists see patterns in the data that are not really there. They can do this because they have been taught all these important relations and their minds automatically connect all the dots with arrows.

14.7.1 CD10 Three-Level Parameter Profile

Let us now examine CD10 (Figure 14.10a). As soon as we examine the dot patterns for CD10, we observe that CD10 is positive early in B-cell differentiation. The crossword puzzle is still a great analogy for what is happening. With a crossword puzzle, when you put down a word, you are providing additional information that is important in finding the answers to the other, more complicated clues. TdT is providing us with information about early CD10. Even with this information, however, we are faced with another decision. After the early B-cell state, CD10 forms two clusters with intermediate and negative intensities. Again, we are faced with a lack of information in the data set and must resort to the literature to help us decide what would be the correct parameter profile to use. Loken and Terstappen showed that CD10 has high intensity early in the lineage, drops to an intermediate intensity, and then becomes negative.

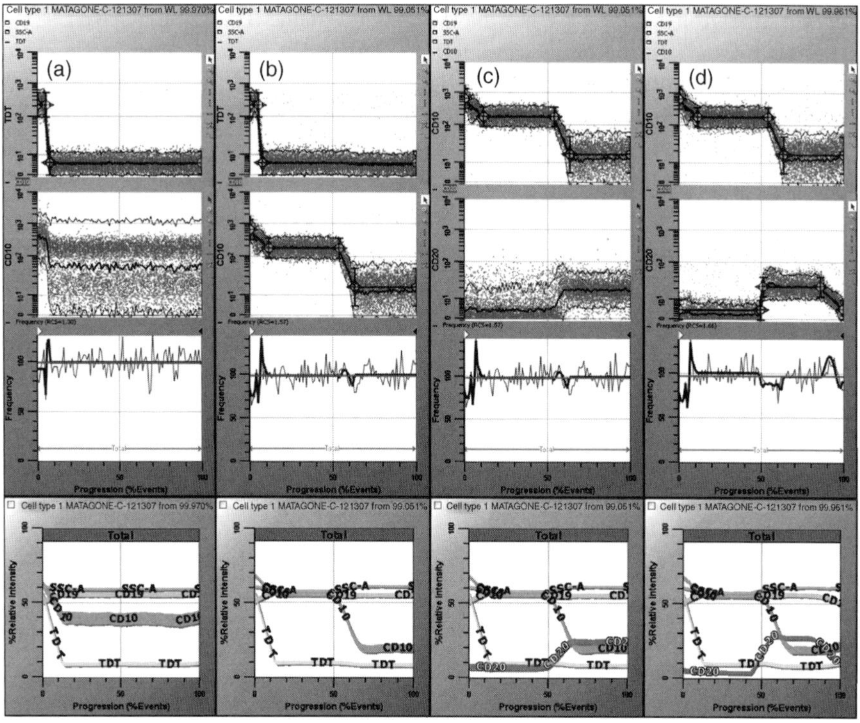

FIGURE 14.10 Sequence of analysis steps for finding the normal bone marrow B-cell lineage in a 12-parameter listmode file. This figure shows the next four analysis steps (data kindly provided by Fred Preffer, Harvard University). See text for details. (See the color version of the figure in the Color Plates section.)

When we choose a three-level parameter profile that follows the progression defined by the literature and let the program find the optimal placements of the control definition points, we find a solution (Figure 14.10b). As soon as CD10 is added to the model, we can see additional structure appear in both the CD19 and the SSC parameters.

You might ask at this point whether we could have started with CD10 instead of TdT. The answer is yes, we could have, but it would not have been as easy as it sounds because CD10 is more complicated than TdT. The slight decrease in CD10 in the early B-cell stage was easier to see with TdT already defined. You are always best off starting with simple parameter profiles and moving toward more complex ones. Could we have started with CD34? As you will soon see, in this data set CD34 did not have quite the dynamic range that TdT did. Therefore, starting with TdT was our best choice for this data set, but we could have started differently and achieved the same solution.

14.7.2 CD20 Step-Up and Step-Down Parameter Profiles

Let us now look at CD20 (Figure 14.10c). The model thus far shows us that CD20 is upregulated when CD10 is downregulated. We do not need the metainformation to know this. The analysis system is showing this to be true, and that serves as another confirmation that we are on the right track in modeling the B-cell lineage. However, we can use the literature to squeeze even more information from the data. We know that when B cells become plasma cells, they lose CD20. Thus, we can have our CD20 parameter profile drop to negative at the end to try to catch these plasma cells when they repopulated the marrow (Figure 14.10d).

So far, we have examined five parameters. We can see all the correlations among these five parameters in our overlay plot (Figure 14.10d, bottom plot). As discussed earlier, we can also read off all the subpopulation percentages because the state frequencies are uniform. We can directly estimate that TdT + cells are around 10%, CD20− is 42%, and CD20$^+$ is 27%. As shown earlier, these percentages are accounting for measurement overlap. The widths of the bands shown in all the overlays are related to the population's confidence limits at each stage in the B-cell progression.

14.7.3 The B-Cell Scaffold

These five parameters are enough to define all the important stages of B-cell development. They form a kind of B-cell scaffold that other parameters can be compared against. Figure 14.11 shows how CD45, CD43, CD23, CD34, and FSC relate to this B-cell scaffold.

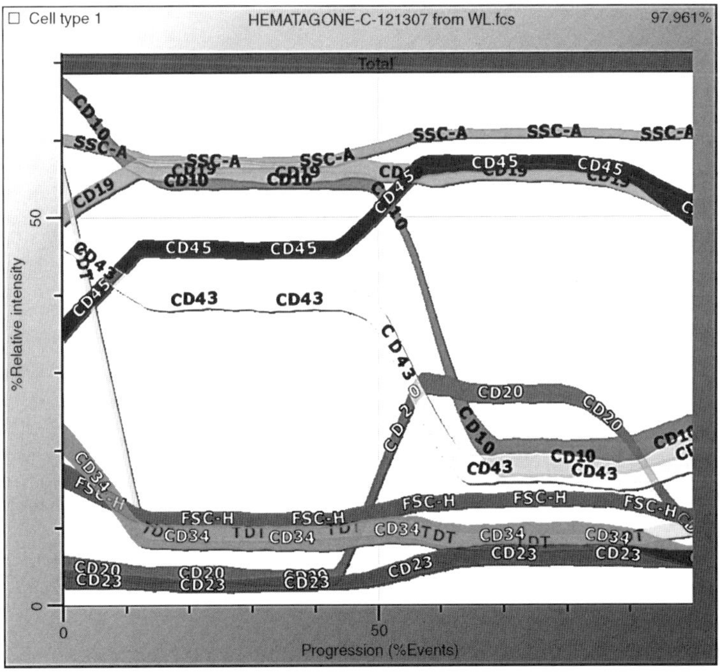

FIGURE 14.11 B-cell scaffold determines the profiles of CD45, CD43, CD23, CD34, and FSC. (See the color version of the figure in the Color Plates section.)

Even though we are not modeling these additional parameters, we can see how CD45, CD43, CD34, CD23, and FSC correlate with each other and the parameters that make up our B-cell scaffold. Once we have a good scaffold, we can use these markers as common parameters and find out how additional panels of markers relate to B-cell development. There really is no limitation to the number of parameters that can be examined in this manner.

14.8 DECREASING NEED FOR METAINFORMATION

Before discussing the importance of the lineage scaffold, let us come back to the question we asked a while ago. Do we need to know what these parameter profiles look like for every single parameter we measure? The answer is no. We need to be pretty sure of the first parameter, but as we go along, we need less and less *a priori* information to build our parameter profiles. As soon as we have defined the skeleton of the progression, we no longer need to know anything about the parameters. Again, it is like the crossword puzzle. There is a point when solving a crossword puzzle you know the answer before you look at the clue.

14.8.1 From Simple and Known to Complex and Unknown

The other important strategy in model building is that you begin with simple parameter profiles such as constant progressions and then slowly move to more complicated ones. For example, if we tried to find the optimal position of the three-level CD20 parameter profile before defining the two-level TdT profile, the computer algorithms would not have been able to position the CD20 control definition points into their optimal locations. The reason is because events are negative for CD20 both in the beginning of the B-cell progression and at the end. Since we have not provided the model information that can distinguish between the beginning and the end, it would undoubtedly find the wrong solution. But if we first did TdT or CD34, it would have no problem in finding the correct solution.

14.8.2 Importance of the Progression-Specific Scaffold

Now that we introduced the concept of a lineage or progression-specific scaffold or skeleton, let us continue the discussion on how important it is and will be in the future. In the B-cell example, once we had our scaffold determined, all the other markers (CD45, CD43, CD23, CD34, and FSC) had their profiles determined by the data and model and therefore were redundant. They may be redundant, but they are still very useful when investigating abnormal lineages. Tumors are typically identified by a disregulation of marker patterns. Therefore, having multiple redundant parameters can help us to distinguish normal patterns from abnormal ones. If these redundant parameters are added to the model, then the potential sensitivity for classifying abnormal from normal events is likely to be drastically improved.

14.8.3 Studying Perturbation Effects of Drugs

Progression-specific scaffolds will also undoubtedly play an important role in the pharmaceutical industry. Although there are no studies to date using probability state models to examine how drugs perturb specific aspects of *in vivo* or *in vitro* cell populations, there are some characteristics of this system that make it quite attractive. Since there are a fixed number of states in a probability state model, the number of cells necessary to achieve statistical significance does not increase geometrically with increasing number of parameters. Partitioning schemes using parameters as their boundaries will always have this unfortunate side effect. The other advantage is that any change in population percentage or expression pattern will perturb the state frequencies. The reduced chi-square (RCS) is typically used to measure this degree of nonuniformity and since RCS can be directly related to a probability value, significance can be estimated from a single sample as opposed to a set of replicates.

14.9 OPTIMAL PANEL DESIGN

In the future, panel design will evolve to carefully constructing tubes with common markers that defined one or more progression-specific scaffolds. These scaffolds will play an increasingly important role in diagnostic cytometry. These scaffolds will allow probability state models to assess all the normal populations and identify specific abnormal populations or abnormal shifts in populations.

14.9.1 Avoiding the Dimensionality Barrier

In order to get some sense of how different probability state modeling is from conventional cytometry, you should realize that it would take 45 two-parameter histograms ($10 \times 9/2$) convey the same amount of correlation as shown in Figure 14.11. Even if we only select those parameters necessary for the scaffold, a total of 10 histograms would be necessary ($5 \times 4/2$). Thus, the dimensionality barrier that has been limiting cytometry for the past few years is largely eliminated by these parametric plots. As soon as we decided to look at one measured parameter versus another, we started down this road that had a seemingly insurmountable barrier at the end.

14.10 THE MUSIC ANALOGY

We can use another analogy to better make this point. Imagine if musical scores were written with one musician's notes on one axis and another musician's notes on the other. Could the musicians play their notes with this type of representation? What happens when there are 10 musicians or 100? I began this chapter showing some simple parametric equations involving the common variable, t. These simple equations can create surprisingly complex figures when the data are graphed with one variable plotted against the other. The same situation is true for our cytometry data. The actual lineage lines for B cells can be quite puzzling when one tries to trace them in some of the two-parameter histograms. These same relations are normally very simple when presented as parameter profiles in a probability state model.

14.11 THE FUTURE OF CYTOMETRIC ANALYSIS

At the outset, I wanted to convince you that this method represents the future of cytometry analysis. We have discussed how it solves the dimensionality barrier and how it accounts for population overlap. Once a progression-specific scaffold is defined by a set of common parameters, any number of tubes can be analyzed and added to the model. There is no limit to the number of parameters that can be examined in this manner. The final output is a graph that merges intensity patterns, percentages, and population variances in an understandable format.

These advantages over conventional techniques are just the beginning of what is possible. These advantages will ultimately move cytometry from being gate-based to model-based, but the paradigm change is likely to be a slow process if these were the only advantages. The reason for the slow acceptance is that we have conventional systems in place that are working and there will be a reluctance to fix these if they are not perceived as broken.

14.11.1 Automation

What we have not discussed yet is the real power of the model-based analysis system – automation. We are living in a time where we are becoming more and more resource limited, especially in the area of medical diagnostics. There is going to be increasing pressure to develop low-cost, high-speed screening systems. These systems will need to be robust and accurate.

The modeling system described in this chapter allows us to engineer model systems that have attributes not attainable with gate-based systems. Each control definition point that defines the parameter profiles can have a defined relation with any other control definition point. We now have the means to create models that represent normalcy where abnormal events are either excluded or captured by additional cell types. These systems can be engineered to be completely automatic.

Initially, these automated systems will handle relatively simple tests, but over time, more and more complex models will become available through the Internet and will allow laboratories to achieve a higher degree of standardization than has been possible so far.

While these automated systems take hold, scientists will use the models as a means to better understand how biological systems work. The models will allow them to concretely define what they think is going on in the cell and test new assumptions. These models can be used to synthesize data sets and therefore be predictive. The art of cytometry will slowly evolve to the science of cytometry.

14.12 BACK TO PARAMETRIC EQUATIONS

We began this chapter with a set of parametric equations that defined a line in n-dimensional space. Most of us initially learned that a line had the following simple equation:

$$y = mx + b$$

A slope, m, multiplies the independent variable, x, and the intercept, b, is added to the product to form the dependent variable, y. A line represents a linear relationship between x and y. What we learned was valuable and we have all used this relation at some time.

What happens to this simple equation if we are interested in a line defined in three-dimensional space? If you do not have a background in mathematics, you might be tempted to think about the solution in terms of the three projections to two-dimensional space.

$$y = m_1 x + b_1$$
$$z = m_2 x + b_2$$
$$z = m_3 y + b_3$$

But it should be clear that as we increase the number of dimensions, the number of 2d projections will increase geometrically. What seemed simple and intuitive in two-dimensional space becomes a nightmare when we dramatically increase the number of dimensions. The parallel with cytometry's history is striking. Parametric equations have been used for over 200 years for a variety of applications, including reducing the complexity of high dimensional mathematical constructs.

I think it is time for cytometry to make the same leap.

REFERENCES

1. Protter MH, Morrey CB. *College Calculus with Analytic Geometry*, Loomis LH, editor, Addison-Wesley Series in Mathematics, Addison-Wesley, 1967, pp. 379–384.
2. Loken MR, Wells DA. Normal antigen expression in hematopoiesis. In: Stewart CC, Nicholson JK, editors, *Immunophenotyping*, Wiley-Liss, 2000, pp. 38–142.
3. Loken MR, Shah VO, Danttilio KL, Civin CI. Flow cytometric analysis of human bone marrow. II. Normal B lymphocyte development. *Blood* 1987;70:1316–1324.
4. Sanchez L, Thieffy D. A logical analysis of the Drosophila gap-gene system. *J. Theor. Biol.* 2001;211:115–141.
5. Mendenhall MD, Hodge AE. Regulation of Cdc28 cyclin-dependent protein kinase activity during the cell cycle of the yeast *Saccharomyces cerevisiae*. *Microbiol. Mol. Biol. Rev.* 1998;62:1191–1243.

6. Tyson JJ, Chen, KC, Novak, B. Sniffers, buzzers, toggles and blinkers: dynamics of regulatory and signaling pathways in the cell, *Curr. Opin. Cell Biol.* 2003;15:221–231.
7. Jacobberger J. Finite state machine and cellular life. Bowdoin Course Lecture, Bowdoin College, June 10–16 2006.
8. Bagwell CB. Theory and Application of DNA Histogram Analysis. PhD Thesis, University of Miami, 1979.
9. Bagwell CB.Probability State Models. Utility Application No. US11/897,148, 19 Sep., 2007.
10. Bevington PR. *Data Reduction and Error Analysis for the Physical Sciences*, McGraw-Hill, New York, 1969 p. 89.
11. Bevington PR. *Data Reduction and Error Analysis for the Physical Sciences*, McGraw-Hill, New York, 1969, p. 245.
12. Gentle JE. Transformations of uniform deviates: general methods. In *Random Number Generation and Monte Carlo Methods*, 2nd ed., Springer Science + Businesss Media, LLC, 2003, pp. 101–109.
13. ModFitLT, Verity Software House, Topsham, Maine, USA.

15

PHOSPHO FLOW CYTOMETRY: SINGLE-CELL SIGNALING NETWORKS IN NEXT-GENERATION DRUG DISCOVERY AND PATIENT STRATIFICATION

Peter O. Krutzik, Sean C. Bendall, Matthew B. Hale, Jonathan M. Irish, and Garry P. Nolan

15.1 INTRODUCTION

Flow cytometry is a unique platform for cellular analysis given its ability to measure the level of 10 or more antigens per cell, including cell surface markers, cell cycle proteins, calcium levels, DNA content, and intracellular cytokines, at the rate of up to 5000 cells per second. Recently, phosphoproteins have been added to the list of molecules that can be measured by flow cytometry with a method called phospho-specific flow cytometry or phospho flow (see Refs [1–5] for reviews). In phospho flow, antibodies that are specific only to the phosphorylated form of intracellular signaling proteins are used to measure the activities of various kinases and their target proteins, often transcription factors [6]. By simultaneously measuring phosphoprotein levels along with levels of other antigens, such as cell surface markers, phospho flow enables complex biochemical events to be measured in heterogeneous populations of cells such as peripheral blood and immunological tissues. This has led our group to apply phospho flow to all stages of drug discovery, from understanding disease mechanism and initial target identification, through high-throughput drug screening and pre-clinical testing, to patient stratification and diagnostics. A major benefit of phospho

Flow Cytometry in Drug Discovery and Development, Edited by Virginia Litwin and Philip Marder

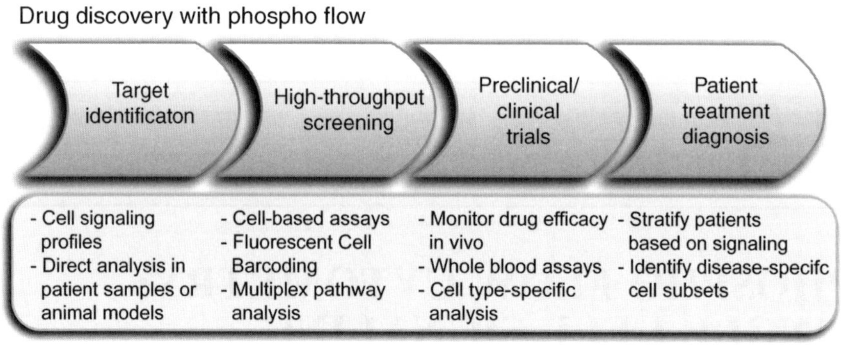

FIGURE 15.1 Phospho flow is an enabling platform in drug discovery. The phospho flow method is unique in that it can be used throughout the drug discovery process. At the target identification stage, phospho flow can be used to generate signaling profiles to characterize disease mechanism directly in patient samples or from animal models. For high-throughput screening, the fluorescent cell barcoding (FCB) method and the multiparameter capabilities of phospho flow provide novel cell-based assays to measure drug effects. At the preclinical and clinical trial stages, the platform enables measurement of drug efficacy directly in whole blood samples from patients. Finally, for treatment and diagnosis of patients, phospho flow can stratify patients based on signaling profiles for more accurate molecular/cellular classification of individual patient's disease states and prognosis.

flow is that the same platform can be used throughout drug discovery, such that the same assay that was used to initially identify a lead compound can be used in preclinical testing as well as monitoring patient response to therapy. This reduces the need to later develop secondary assays to monitor drug action or to determine efficacy *in vivo*, greatly streamlining the discovery process. In this chapter, we will explore all these areas and the unique capabilities of the flow cytometer in advancing our understanding of signaling networks at the single-cell level (Figure 15.1). In addition, we will discuss a novel hardware platform for flow cytometric analysis that is based on mass spectrometry analysis of elemental mass tags rather than fluorescence. This platform may push flow cytometry into next-generation diagnostics and applications, enabling measurement of 30–50 parameters per cell.

The chapter will be divided into sections that are summarized below.

Understanding Disease Mechanism and Target Identification Perhaps the most important aspects of effective drug discovery are determining the molecular causes of the disease of interest and finding appropriate drug targets. Phospho flow enables researchers to measure the complex changes that occur in cellular signaling networks directly from primary tissues, such as splenocytes from murine models of disease, or directly from patient blood samples. To characterize the disease, we probe the network by adding exogenous stimuli to the cells, forcing them to respond. In this way, changes such as negative regulation or mutations in signaling proteins can be observed and understood. These "signaling profiles" provide a more complete view of disease pathogenesis than does evaluation of the constitutive state of the network and a rich source of potential drug targets. In this chapter, we describe the application of phospho flow to analysis of samples from murine models of systemic lupus erythematosus (SLE) and from human patients.

High-Throughput Screening Flow cytometry is often used for in-depth analysis of a small number of samples, perhaps for secondary screening. Here we will discuss fluorescent cell barcoding (FCB), a technique that multiplexes samples to dramatically improve sample throughput and reduce reagent consumption. This technique, coupled with modern hardware for sample acquisition, makes it feasible to run thousands of samples per day, providing a platform for screening of focused libraries of tens of thousands of compounds.

Secondary Screening/Preclinical Testing Once lead compounds are identified, they must be tested in animal models. Here, the same assays that were used to first understand the disease and identify the target can again be used to determine if the compound is selective for the pathway and cell type of interest. In this way, no new assays must be developed to study the mechanism of drug action. In this chapter, we will discuss the search for inhibitors of cytokine-induced Janus kinase–signal transducer and activator of transcription (Jak–Stat) signaling pathways and the testing of these molecules *in vivo*. Importantly, analysis of multiple cell types allowed us to identify potent drugs that would otherwise have been overlooked because they affect only a small proportion of the total peripheral blood cells. In addition, we will show how a "druggability landscape" can be developed based on observations of how multiple compounds inhibit particular cellular subsets and signaling pathways.

Patient Stratification and Diagnostics Drug therapies can only be effective if a patient has the appropriate drug target or modifications to their intracellular signaling networks. Examples from cancer research have shown that certain therapies should only be used in patients who overexpress particular antigens. In this chapter, we will discuss patient stratification in acute myeloid leukemia (AML). By analysis of signaling networks of individual patients, it was possible to stratify patients into groups that were either responsive or nonresponsive to standard treatments. This type of analysis allows important diagnostic decisions to be made and treatments to be tailored to individual patients. The ability to monitor changes in multiple cellular subsets enables a detailed view of disease and allows early identification of patients who will not respond or who will relapse after treatment.

Next-Generation Flow Cytometry The multiparameter capabilities of the flow cytometer enable a unique view of drug action and disease analysis. Current cytometers, based on fluorescence, can measure up to 17 parameters per cell. However, in practice, making this many measurements is limited by the presence of fluorescence emission compensation, where signals from one fluorophore emit into multiple detector channels, convoluting analysis. Here we discuss a new platform, called mass cytometry, which is based not on fluorescently tagged antibodies but on metal isotope-tagged antibodies. Cells are introduced into a mass spectrometer and the quantity of each of the metal isotopes is measured on a per cell basis. The mass cytometer has two major advantages relative to standard fluorescence cytometry: There are 30–50 metal isotopes that can be used to tag antibodies and there is essentially no compensation required, enabling simple experimental setup and reducing the "art" behind high dimensional flow cytometry.

15.2 INSIGHT GAINED THROUGH MULTIPARAMETER MEASUREMENTS

Diseases affect many cell types and cellular processes. Although this might seem obvious, its value is often underestimated in drug discovery paradigms. Current models attempt to distill diseases down to a single drug target. Massive compound libraries are then screened against this one target to find lead compounds. Once a drug is identified, its effects on cellular components other than the initial target are often poorly defined due to the complexity of cellular biochemistry and the inability to measure drug action on the many cell types of the body.

The multiparameter nature of flow cytometry, and the unique capabilities of phospho flow to interrogate signaling networks, provides solutions to these difficulties. First, by measuring 10 or more parameters simultaneously, phospho flow enables us to analyze the phosphorylation levels of multiple signaling proteins to yield a clear picture of how the relevant signaling network is structured and how it may be perturbed. In the context of kinase drug discovery, phospho flow can be used to determine how selective a drug is for a particular pathway of interest. For instance, one may be interested in inhibiting signaling via PI3 kinase and Akt without affecting MAP kinase or Jak–Stat signaling. With phospho flow, major players from each of these signaling pathways can be measured simultaneously in the presence of the drug to determine how selective the compound is in the context of the cellular system. Nonspecific effects of potential drugs on metabolic state or ATP metabolism, for instance, can quickly be identified. In our experience, analyzing more signaling pathways yields fewer lead compounds that are selective for the signaling pathway of interest. Therefore, it is of critical importance to analyze drug action in the context of the cell and in the context of active signaling, not as isolated proteins devoid of cellular machinery.

Second, measurement of 10 or more parameters enables us to measure multiple cell surface proteins in addition to multiple components of the cell signaling network. This allows work to be done directly in patient samples, effectively describing the landscape of signaling abnormalities or changes that occur during disease. For instance, a blood sample from a cancer patient can be taken and signaling via MAP kinase and Akt pathways can be measured in B cells, T cells, monocytes, neutrophils, or known disease-specific cell subsets. These "signaling profiles" obtained after different extracellular ligands are added to the patient samples to determine how the signaling system responds can be used to identify potential drug targets and to stratify patients into unique groups. Importantly, such signaling profiles describe both the cell types that play a role in the disease and the signaling pathways that are affected. The former point is one that previously has been difficult to characterize because no method was available that allowed the different cell types of peripheral blood to be analyzed individually. Measurements were typically performed on the blood as a whole, lysing and mixing all the different cellular subsets together. Information regarding the complexity of the disease was lost and large changes in small cell subsets were hidden due to lack of change in predominant cell subsets. With phospho flow, disease complexity can be described, and harnessed, for more effective drug development.

15.3 UNDERSTANDING DISEASE MECHANISM: AUTOIMMUNITY AS CASE STUDY

Drug discovery has recently focused much attention on intracellular signaling in the pursuit of rational drug design. Even as cell line assays reach new heights, there is a growing realization that cell lines do not perfectly recapitulate the signaling biology present in human disease, a difficulty compounded by the limitations of existing pharmacodynamic assays. These factors contribute to the high rate of failure of drug candidates during clinical trials that had looked promising early in the development pipeline. But the difficulty transitioning from high-throughput cell line assays to working with heterogeneous primary tissues requires new tools and a new perspective.

15.3.1 Dynamic Signaling Profiles in Primary Samples

Flow cytometry is an ideal platform for resolving heterogeneous samples. The technology has been embraced by immunologists for its ability to resolve all the major populations of interest in a complex mixture of cells, even cell populations that make up less than 0.1% of the sample. Only recently, however, has the flow cytometer been recognized as a uniquely powerful proteomics platform for investigating signaling biology. Signaling networks in eukaryotic cells make extensive use of phosphorylation and dephosphorylation to control the activity of signaling kinases and other classes of molecules used in signaling cascades. From our studies and those of other research groups, we have learned that when cells undergo profound changes, such as differentiation, activation, or transformation, there are underlying alterations to intracellular signaling pathways that allow the cell to adapt its functional responses to environmental stimuli. The most important modifications do not appear to be in the basal activity of signaling networks, but rather involve changes in transduction of incoming signals from the extracellular environment. These changes are revealed when the cells are probed with extracellular stimuli such as cytokines and growth factors.

The ability to resolve heterogeneous populations is of critical value as disease samples often appear to respond normally when analyzed in bulk, for example, in whole blood lysates. This result is an artifact of averaging, because only a small subset of cells in the sample might be responding in a very aberrant manner. In our laboratory, a study of juvenile myelomonocytic leukemia (JMML) revealed that a rare population of cancer cells, otherwise masked by a preponderance of normal myeloid cells, responded to a growth factor with very aggressive signaling [7]. Similarly, in autoimmune disease, analysis of naïve T cells and B cells may give an incomplete picture of what is driving disease, whereas analysis of specific populations of effector lymphocytes and regulatory T cells will provide a mechanistic understanding of the disease.

15.3.2 Mechanistic Complexity of Signaling Changes

When analyzing changes in signaling responses to cytokines, a common question is whether the level of the cytokine receptor at the cell surface correlates with the magnitude of signaling observed. Although in some cases there is a measurable correlation, intracellular factors also play roles in modulating the signaling transduced from the cell surface. Strong cellular responses can come from cells that exhibit barely detectable levels of receptor on their surface and cells that have high levels of receptor expression often show muted or limited signaling responses. Therefore, it is of critical importance to evaluate the actual signaling induced by a particular cytokines and not to exclusively use receptor expression as a surrogate marker. Indeed, our investigations of systemic lupus erythematosus show that receptor expression levels do not directly correlate with the ability of cells to strongly or weakly transduce cytokine signals from their environment [8]. Starting early in disease progression, the Stat1 signaling response to IL-6 was greatly reduced in both $CD4^+$ and $CD8^+$ T cells [8]. While downregulation of the IL-6 receptor partially explained this reduced sensitivity to IL-6 in $CD4^+$ T cells, expression of the receptor was unchanged in the $CD8^+$ T cells and the Stat3 response to this cytokine remained robust in the $CD8^+$ subset. This differential activation of Stat1 and Stat3 was evident in individual cells. We found that the Stat1 arm of the signaling response was greatly inhibited by the upregulation of SOCS1 protein levels, a molecule known to inhibit the ability of Jak2 to phosphorylate Stat1. A similar pathway-specific inhibition of Stat1 relative to Stat3 was observed when T cells were stimulated with IL-21. As expected, the disease-associated inhibition of Jak2 activity was found to reduce the induction of a number of Stat1 target genes. Therefore, both receptor expression and intracellular factors played roles in modulating the IL-6 and IL-21 responses, yielding multiple potential drug targets.

In the MRL/*lpr* model of SLE, mice progress with great consistency from a relatively normal immune profile (at 5 weeks of age) through an early phase of autoimmunity characterized by relatively mild symptoms (at 10–12 weeks of age) on to more severe pathology, including lupus nephritis (at 15–20 weeks of age). We examined cytokine signal transduction through Jak–Stat signaling pathways at four different stages of lupus progression in these mice. At 5 weeks of age, we found that signaling was relatively normal across all cell types in the study and there was no significant difference from the control MRL strain. However, starting at 10 weeks of age, we observed robust changes in the ability of multiple immune cell types from the *lpr* mice to transduce cytokine signals, in particular an ablation of signal transduction through Stat1 for several cytokines across many of the cell types [8]. In contrast, the Stat6 response to IL-4 in B cells was completely preserved despite nearly complete loss in all T cell subsets. IL-4 signaling is required for high-affinity antibody responses, so this preservation of Stat6 signaling in B cells may contribute to pathogenic autoantibody production.

Interestingly, not all signaling pathways showed decreased activity in the mouse model. As an example, the Stat5 signaling response to IL-15 was higher in $CD4^+$ T cells from the *lpr* mice than from the control mice. Surprisingly, this pathway was downregulated in $CD8^+$ T cells from *lpr* mice and this response decreased further during disease progression. Stat3 signal transduction varied greatly, with some responses being greatly reduced as early as 10 weeks, while others such as the Stat3 response to IL-10 were preserved at all stages of SLE. These studies revealed the complexity of SLE disease mechanism in a murine model and showed how broad signaling profiling can yield a more complete picture of the immune signaling network in the context of disease. By more clearly defining the perturbed signaling network, we were able to identify many signaling pathways that potentially play causative roles in the disease and may be excellent drug targets.

15.3.3 Signaling Ratios

There are a number of cytokines that have been shown to have tolerogenic actions in one context and proinflammatory actions in another context. In addition, some cytokines act on many different cell types, potentially inducing very different functional responses in each (Figure 15.2). The contribution to the immune system as a whole that comes from each of these cell types when stimulated by the cytokine is determined, in part, by the different signaling responses activated by the cytokine. A striking example is IFNα, which in $CD4^+$ or $CD8^+$ T cells sends an apoptotic signal through Stat1, a weaker mitogenic signal through Stat3, and an even lesser signal through Stat5 [9]. Under most conditions, the mitogenic signal through Stat3 is overwhelmed by strong Stat1 activation. When the ratio of Stat3 activation to Stat1 activation reaches a high enough level, the cellular program induced by IFNα switches from death to proliferation. There are a number of intracellular mechanisms, including those acting through the SOCS family, that strongly and specifically inhibit different Jak isoforms upstream of Stat pathways and shift the signaling ratio of a number of cytokine receptors. An example of this shift in the signaling ratio was observed in the aforementioned Stat1 and Stat3 responses to IL-6 in the T cell populations of *lpr* mice.

15.3.4 Analysis of Human SLE Patient Samples

Given the findings with the mouse model, we hypothesized that similar changes in signaling ratios would be proximal to an induction mechanism responsible for flare in human patients. To clarify, we were focusing not on signal transduction regulation that correlated with static symptoms, but rather on signaling phenotypes that indicated

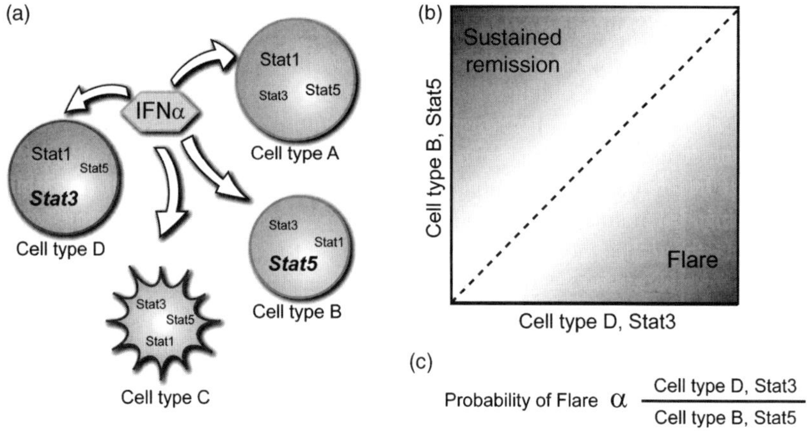

FIGURE 15.2 Understanding immune system network function during SLE disease progression. (a) IFNα acts on multiple cell types by activating Stat1, Stat3, and Stat5 in multiple cell types. The relative activation of each of these signaling molecules, represented as the different font size of the Stat labels, varies between cell types. The functional response induced by the cytokine in a particular cell type is determined in part by the relative activation of the Stats. In turn, the influence of a cytokine on overall immune system function is determined by the relative activation of the Stats in two or more cell types. (b) For each SLE patient sample, the activation of Stat5 in cell type B in response to IFNα was plotted against the activation of Stat3 in cell type D. Most patients that entered flare within 90 days of the blood drawn fell into the lower right quadrant. The probability of a patient entering flare within 90 days can be expressed by the equation shown in (c), which is a ratio of the Stat3 phosphorylation induced in cell type D divided by the Stat5 phosphorylation induced in cell type B. Although the reciprocal of this expression does approximate the probability of staying in remission, other signaling pathways enable more accurate prediction of sustained remission.

increased immune activity and upcoming flare episodes. Human SLE is a cyclic disease for many patients, characterized by long periods of stable, low disease activity or even complete remission, punctuated by episodes of increased pathology. When an increase in symptoms is observed, quantified by an increase in the score assigned by assessment systems such as the SLE Disease Activity Index (SLEDAI), the patient is said to be experiencing a flare. Increases in the SLEDAI score, corresponding to new and/or worsening symptoms, are the product of active autoimmune processes that may require days or weeks to become manifest and weeks or months to completely subside once these processes abate. Episodes of flare are thought to be initiated by an event that induces increased autoimmune activity, which in a matter of time yields an increase in symptoms. The initiating event for most flares is unknown, but we can refer to the immediately upstream induction event of autoimmune activity as the induction mechanism. The specifics of this induction mechanism are not well known, but the resulting increase in autoimmune activity is thought to precede any increase in symptoms (Figure 15.3a). Our experience with the mouse model suggested that changes in cytokine signal transduction would accompany any shift in immune system function and predict flare weeks in advance of worsening symptoms.

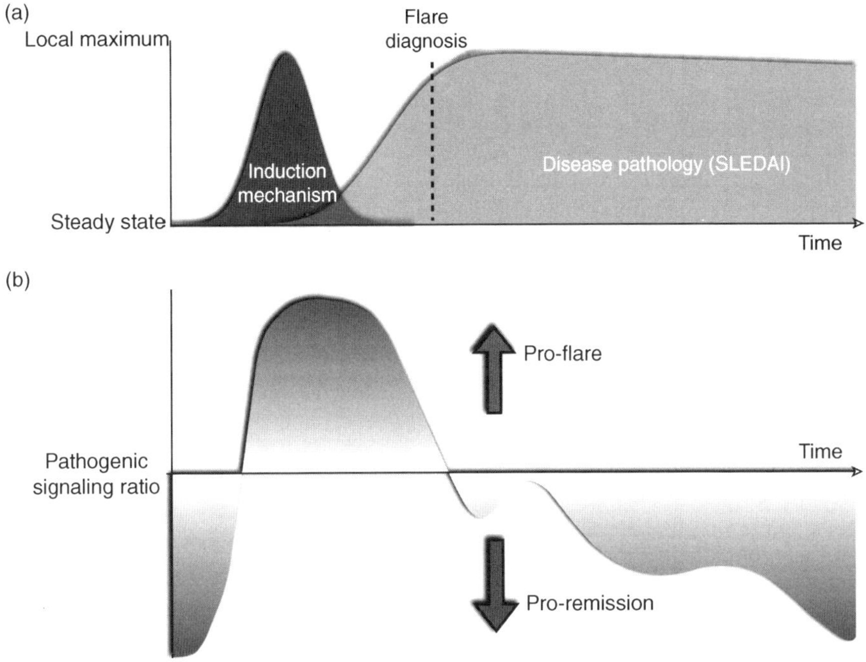

FIGURE 15.3 Prediction of SLE disease flare based on phospho flow analysis of critical signaling pathways in patient blood samples. (a) Flare is defined as a significant increase in SLE symptoms that corresponds to a numerically significant rise in SLEDAI score. This rise is due to a higher level of autoimmune activity that is initiated by an induction mechanism that may precede the observed change in symptoms by days or even weeks. A reduction or plateau in pathology may indicate that no further damage is occurring and that the autoimmune activity responsible for the flare has concluded or been reduced in magnitude. (b) The pathogenic signaling ratio (PSR) is a numerical value comparing the ratio of two or more signaling pathways in specific cell types in SLE patients. From our studies in SLE patient samples, PSR values below 1 correlated with remission or static disease. PSR values significantly above 1 strongly predicted flare within 90 days, but did not correlate well with static scores of disease pathology or past symptoms. PSR changes are temporally concordant with the induction mechanism and may be one of the earliest detectable changes in immune system function preceding flare.

To test this hypothesis, we had to be able to calculate the rate of change of patient symptoms through time. To have the appropriate data set, we scheduled multiple patient visits over the course of a year, scoring symptoms at each clinical visit with both SLEDAI and Physician's Global Assessment (PGA). Patient enrollment in three successive cohorts was randomized to make the results broadly representative. Blood samples were treated with a set of cytokines and multiple cell types were analyzed for their phosphoprotein response profiles.

The resulting signaling profile consisted of hundreds of signaling nodes, which represent the signaling response through a specific signaling molecule, in a particular cell type, in response to a specific cytokine. For example, one signaling node was the Stat1 response in B cells to stimulation with IFNα. We found that single signaling nodes in isolation were imperfect predictors of flare. However, when we used an algorithm that looked at each pair of nodes in the data and determined whether they accurately distinguished patients who would experience a flare within 90 days, we found several pairs of nodes that appeared highly accurate at predicting flare. Importantly, the classifiers most predictive of flare paired a decrease in a pro-tolerance response in one leukocyte subset with a pro-inflammatory or pro-proliferative response in another leukocyte population. The relative activation of the nodes making up each two-dimensional classifier can be quantified as a signaling ratio. For many of the classifiers, patients who stayed in remission had a positive signaling ratio such that the protolerance signaling response balanced the proinflammatory signaling response. The patients who went on to experience flare within 90 days had a signaling ratio significantly below a value of 1. Inverted signaling ratios that predicted flare were termed pathogenic signaling ratios (Figure 15.3b). The observed balance of cell type-specific signaling highlights the importance of designing drugs, such as kinase inhibitors, that are not only pathway specific but also act on the appropriate cell type(s).

15.4 HIGH-THROUGHPUT SCREENING

Flow cytometry is not often associated with high-throughput screening of large (>100,000 molecule) compound libraries. Instead, along with microscopy platforms, flow cytometry is typically used for "high-content" screening applications. In high-content screening, many parameters are measured simultaneously to obtain a richer understanding of a drug candidate's effects on the cell as a whole, not just on the particular target molecule. However, recent advances in both assays and instrumentation are enabling the flow cytometer to simultaneously be used as a high-throughput and a high-content platform. To improve throughput, we recently developed a technique called fluorescent cell barcoding or FCB, which enables one to combine many different samples together into one tube prior to adding antibodies or other measurement reagents [10]. This dramatically reduces reagent consumption and improves data robustness. On the hardware side, an autosampler has been developed that enables data acquisition from 384-well plates in approximately 15 min. The combination of these two advances will push flow cytometry-based high-content, multiparameter analyses to staggering throughput rates.

15.4.1 Fluorescent Cell Barcoding

One of the major limitations of many high-content analysis platforms is the cost associated with the reagents that are used to make the measurements. While the typical high-throughput analysis uses fluorescent dyes or genetically encoded reporter constructs that are extremely inexpensive on a per well basis, high-content assays typically use antibodies for staining of cellular antigens. Antibody costs can be overwhelming, particularly when using proprietary reagents developed by a particular vendor. Even with reductions in sample volume and optimization of reagent amounts, antibodies can cost on the order of $1 per well. When we first set out to perform flow cytometry experiments on hundreds or thousands of samples, we quickly ran into the limitations of cost. FCB minimizes reagent consumption, enabling larger screening efforts, disease signaling profiling experiments, and patient stratification methods discussed elsewhere in this chapter.

15.4.1.1 The FCB Technique The FCB technique is simple in practice [10]. Different samples are labeled with different amounts of various fluorescent dyes, washed, and then mixed together before staining with antibodies or other reagents. The combined sample is then run on the flow cytometer (Figure 15.4). For instance, four wells in a 96-well plate can be labeled with none, low, medium, or high amounts

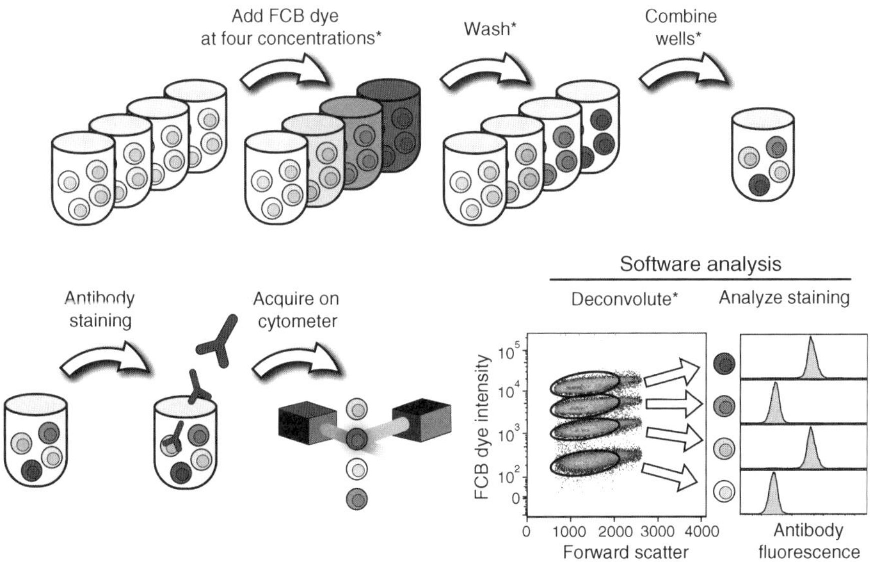

FIGURE 15.4 The fluorescent cell barcoding (FCB) method. Shown is an example where four different samples are barcoded. In the first step, the FCB dye is added at a different concentration to each of the four samples: none, low, medium, or high. After the cells are covalently labeled with the dye, they are washed to eliminate free dye. At this point, the cells are irreversibly labeled with dye, and can be combined together into one tube or well. Once combined, the sample is treated like an ordinary sample and stained with antibodies before acquiring on the flow cytometer. During software analysis, the four original samples are deconvoluted, or separated, based on the intensity of fluorescence in the FCB dye channel. Here, the four populations are gated and then analyzed for the fluorescence intensity of the antibody staining. Expansion of the number of FCB dyes and/or dye levels enables entire 96-well plates to be combined into one sample for staining and acquisition on the cytometer, dramatically reducing antibody consumption and increasing acquisition rates. Steps with an asterisk (*) indicate steps unique to the FCB method relative to a standard flow cytometry experiment.

of an amine-reactive fluorescent dye. The cells are labeled for a short period of time and then washed. The dye is covalently bound to the cells, and the cells that received the most dye have the highest fluorescent signal, whereas those that received less dye have less fluorescent signal. These four wells are then combined into one well and stained together. When run on the flow cytometer, one can deconvolute, or separate, the combined sample back into the constituent wells based on their fluorescent signal in the barcoding channel.

Of course, because the flow cytometer has many parameters available (many of which are not used in every experiment on modern instruments with 10–15 parameters), more than one fluorescent dye may be used for barcoding purposes. The dyes can be added in different ratios to create unique fluorescent signatures for each sample/well; combining two dyes at four concentrations yields 16 (4×4) unique signatures. Combining three colors yields 64 ($4 \times 4 \times 4$) signatures, and so on. We have been able to combine an entire 96-well plate into 1 tube by using a $4 \times 4 \times 6$ matrix of dyes [10]. It is possible to barcode up to 10 different samples using a single fluorescent dye; however, encoding 4–6 channels is most robust.

FCB can be used in nearly all flow cytometry cell-based assays. Cells can be barcoded when they are viable or once they have been fixed and/or permeabilized. Higher signal intensities are obtained, and therefore lower amounts of dye are needed, in permeabilized cells, but the method works in live cells as well.

15.4.1.2 Benefits of Applying FCB to Flow Cytometry

Reduced Antibody Consumption As mentioned above, a major driving force for developing FCB was to enable us to perform larger screens using minimal antibody reagent. In the example of a 96-well plate being combined into 1-well using FCB, it is possible to reduce antibody consumption nearly 100-fold. That is, instead of staining 96 individual samples, only one sample is stained. Obviously, if the barcoding was expanded to four dyes, one could multiplex an entire 384-well plate. This could reduce the cost per well down to a few pennies, making high-content antibody-based screens viable on 10,000 or more compounds.

It should be noted, however, that FCB cannot reduce antibody consumption in all cases. The method relies on the fact that most experiments require only 1000–100,000 cells per sample to be acquired on the flow cytometer. When 10–100 samples are combined, one can still stain a reasonable cell number (about 10,000–10 million cells) in a single well. However, when studying extremely rare cell types, it may be necessary to acquire data on two million cells. In this case, because the mass of antibody may be limiting if more than 10 million cells are stained in one tube (and require more antibody to be added), the benefits of FCB are slightly reduced. Even in this "worst case" scenario, reagent consumption is cut two to five-fold. Also, these types of analyses are typically limited to basic research applications and not drug screening efforts.

Improved Assay Performance When designing assays for cell-based analysis, there are many experimental components that can lead to variability in data, including temperature fluctuations, cell density, timing of reagent addition, and for antibody-based assays, staining volume, time, and temperature. In addition, when acquiring samples on the flow cytometer or any analytical instrument for that matter, changes in detection sensitivity, optics, or flow rates may affect results. With the FCB method, many of these variables are minimized. By staining many samples together, one eliminates staining variability. In addition, because positive and negative controls can be stained with drug-treated samples, complex normalization procedures are not needed. Each barcoded sample can be normalized to itself. Another advantage is that data on many samples are acquired simultaneously on the instrument. Therefore, if there is a mechanical error, all the samples in that barcoded sample will be affected equally. If the samples had been run in sequence, only some wells would have been affected, leading to a large drift in data. Thus, FCB eliminates variability on the instrument acquisition end, leaving the major sources of variability at the front end of the experiment, with cell preparation, distribution, drug treatment, and reagent addition. In our studies, FCB assays are superior to singly stained assays with Z' values of 0.8 or greater in most cases [11].

Increased Acquisition Speed In typical cytometers, the largest bottleneck in acquisition rate is the switching of samples from one tube to the next. Because FCB combines multiple samples into one, no switching of samples is required. Even without plate loading hardware, one can easily run the equivalent of a 96-well plate in less than 5 min using FCB, because the acquisition is from one continuous tube. Therefore, on any standard cytometer setup, a researcher can load and acquire 10 or more 96-well plates worth of samples in 1 h, dramatically improving throughput.

15.4.2 High-Throughput Autosamplers

In addition to the FCB assay, recent advances in hardware have made it possible to analyze a 384-well plate in less than 15 min. Therefore, screening 10,000 wells in one day is entirely feasible, eliminating sample throughput as a major concern in flow cytometry. This platform, called plug flow cytometry, uses a HPLC-style autosampler and peristaltic pump to serially sample from the assay plate (for review, see Refs [12–14]). Each sample is separated by a small air gap, or plug, so that the samples do not mix as they travel up the autosampler tubing and into the cytometer. This was a counterintuitive innovation because most researchers are taught to never introduce air bubbles into the flow cell of the flow cytometer. However, because most of the fluid entering the flow cell at any given time is sheath fluid, the small air bubbles that are introduced in the sample line do not cause any major perturbations. The plug flow cytometry platform has been used extensively for drug screening as part of the NIH Roadmap initiative. Millions of wells have been screened in assays on beads, in yeast cells, and mammalian cells. These include receptor binding assays, enzyme modifications, proteases, GTPases, and GPCRs [15]. This work is pioneering in its application of flow cytometry to high-throughput, high-content screening.

15.4.3 Combining FCB and High-Throughput Autosamplers

Moving forward, we anticipate that FCB will be used together with high-throughput autosamplers, such as the plug flow platform, to enable ultrahigh throughput screening by flow cytometry. Combining a relatively "simple" FCB matrix that encodes 9 samples (a 3×3 matrix) with a high-throughput autosampler system would allow nearly 100,000 compounds to be screened in a day. At this point, the bottleneck would likely rest not on sample acquisition on the cytometer, but rather with drug treatment, sample preparation, and staining. Therefore, flow cytometers can clearly manage the sample throughput necessary to screen large compound libraries. What is perhaps more important is development of cell-based assays that are robust at large scales and minimization of antibody costs. The combination of assay optimization and hardware configurations makes it feasible to perform advanced cell-based assays on a completely new scale.

15.5 SECONDARY SCREENING AND PRECLINICAL TESTING

Once lead compounds have been identified via high-throughput screens such as *in vitro* kinase assays, translocation, or gene reporter assays, flow cytometry provides an excellent platform for more fully elucidating compound selectivity and toxicity, as well as monitoring compound effects *in vivo* in preclinical animal models. In particular, phospho flow is perfectly suited to drug discovery targeting signaling proteins such as kinases, phosphatases, and transcription factors [2, 16]. The multiparameter capabilities of phospho flow enable the researcher to evaluate multiple components of the same signaling pathway or key members of multiple different signaling pathways to characterize selectivity in cell-based systems. Then, once a suitably selective compound has been found, phospho flow provides a platform for analyzing compound efficacy directly in peripheral blood from preclinical models of disease. This ability to use the same assay platform through multiple steps of the discovery process reduces the time needed to validate compounds and eliminates the requirement for completely distinct metrics to monitor drug efficacy *in vivo*. In this section, we will discuss the use of phospho flow in kinase drug screening and how valuable insight is gained by analyzing drugs in the context of a signaling network, not individual proteins.

15.5.1 Cell-Based Screening

Our laboratory has a strong interest in understanding how the immune system functions in the context of inflammation, bacterial challenge, antigen challenge, and autoimmune diseases such as rheumatoid arthritis and SLE [7, 8, 17–19]. In multiple disease states, such as cancer and SLE, we have found that modulation of cytokine signaling plays a central role in the disease and correlates with disease severity and patient outcome.

Cytokines, small proteins that bind to specific cell-surface receptors, play a central role in nearly all immunological system challenges as the major means of communication between cells. When cytokines bind to their cognate receptors on the surface of a particular cell, they induce activation of Jaks, which in turn phosphorylate Stats. The Stats act as transcription factors and lead to transcription of target genes necessary for cell differentiation, maturation, or activation. In signaling profiling studies, as well as in drug screening efforts, we measure the phosphorylation state of the Stat proteins as readouts of cytokine activity. Because the Stat proteins are relatively abundant transcription factors, signal-to-noise and Z' characteristics of phospho flow assays for Stat phosphorylation are excellent.

15.5.2 Analyzing Multiple Pathways Improves Selectivity

In our efforts to identify inhibitors of Jak–Stat signaling, we began by testing a natural product compound library in a myeloid cell line [16]. In these cell line assays, cells were treated with compounds, incubated, and then activated by adding three different stimuli to the cells simultaneously. The stimuli were chosen to induce three different signaling pathways: IFNγ to induce Stat1, GM-CSF to induce Stat5, and anisomycin to induce p38 MAP kinase. The goal of the screen was to find compounds that modulated Stat1 and/or Stat5 signaling, but not p38 (i.e., we sought compounds selective for Jak–Stat signaling relative to MAP kinase signaling pathways).

What was striking from this initial screen was the importance of measuring more than one signaling pathway. If only one pathway had been examined, rather than three at once in individual cells, we would have identified several potential lead compounds that inhibited each of the three pathways. Yet, as we looked at the effects of these inhibitors on the other pathways measured, the number of selective hits dwindled. For instance, although eight compounds inhibited GM-CSF-induced Stat5 phosphorylation, only one of these compounds was selective for this pathway. Six compounds inhibited IFNγ-induced Stat1 phosphorylation, but none were specific to only this pathway [16]. Therefore, by examining only three pathways simultaneously, we were able to rapidly reduce the number of false leads and focus on compounds with desired selectivity profiles. One can imagine that if the number of pathways examined were expanded to 10 or 20, even fewer selective compounds would have been identified, but each lead compound would have a good chance of success in preclinical and toxicology testing.

These results highlight the difficulty of modulating signaling in the context of the multitude of processes constantly occurring during cellular biochemistry. Therefore, although inhibitors can be identified via assays with purified proteins, the complexity of cellular metabolism cannot be appreciated until the compound is added directly to cells. For example, many compounds bind nonselectively to critical enzymes in ATP metabolism or glycolysis leading to toxicity. Clearly, the ideal situation is to find compounds that are selective in *in vitro* assays and then to test these using phospho flow to find compounds that are not only effective in the cellular context but also show selectivity toward the pathway of interest.

15.5.3 Which Cell Type Is Inhibited?

One question that is not often asked during drug discovery is "Does the lead compound target the correct cell type?" One reason that this question is not often asked is that it is difficult to measure the effects of a compound on a particular cell type when that cell type is mixed with many others. One of the best attributes of the flow cytometry platform is that it can be used to identify cellular subsets present in patient samples like peripheral blood (e.g., T cells, B cells, monocytes, and neutrophils). Well-characterized cellular markers are available to identify these cell subsets and by coupling these markers to phospho-specific antibodies, we are able to measure intracellular signaling pathways directly in complex tissue samples such as the murine spleen, lymph nodes, and peripheral blood [20, 21]. This enables us to answer the question of whether a compound inhibits the target pathway in the target cell type when the cell type is a part of a mixture of cell types.

A compound displays one of four general patterns of selectivity when analyzed in a mixture of cells: (1) It inhibits all signaling in all cell types. (2) It inhibits one signaling pathway, but in all cell types. (3) It inhibits all signaling in one cell type. (4) It inhibits one signaling pathway in one cell type (Figure 15.5). The last option provides the ideal drug candidate, as it would presumably have minimal side effects and optimal efficacy.

However, as we began to examine the effects of some of the lead compounds from our screen of natural products, as well as commercially available Jak–Stat inhibitors, we quickly discovered that inhibiting one pathway in one cell type was not common. What we found were compounds that displayed strong patterns of selectivity for certain signaling pathways, and one compound that showed potent selectivity toward signaling in only one cell type (Figure 15.5). However, none of the compounds tested showed selectivity for only one pathway in one cell type. Indeed, it has proven difficult to incorporate moieties into drug leads that guide the molecule to a particular cell type. Therefore, hybrid drugs, such as antibody-targeted therapeutics, might be the best approach to achieving the ideal one target, one cell type drug. Importantly, phospho flow provides a unique platform that is capable of quantitatively measuring the inhibitory effects of molecules directly in complex, heterogeneous populations of cells.

15.5.4 Painting a Druggable Landscape

Perhaps more important than understanding any one drug in our screening experiments was the knowledge that we gained about the immune cell signaling network when we clustered the inhibition profiles of all the compounds simultaneously. We were able to identify signaling pathways and cell types that were inhibited at low doses by all the compounds. These patterns of pathway and cell type "druggability" were apparent even with the small number of compounds that we tested. We plan to expand our screening efforts to more fully elucidate this druggable landscape. In our primary cell screens, we performed dose–response experiments with each compound on 24 signaling pathways in 5 cell types [16]. Of the 120 cell type–pathway combinations possible, 27 showed signaling changes greater than twofold upon induction. For these 27 cell type–pathway combinations, we obtained IC_{50} values and were able to

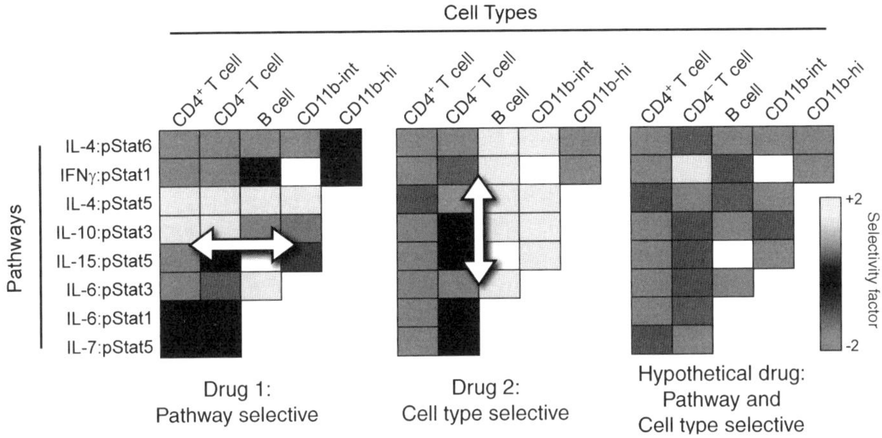

FIGURE 15.5 Phospho flow enables identification of both pathway and cell type selective drugs. Shown are the selectivity fingerprints of two different compounds when analyzed on five cell types and eight signaling pathways in primary immune cell populations. The selectivity factor is a measure of the relative inhibitory effect of a compound on the different pathways or cell types, with yellow indicating potent inhibition (low IC_{50} values) and cyan indicating weak inhibition (high IC_{50} values). Drug 1 (left) displays a pathway selective profile (horizontal pattern), inhibiting IL-4:pStat5 and IL-10:pStat3 strongly, while having relatively less effect on IL-4:pStat6, IL-6:pStat1, and IL-7:pStat5. Drug 2 (center) displays a cell type selective profile (vertical pattern), inhibiting B cells and CD11b-int cells potently across all the signaling pathways measured. Also shown is a hypothetical selectivity fingerprint (right) of an ideal compound that is able to inhibit one pathway, IFNγ:pStat1, in only one cell type, $CD4^-$ T cells. Such a compound could target an aberrant signaling pathway in a cell subset that drives a disease without having negative side effects in other cell populations. Phospho flow is a unique platform for identifying these types of compounds in complex primary cell populations. (See the color version of the figure in the Color Plates section.)

determine a selectivity factor. The selectivity factor is a metric that describes whether a particular signaling pathway or cell type is inhibited more or less easily than the other pathways evaluated for any given drug. Therefore, the selectivity factor is a normalized score for each drug allowing us to compare patterns of inhibition rather than potency alone.

By clustering the 27 cell type–pathway combinations according to their selectivity factors across the 8 drugs tested, we observed two clear trends of druggability (Figure 15.6). First, some signaling pathways were more easily inhibited than others. For instance, although IL-4 induced both Stat5 and Stat6 phosphorylation, across the eight drugs we tested, the Stat5 pathway was inhibited at 10-fold lower doses than the Stat6 pathway. Therefore, even though both Stat5 and Stat6 are induced by the same upstream receptor, the Stat6 pathway is more "strongly" induced and is more resilient to drug inhibition. Stat5, on the other hand, is "weakly" induced and is more druggable. In this particular case, the magnitude of the signaling induced through these two proteins is different and may explain why Stat5 is more easily inhibited.

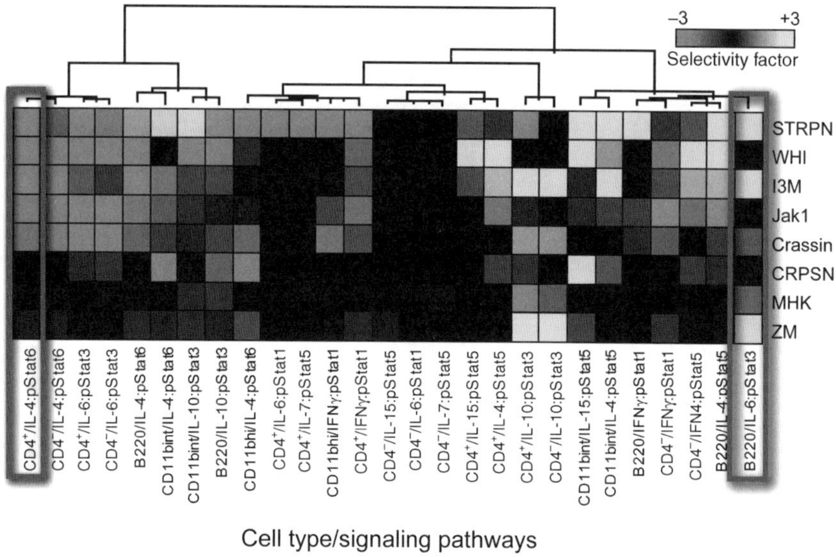

FIGURE 15.6 Characterization of druggable signaling pathways and cell types. After measuring the inhibitory effects of 8 compounds on 27 different cell type/signaling pathway combinations, the resulting selectivity profiles were clustered to determine overall patterns of selectivity or druggability. Cell type/pathway combinations that were inhibited at lower drug concentrations (low IC_{50}) have high selectivity factors (yellow) and cluster together at the right side of the dendrogram. As an example, the IL-6:pStat3 pathway in B cells was inhibited strongly by the majority of the compounds tested and, therefore, is a druggable cell type/pathway. Conversely, cell type/pathway combinations that were inhibited at high doses of drugs (high IC_{50}) have low selectivity factors (cyan) and cluster to the left side of the dendrogram. The IL-4:pStat6 pathway in $CD4^+$ T cells was particularly resistant to drug inhibition, making it a poorly druggable cell type/pathway combination. Understanding of druggability of both cell types and signaling pathways will enable more efficient choice of drug targets and ultimately more successful drug discovery campaigns. (See the color version of the figure in the Color Plates section.)

However, we did not observe a clear trend between the magnitude of signaling induced and the druggability [16]. A larger panel of drugs must be tested to ensure that these observations hold true; it may still turn out that signaling pathways that are not fully activated by a particular ligand are more susceptible to inhibition.

Second, we observed that some cell types are more easily inhibited than others. For instance, although IL-6 induced Stat3 phosphorylation at similar magnitudes in B cells and $CD4^+$ and $CD8^+$ T cells, the B cells were consistently inhibited at 10-fold lower doses than the two T cell populations [16]. The B cells appear to be more "druggable" for IL-6 signaling, whereas T cells are more resilient. We believe that these insights into how readily particular cell types are inhibited will greatly improve drug discovery by guiding selection of potential drug targets in cell types that appear amenable to treatment. For instance, research may show that a particular signaling pathway is an ideal drug target for the treatment of autoimmune disease. However, if that pathway is difficult to inhibit (i.e., not very druggable), then the possibility of finding a compound that specifically inhibits the target pathway without inhibiting other pathways is low. Similarly, a certain cell subset might be found to act aberrantly in a disease state. However, targeting that cell type might also invariably lead to adverse side effects in other cell types that are more druggable. Therefore, although target identification is perhaps the most important part of drug discovery, prudent selection of targets requires prior knowledge of the landscape of druggable signaling pathways and cell types, so that the chances of successfully identifying an efficacious drug are maximized.

15.5.5 Needle in a Haystack: Drug Discovery in Rare Cell Types

One of the inherent challenges of drug discovery is attempting to target a cell type that is rare in the body, such as a hematopoetic stem cell, a cancer stem cell, or an antigen-specific T cell. These types of cells comprise 0.1% or less of total circulating blood cells. However, these might be important drug targets, particularly with the potential of stem cells to ameliorate many disparate diseases. These cell types are difficult to analyze due to their rarity and a lack of *in vitro* model systems of their behavior. Here again, the phospho flow platform provides a method for measuring the effects of compounds directly on these cell types in the mixture of primary cell types in the blood.

As an example of this capability, one of the natural product molecules we identified in initial screens, streptonigrin, proved to have potent inhibitory effects on B cells relative to T cells and neutrophils when we analyzed cell signaling in murine splenocytes [16]. In the murine spleen, B cells comprise approximately 60% of the total white blood cells. Therefore, even if we had measured the entire leukocyte population as a whole, we would have identified streptonigrin as an inhibitor of Jak–Stat signaling. However, what was more illustrative was when we treated mice with streptonigrin and measured its inhibitory effects on cytokine signaling directly in the blood. In the blood, B cells comprise only 5% of the total population. Therefore, when we analyzed the entire white blood cell population present in the blood, streptonigrin did not have an appreciable effect on signaling. However, by using surface markers to identify the various cell subsets, we observed a clear and potent inhibition of signaling in B cells, but not in any of the other cell subsets.

This example shows how the proportion of cell types in the blood can dramatically affect the apparent activity of a drug. If the drug has selectivity for one cell type relative to another, it will appear to be more potent/effective if it inhibits a more prevalent cell type. If it inhibits a rare cell type, it will appear to have no effect on the general population. This can lead to misleading conclusions about drug action and toxicity.

Because many diseases, especially immunological diseases, are restricted to particular cell types, we believe that drug discovery must take into account the effects of compounds on one cell type relative to others. It cannot be assumed that because a lead compound inhibits a kinase in enzymatic assays and appears to function in a cell line model, it will equally affect all cell subsets in the blood. In particular, different cell subsets have different rates of endocytosis, different life spans, and different rates of differentiation. This makes it imperative to measure drug action directly in the blood, where the drug is ultimately acting. Phospho flow is one of the few platforms that allows this type of analysis.

15.6 PATIENT STRATIFICATION AND DIAGNOSTICS

Assuming that an effective drug has been identified, flow cytometry holds significant promise for the growing field of personalized medicine. Appropriate protein expression and signal transduction are essential in the development of healthy immune cells and remodeling of cell communication networks drives proliferation of diseased cells in cancer, autoimmunity, and many other human diseases. A network-level single-cell view of signaling in healthy and diseased cells is needed to identify shared features of malignant signaling networks and identify new therapies that target signaling.

15.6.1 Signaling Profiles of Human Cancer

As mentioned previously, one advantage of flow cytometry is that primary samples can be studied without the need to physically isolate the cell types of interest from patient samples. This technique for dissecting signaling in heterogeneous cell populations within a primary tumor was demonstrated in studies where single-cell signaling profiles were used to map acute myeloid leukemia cell signaling and identify patients with a high risk of not responding to chemotherapy [22]. By seeking disease-specific signaling features and then comparing them to clinical outcome, a profile of signaling that stratified patient risk was obtained (Figure 15.7). As the notion of "cancer stem cells" has moved into the mindset of molecular oncology, there are some tantalizing hints that the signaling profile of the cell subset that appears to contain the "killer" cells for the AML patients has stem cell-like attributes.Since the original analysis of AML, single-cell signaling profiles have been used to determine the effect of oncogene mutations, such as G12D in murine *K-Ras* and Y591 duplication in human *FLT-3*, on cell signaling networks and to diagnose JMML [7, 23, 24]. The JMML study provided a flow cytometry signaling profile that can be obtained in an afternoon (instead of the standard 3 week colony formation assay) for JMML diagnosis. When a profile of cell signaling is closely associated with patient outcome, it suggests that the measured signaling events might be responsible for that outcome [4]. If so, targeting signaling events associated with poor clinical outcomes might improve patient survival.

15.6.2 Perturbing Signaling Networks for Research and Therapy

An important aspect of the signaling profile approach is the value of evoking signaling in the signal transduction network (Figure 15.8). By perturbing cells, through

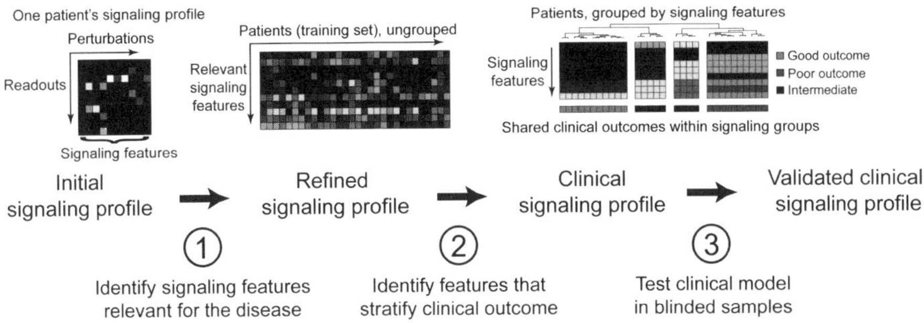

FIGURE 15.7 Creating a validated clinical signaling profile. An initial signaling profile is obtained for each patient by measuring numerous phosphoprotein signaling events following perturbations (stimuli and/or inhibitors) that evoke responses throughout the signaling network. The relevance of signaling responses to the disease is judged by looking for variance in the disease that is greater than that observed among healthy individuals. Here, a signaling feature is the fold induction of a phosphoprotein under one of the stimulation conditions or the basal level of the phosphoprotein. Next, features that stratify clinical outcome are identified. This can be done in an unsupervised manner, where all relevant parameters are used, or in a supervised manner, where features are selected based on their ability to stratify a given clinical outcome. Finally, a simplified model using the signaling features that best stratify outcome is finalized and then tested in a new set of patients where researchers are blinded to clinical outcome. In the end, this approach produces a validated clinical signaling profile. (See the color version of the figure in the Color Plates section.)

stimulation and/or inhibition, researchers vastly expand the capability to understand and map the cell signaling network. For example, loss of signaling pathways that control cell death, cell cycle checkpoints, and immune presentation is common in cancer and can be a critical component of signaling pathways that stratify clinical outcome [4]. By stimulating cells and looking for signaling events "missing" in a patient subpopulation, one can identify the signaling equivalent of "loss of a tumor suppressor" function [25]. In addition to revealing losses of signaling, perturbing the signaling network can also reveal when cancer signaling networks have gained abnormal signaling responses, such as hypersensitivity to a growth factor. In the example of JMML, stimulation of the signaling network revealed that hypersensitivity to a low dose of the cytokine GM-CSF was diagnostic for the disease [7]. Flow cytometry measurements of phosphorylated Stat5 revealed a small population of GM-CSF hypersensitive cells only present in patients with JMML. Combinations of stimuli and inhibitors can also be used to map the structures of signaling networks computationally [24]. Thus, while the resting or basal state of the signaling network can provide valuable information, much more information is obtained through perturbational mapping (Figure 15.8).Moving forward, signaling profiles that are diagnostic or prognostic for a disease can be applied in preclinical or early clinical settings. When designing a therapy based on signaling perturbation, such as inhibition of malignant signaling, a key issue is whether targeting any one molecule is a better strategy than simultaneously targeting multiple signaling events identified as negative

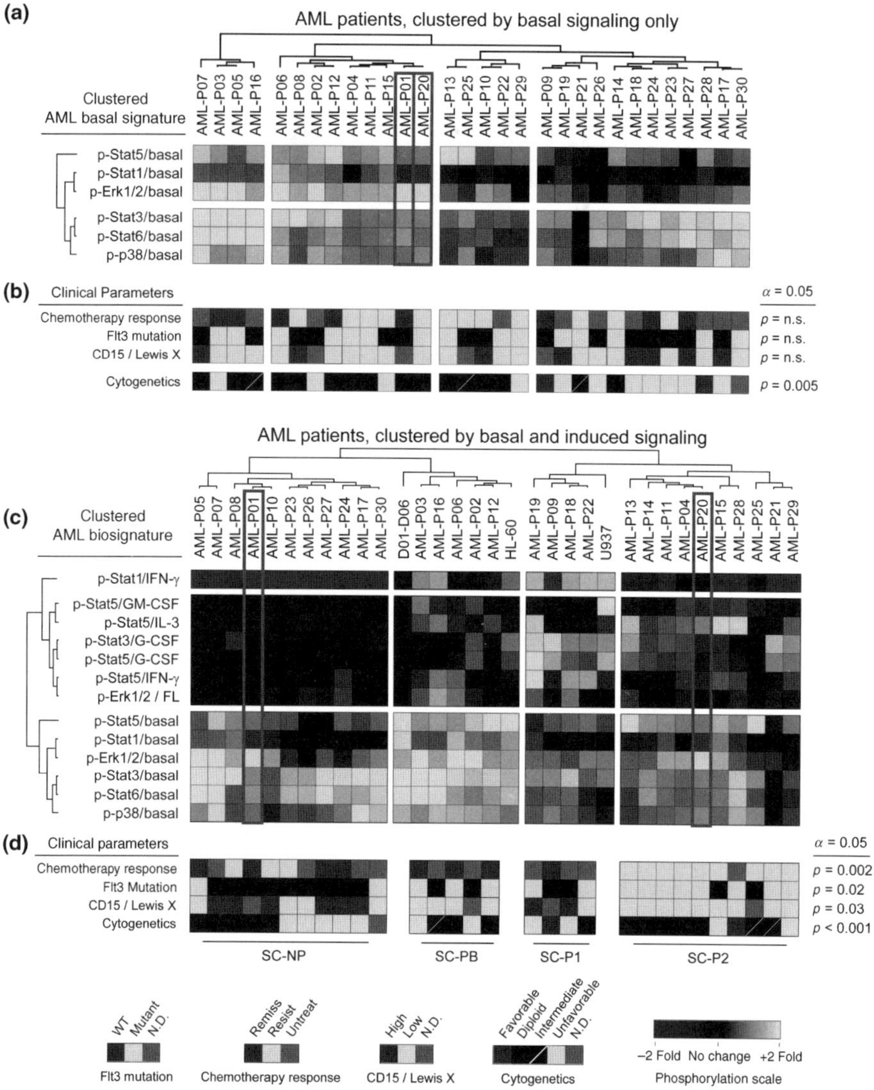

FIGURE 15.8 Measuring induced signaling responses provides clinically useful information on disease signaling networks. (a) Shown are 30 AML cancer patients clustered by the basal level of phosphorylation measured in the AML cell population. Large differences in basal phosphorylation were observed among the patient samples. Four groups of patients were defined using this profile of basal signaling only. Note patients P01 and P20, whose AML cells displayed extremely similar basal signaling. (b) Key clinical features, including the initial response to induction chemotherapy, FLT3 mutation status, CD15 expression, and cytogenetics of the 30 patients from (a). Basal signaling stratified the cytogenetics of the patients, but not initial chemotherapy response, CD15 expression, or FLT3 mutation status. (c) Additional data on the induced signaling response of the same 30 patients are shown. Strikingly, the addition of signaling responses reveals differences between AML-P01 and AML-P20. Many induced signaling events were potentiated in AML-P01. (d) The same clinical parameters as in (b) are now shown for the patients, as grouped in (c), according to both basal and induced signaling. Potentiated signaling responses now significantly stratify all four clinical parameters: chemotherapy response, FLT3 mutation status, CD15 expression, and cytogenetics. (See the color version of the figure in the Color Plates section.)

prognostic in a disease. In a preclinical research setting, a set of inhibitors chosen to target a profile of signaling events specifically observed in a patient's cancer cells could be studied in primary cancer specimens. A tailored therapy may be no better than a single-agent inhibitor. However, the potential exists that a tailored therapy will be far more effective at killing cancer cells—and only cancer cells—if based on multiple unique signaling features of each disease. Single-cell assays for cell death can help determine the cell-type specificity of therapies in primary patient samples. Flow cytometry reveals whether all the disease cells are being killed or only a subpopulation. Finally, a validated clinical signaling profile can be used when designing a clinical trial to initially stratify patient risk, guide patients toward the trial, balance treatment arms of the trial, and measure, as a corollary end point, changes in the signaling profile over time.

15.6.3 Measuring and Targeting Cell Subsets

As discussed in Section 15.5, a key advantage of flow cytometry is the ability to track multiple populations of cells and to know which cell types are inhibited, when conducting assays. In primary tumor samples from cancer patients, tumors have variable, but significant, nonmalignant immune cell infiltrate [23]. Flow cytometry analysis of live cells from cryopreserved lymphoma tumor samples allows simultaneous study of both cancer cells and the patient's immune system cells. We used single-cell signaling profiles to identify cell-intrinsic changes to BCR signaling kinetics that distinguished lymphoma B cells from nonmalignant B cells within a patient's tumor [26]. The same approach can also be applied to dissect signaling in developmental subsets of healthy cells, providing a valuable reference point for characterizing abnormal signaling in tumor cells.

In studies of clinical therapeutics, the ability to measure the effects of a therapy in all patient sample cells is significant advantage of single-cell assays. Signaling inhibitors may have greater potency in killing primary human cancer cells while remaining below the threshold of toxicity to adjacent nonmalignant cells that are less dependent on abnormal signaling events. For example, for a B cell cancer, the effects of a drug on lymphoma B cells, tumor-infiltrating T cells, and nonmalignant B cells might be compared. An ideal drug might induce apoptosis specifically in the cancer B cells, while having minimal toxicity against the healthy cells. With flow cytometry, researchers can identify control cell populations within a single well of a plate, providing confidence that all cells were treated with equivalent conditions during an experiment.

In addition to comparing tumor and nonmalignant subsets, flow cytometry can identify subpopulations within the tumor that have distinct signaling profiles and these subsets can be related to the patients' clinical outcomes [19]. Defining tumor heterogeneity in terms of signaling cell subsets gives us insight into the complexity of individual patient's disease. Populations such as cancer stem cells, evolving tumor subclones, or therapy-resistant cell subsets could be identified by flow cytometry and tracked over time.

15.7 NEXT-GENERATION CYTOMETRY

Current cytometers, based on fluorescence, can measure up to 17 parameters per cell simultaneously [27]. However, even through the use of time-delayed excitation and detection, novel fluorophores such as quantum dots, and upconversion, extension to more than 20 parameters seems to be unlikely. In addition, more than 10–12 parameter experiments, possible on some commercial instruments, require skilled experimental design to account and correct for spectral overlap. Often called "compensation," spectral overlap causes emission from one fluorophore to be measured in multiple detector channels, convoluting analysis and reducing the quantitative nature of the experiments (Figure 15.9). Now a new platform, mass cytometry, offers single-cell analysis capabilities potentially approaching 100 parameters without fluorescent agents or interference from spectral overlap (Figure 15.9). This platform utilizes nonbiological, rare earth metal isotopes as reporters. By exploiting the resolution, sensitivity, and dynamic range of elemental mass spectrometry, but on a timescale that allows the measurement of 1000 individual cells per second, this device potentially offers a much simplified alternative for ultrahigh content cytometric analysis.

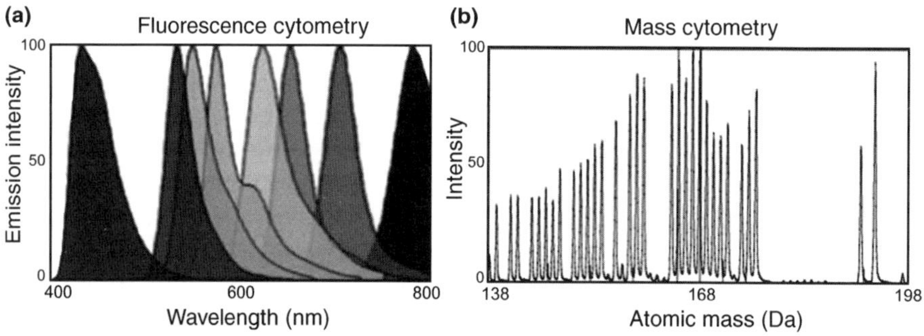

FIGURE 15.9 Comparison of traditional fluorescence-based cytometry and mass-based cytometry. In fluorescence-based cytometry (a), the emission spectra of the various fluorophores overlap one another significantly. Although bandpass filters can be used to minimize compensation between fluorophores, they are not able to eliminate fluorescence spillover completely. The breadth of the emission spectra limits the number of fluorophores that can realistically be used in any one experiment. In the mass-based cytometry on the other hand (b), the mass labels appear at only one discrete mass value when measured. With a well-tuned mass cytometer, compensation is not required. Signals from one mass label will only interfere with another mass label in cases of oxidation (+16 Da) or from impurities in the isotopic labels themselves. The 30 elemental metal isotope labels plus 2 DNA intercalators (Ir-191 and Ir-193) shown in this plot are available at extremely high purity and do not produce overlapping oxides, essentially enabling compensation-free analysis of all 30 parameters plus DNA content. (See the color version of the figure in the Color Plates section.)

15.7.1 Elemental Mass Spectrometry

Inductively coupled plasma mass spectrometry (ICP-MS) is the most advanced and sensitive means of determining the elemental composition of materials. Classically, it has been used for ultratrace (as low as parts per quadrillion) detection of metals and other elements in both environmental (water, soil, and air) and clinical (blood and urine) samples. The key component of the analysis is high-temperature plasma (approximately 7500K) that is created through the coupling of high radio frequency energy into a flowing inert gas, such as argon. Material introduced into the plasma is completely vaporized into its component elemental ions. The elemental ions representative of the sample are extracted and introduced into the mass analyzer, a quadrupole mass filter for example, which can provide approximately seven orders of magnitude abundance sensitivity (resolution between adjacent mass channels) and nine orders of dynamic range. It is these characteristics that make elemental mass spectrometry a logical alternative to fluorescence as a detection method in bioanalysis [28–31].

15.7.2 Mass Cytometry

Mass cytometry uses single isotopes of pure elements in the form of soluble metal ions as tags bound to antibodies instead of fluorophores for measuring target epitopes in single cells [28]. Notably, mass cytometry has essentially the same workflow as conventional fluorescence cytometry. Once labeled, single cells are individually sampled, atomized and ionized by ICP injection, and the resulting ions are mass measured. The signal from each unique mass channel is then correlated back to a particular isotopic element and thus to the expression of a given epitope on each cell (Figure 15.10). This instrument was originally conceived in Scott Tanner's group at the University of Toronto and is now available under the commercial name of CyTOF, made by DVS Sciences. The general workflow for the mass cytometer data acquisition is as follows:

- Labeled cells are nebulized into single-cell droplets and enter the plasma at a rate of up to 1000 cells/s. The cells are completely destroyed and reduced to a "cloud"

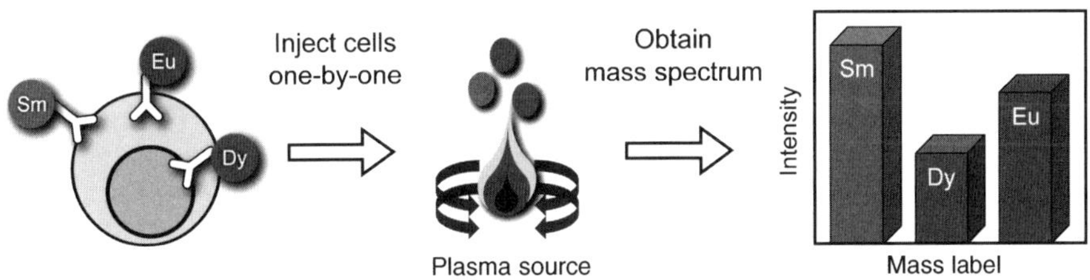

FIGURE 15.10 Analysis of cellular markers with ICP-based mass cytometry. An affinity product (e.g., antibody) tagged with a specific elemental isotope binds to the protein or molecule of interest. The cell is introduced into the plasma source, atomized, ionized, and then the elemental composition is mass measured. Signals corresponding to each elemental isotope tag are quantified and correlated with the abundance of the target of interest.

of the individual atoms that made up the cell, including the elemental tags used to label the antibody.

- The resulting elemental ions are transferred into a high-vacuum region, neutral species are removed, and the cloud becomes focused into a beam that is transmitted further into the mass analysis region.
- The beam of elemental ions then passes through a radio frequency-only quadrupole mass filter that removes the highly abundant, low molecular weight elemental ions (i.e., C, N, O, S, Na, K, Cl, and Ca) that compose the bulk of biomass. This is a necessary step, as these elements would overwhelm the detector hindering the analysis of the low-abundance heavy metal isotope tags.
- The heavier elemental ions (approximately 100 Da and higher) are then transmitted into the time-of-flight (TOF) mass analysis region where the beam is sampled and mass measured (scanned) approximately 80,000 times per second. The very fast TOF scan rate is the component of the platform that allows single-cell measurement on the order of 100–1000 times per second.
- The average ion cloud corresponding to a single cell event has a duration of 25–30 scans. Moreover, this ion cloud has a signal intensity distribution that is mainly Gaussian. Using these principles, the single-cell signal is extracted from the raw data by integrating the signal from each parameter measured, reporting it as a single value for each cellular event measured. This is quite similar to the protocols in modern fluorescent cytometric analyzers.
- The data are saved as text and in FCS3.0 formats, the latter being compatible with conventional flow cytometry data processing software.

On the CyTOF instrument, the TOF mass analyzer is optimized to measure elements approximately between 100 and 250 Da. There is currently a panel of 30 nonbiological, rare elemental metal isotopes commercially available for use as reporters in mass cytometry analysis. The metal isotopes are listed in Table 15.1. The majority of these metals belong to the lanthanide series of elements. These metal

TABLE 15.1 Common Metal Isotopes Used in Compensation-Free Mass Cytometry Analysis

Isotope/Elemental Symbol			
139 La	150 Nd	160 Gd	170 Er
141 Pr	151 Eu	164 Dy	171 Yb
142 Nd	152 Sm	165 Ho	172 Yb
144 Nd	153 Eu	166 Er	174 Yb
145 Nd	154 Sm	167 Er	175 Lu
146 Nd	156 Gd	168 Er	176 Yb
147 Sm	158 Gd	169 Tm	197 Au
148 Nd	159 Tb		

isotopes can be detected simultaneously without any overlap between their resulting mass signatures. In addition, their signal response in the CyTOF platform is within a twofold difference from the best channel to the worst channel suggesting that they could all be used relatively interchangeably.

15.7.2.1 Reagents The affinity reagents used (e.g., antibodies) are labeled with chelating polymers that can then be loaded with the desired elemental salt for mass measurement [32]. The prototype tags that are currently available employ an acrylate polymer backbone functionalized with chelators (e.g., diethyl-triamine pentacetic acid (DTPA)) that specifically bind trivalent heavy metal cations tightly ($K_d \sim 10^{-16}$ M). The current generation of tag is designed for attachment to antibodies, linking to the reduced disulfide bridges of the antibody Fc chains. For this purpose, current commercial reagents are maleimide derivatized and, consequently, the labeling procedures are similar to those used in labeling antibodies with fluorophores. Therefore, it is anticipated that all affinity reagents compatible with conventional cytometry platforms should translate into the mass cytometry system.

It has already been demonstrated that the approach of using polymer-coupled metal ligands can be used to effectively label antibodies [28, 31, 33]. Furthermore, the signal generated by each metal isotope tag is highly and linearly correlated with the amount of antigen detected, making these reagents at least as sensitive and quantitative as radio-immuno or other colorimetric assays with the added benefit of high multiplexing capacity [34].

With respect to sensitivity, the polymers produced to date bind approximately 30 atoms of metal per polymer label and 4–8 polymer tags can be linked to each antibody without loss of function. This results in approximately 120–240 metal isotope atoms per antibody. With the current mass cytometry efficiency of approximately 1 ion detected per 4×10^4 atoms introduced to the plasma, the current limit of detection is as low as 100–400 copies of antigen per cell. Certainly, the number of metal atoms attached to each detection reagent will be a driving force behind the mass cytometry platform sensitivity. Increasing the number of metal ions per tag will likely involve synthesizing either longer polymers than those used now or polymers that are more densely functionalized with metal atoms.

The other major reagent, which has to be considered in the case of cell analysis, is a general cellular stain. Because of the lack of parameters such as forward and side light scatter used in conventional fluorescence-based cytometry and the fact that the majority of elemental ions inherent in biological substances are filtered out prior to analysis, a cell negative for the antigens of interest in an experiment would appear "invisible" in mass cytometry analysis. To address this, Rh (103 Da) and Ir (191 and 193 Da) DNA intercalators have been developed for the baseline detection of a cell in the instrument, as well as to provide some information about DNA content [33]. The algorithm currently used for the detection and extraction of cell-induced transients from the mass analysis includes summation of the Ir responses allowing determination of single cell events.

15.7.3 Advantages of the Mass Cytometry Approach

15.7.3.1 High Dimensionality Using the current set of reagents, the ability to measure 30 parameters simultaneously, as compared to the maximum of 17 reported in fluorescence, offers a number of new avenues in cytometry analysis. More detailed characterization based on cell surface antigen expression of immune and other hematopoetic cell types are some of the first obvious experiments. In addition, simultaneous measurement of intracellular antigens (e.g., proteins) and their modifications (e.g., phosphorylation) that regulate cell cycle, apoptosis, and gene expression in rare cell populations (e.g., hematopoetic stem cells) becomes feasible. This type of measurement is particularly difficult to do in greater than 10 dimensions by fluorescence because many of the changes detected are subtle and can suffer heavy interference from spectral overlap. The lack of compensation combined with high dimensionality in mass cytometry should enable the measurement of more than 10 intracellular parameters while still allowing simultaneous measurement of surface antigen expression. In preliminary experiments, we have already been able to use the CyTOF mass cytometer to perform phosphorylation analysis on 17 parameters simultaneously (Figure 15.11). This high-dimensional phosphorylation analysis will also aid in the creation of better profiles for clinical diagnostics and will further our ability to evaluate target and off-target effects simultaneously during drug screening.

15.7.3.2 Simplicity/Versatility Because the mass spectrometer provides at least three orders of magnitude resolution between adjacent detection channels, meaning that two adjacent metal isotopes can differ in abundance by 10^3 before spectral overlap becomes an issue, compensation is only minimally required, and in most instances it is not required at all. In addition, the signal responses for the current commercial metal isotope tags are approximately twofold of one another on the CyTOF mass cytometer. Taken together, these attributes indicate that tag-panel selection can essentially be arbitrary, making experimental design much simpler. Consequently, even if a user does not have 30 parameters to interrogate, the design of staining panels for as few as 10 parameters could be accomplished more rapidly than it can for fluorescence-based experiments.

15.7.3.3 Stability The elemental tags have an added advantage over many fluorescent reagents in that they are extremely stable chemically. So unlike several fluorophores, such as the phycobiliproteins (PE and APC), they remain unaffected during procedures such as cell permeabilization with methanol or other strong denaturants. Elemental tags will be stable under these conditions. Characteristics such as this are essential in intracellular phosphoprotein analysis.

15.7.3.4 Speed/Throughput Although the current throughput is limited to about 1000 cells per second, the large number of parameters that can be measured simultaneously by mass cytometry provides a dramatic increase in the data acquisition rate. For example, a simultaneous 30 parameter single-cell analysis would require multiple samples with overlapping parameters on a fluorescence

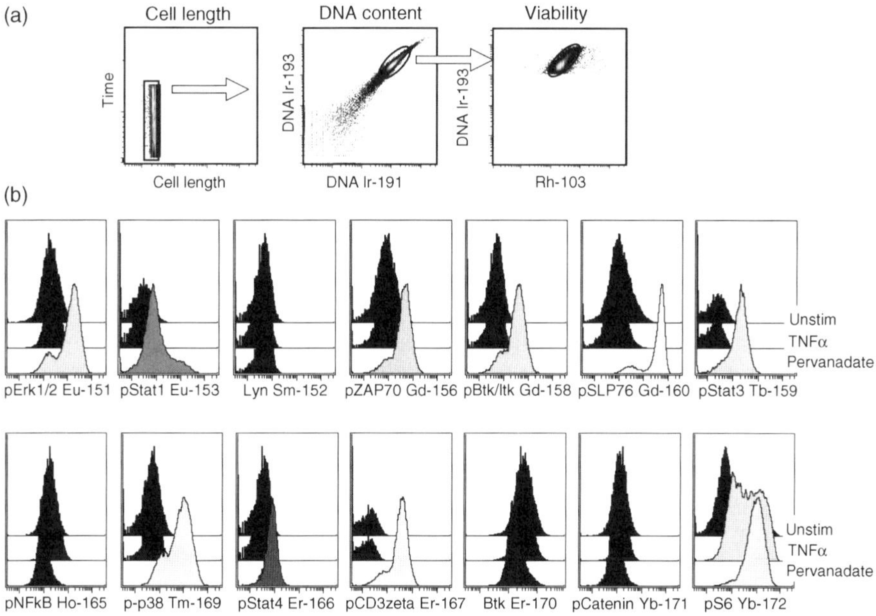

FIGURE 15.11 Mass cytometry experiment analyzing 17 parameters simultaneously, including 14 intracellular parameters. KG1 cells were left untreated or stimulated with TNFα or pervanadate. Fixed and permeabilized cells were stained for the indicated intracellular antigens and analyzed by mass cytometry on a CyTOF instrument. (a) Cellular events were identified by gating based on cell event length, DNA content, and viability before fixation. (b) Histogram overlay representation of the data. Histograms are colored according to the fold change in phosphorylation or protein level upon induction, relative to the untreated control. Data analysis was performed in Cytobank software [35].

platform. Even then, information would be lost due to the lack of simultaneous measurement. The other aspect of increased throughput comes from barcoding approaches. As mentioned, FCB provides the ability to mix cells from multiple conditions into a single tube where their individual profiles can be reconstructed postanalysis, thereby saving analysis time and reagents. In the case of mass cytometry, it is also possible to label cells with metal isotopes using reactive chelators of nanoparticle labeling [36]. Even taking a binary approach with 7 metals (present or absent), one could combine 128 different samples (2^7) into the same tube leaving 23 parameters available for analysis.

15.7.3.5 Accuracy Fluorescence flow cytometry can be highly quantitative. However, as the number of parameters measured simultaneously increases, data become less reliable due to spectral overlap. On the other hand, ICP-MS measurements at the heart of mass cytometry were originally developed for the absolute quantification of elements and stable isotopes in a variety of complex matrices. Consequently, the signal obtained is directly and quantitatively related to the number of atoms of the tag introduced. Accordingly, using elemental standards, it might become routine to measure the absolute number of atoms per cell, regardless of the number of metal isotopes used. Again, such attributes should greatly enhance cytometry-based clinical assays, drug screening panels, and overall assessment of regulatory phosphorylation at the single-cell level.

15.7.4 Future of Mass Cytometry

The traditional three to five color fluorescence flow cytometer helped define the major cell subsets of the immune system that we understand today (T cells, B cells, macrophages, among others). Eight-color machines brought characterization of even more rare immune subsets and stem cells, leading to advances in stem cell biology and cancer stem cell characterization. Merging this capability with intracellular staining, 11–15 color analysis provided by state-of-the-art instruments allowed analysis of regulatory signaling networks, drug design/screening, and patient stratification of clinical and drug response outcomes. Given this, the new generation of mass cytometry instrumentation will likely drive an unparalleled revolution in the understanding of hematologic processes—both in mapping biological mechanisms in normal human development and disease processes that impact the immune system.

15.8 SUMMARY

What has become evident during our work is that drug discovery and patient treatment should best be viewed as a cyclic process. Analyzing diseased patient samples tells us much about the complexity of disease, how it affects particular groups of signaling pathways, sets of cell types, and the time dependence of certain features. Analyzing the effects of many compounds on signaling pathways in primary cell samples shows patterns of druggability. When both these aspects are coupled, a more thorough understanding of signaling networks emerges and more effective therapies can be developed. For instance, a target may be identified from patient samples, but this target may be known to be undruggable based on prior screening efforts. By understanding the repertoire of druggable targets and matching those to potential disease-specific targets, drug discovery will become more effective and rapid and treatments will be individualized.

REFERENCES

1. Krutzik PO, Irish JM, Nolan GP, Perez OD. Analysis of protein phosphorylation and cellular signaling events by flow cytometry: techniques and clinical applications. *Clin. Immunol.* 2004;110:206–221.
2. Clutter MR, Krutzik PO, Nolan GP. Phospho-specific flow cytometry in drug discovery. *Drug Discov. Today* 2005;2:295–302.

3. Hale MB, Nolan GP. Phospho-specific flow cytometry: intersection of immunology and biochemistry at the single-cell level. *Curr. Opin. Mol. Ther.* 2006;8:215–224.
4. Irish JM, Kotecha N, Nolan GP. Mapping normal and cancer cell signalling networks: towards single-cell proteomics. *Nat. Rev. Cancer* 2006;6:146–155.
5. Danna EA, Nolan GP. Transcending the biomarker mindset: deciphering disease mechanisms at the single cell level. *Curr. Opin. Chem. Biol.* 2006;10:20–27.
6. Krutzik PO, Nolan GP. Intracellular phospho-protein staining techniques for flow cytometry: monitoring single cell signaling events. *Cytometry A* 2003;55:61–70.
7. Kotecha N, Flores NJ, Irish JM, Simonds EF, Sakai DS, Archambeault S, Diaz-Flores E, Coram M, Shannon KM, Nolan GP, Loh ML. Single-cell profiling identifies aberrant STAT5 activation in myeloid malignancies with specific clinical and biologic correlates. *Cancer Cell* 2008;14:335–343.
8. Hale MB, Krutzik PO, Samra SS, Crane JM, Nolan GP. Stage dependent aberrant regulation of cytokine-STAT signaling in murine systemic lupus erythematosus. *PLoS One* 2009;4:e6756.
9. Tanabe Y, Nishibori T, Su L, Arduini RM, Baker DP, David M. Cutting edge: role of STAT1, STAT3, and STAT5 in IFN-alpha beta responses in T lymphocytes. *J. Immunol.* 2005;174:609–613.
10. Krutzik PO, Nolan GP. Fluorescent cell barcoding in flow cytometry allows high-throughput drug screening and signaling profiling. *Nat. Methods* 2006;3:361–368.
11. Zhang JH, Chung TD, Oldenburg KR. A simple statistical parameter for use in evaluation and validation of high throughput screening assays. *J. Biomol. Screen.* 1999;4:67–73.
12. Sklar LA, Carter MB, Edwards BS. Flow cytometry for drug discovery, receptor pharmacology and high-throughput screening. *Curr. Opin. Pharmacol.* 2007;7:527–534.
13. Edwards BS, Young SM, Ivnitsky-Steele I, Ye RD, Prossnitz ER, Sklar LA. High-content screening: flow cytometry analysis. *Methods Mol. Biol.* 2009;486:151–165.
14. Edwards BS, Oprea T, Prossnitz ER, Sklar LA. Flow cytometry for high-throughput, high-content screening. *Curr. Opin. Chem. Biol.* 2004;8:392–398.
15. Arterburn JB, Oprea TI, Prossnitz ER, Edwards BS, Sklar LA. Discovery of selective probes and antagonists for G-protein-coupled receptors FPR/FPRL1 and GPR30. *Curr. Top. Med. Chem.* 2009;9:1227–1236.
16. Krutzik PO, Crane JM, Clutter MR, Nolan GP. High-content single-cell drug screening with phosphospecific flow cytometry. *Nat. Chem. Biol.* 2008;4:132–142.
17. O'Gorman WE, Dooms H, Thorne SH, Kuswanto WF, Simonds EF, Krutzik PO, Nolan GP, Abbas AK. The initial phase of an immune response functions to activate regulatory T cells. *J. Immunol.* 2009;183:332–339.
18. Hotson AN, Hardy JW, Hale MB, Contag CH, Nolan GP. The T cell STAT signaling network is reprogrammed within hours of bacteremia via secondary signals. *J. Immunol.* 2009;182:7558–7568.
19. Irish JM, Hovland R, Krutzik PO, Perez OD, Bruserud O, Gjertsen BT, Nolan GP. Single cell profiling of potentiated phospho-protein networks in cancer cells. *Cell* 2004;118: 217–228.
20. Krutzik PO, Clutter MR, Nolan GP. Coordinate analysis of murine immune cell surface markers and intracellular phosphoproteins by flow cytometry. *J. Immunol.* 2005;175: 2357–2365.

21. Krutzik PO, Hale MB, Nolan GP. Characterization of the murine immunological signaling network with phosphospecific flow cytometry. *J. Immunol.* 2005;175:2366–2373.
22. Irish JM, Hovland R, Krutzik PO, Perez OD, Bruserud O, Gjertsen BT, Nolan GP. Single cell profiling of potentiated phospho-protein networks in cancer cells. *Cell* 2004;118: 217–228.
23. Irish JM, Czerwinski DK, Nolan GP, Levy R. Altered B-cell receptor signaling kinetics distinguish human follicular lymphoma B cells from tumor-infiltrating nonmalignant B cells. *Blood* 2006;108:3135–3142.
24. Irish JM, Anensen N, Hovland R, Skavland J, Borresen-Dale AL, Bruserud O, Nolan GP, Gjertsen BT. Flt3 Y591 duplication and Bcl-2 overexpression are detected in acute myeloid leukemia cells with high levels of phosphorylated wild-type p53. *Blood* 2007;109:2589–2596.
25. Hanahan D, Weinberg RA. The hallmarks of cancer. *Cell* 2000;100:57–70.
26. Irish JM, Czerwinski DK, Nolan GP, Levy R. Kinetics of B cell receptor signaling in human B cell subsets mapped by phosphospecific flow cytometry. *J. Immunol.* 2006; 177:1581–1589.
27. Perfetto SP, Chattopadhyay PK, Roederer M. Seventeen-colour flow cytometry: unravelling the immune system. *Nat. Rev. Immunol.* 2004;4:648–655.
28. Bandura DR, Baranov VI, Ornatsky OI, Antonov A, Kinach R, Lou X, Pavlov S, Vorobiev S, Dick JE, Tanner SD. Mass cytometry: technique for real time single cell multitarget immunoassay based on inductively coupled plasma time-of-flight mass spectrometry. *Anal. Chem.* 2009;81(16): 6813–6822.
29. Baranov VI, Quinn Z, Bandura DR, Tanner SD. A sensitive and quantitative element-tagged immunoassay with ICPMS detection. *Anal. Chem.* 2002;74:1629–1636.
30. Ornatsky O, Baranov VI, Bandura DR, Tanner SD, Dick J. Multiple cellular antigen detection by ICP-MS. *J. Immunol. Methods* 2006;308:68–76.
31. Ornatsky OI, Kinach R, Bandura DR, Lou X, Tanner SD, Baranov VI, Nitz M, Winnik MA. Development of analytical methods for multiplex bio-assay with inductively coupled plasma mass spectrometry. *J. Anal. At. Spectrom.* 2008;23:463–469.
32. Lou X, Zhang G, Herrera I, Kinach R, Ornatsky O, Baranov V, Nitz M, Winnik MA. Polymer-based elemental tags for sensitive bioassays. *Angew. Chem. Int. Ed.* 2007;46: 6111–6114.
33. Ornatsky OI, Lou X, Nitz M, Schafer S, Sheldrick WS, Baranov VI, Bandura DR, Tanner SD. Study of cell antigens and intracellular DNA by identification of element-containing labels and metallointercalators using inductively coupled plasma mass spectrometry. *Anal. Chem.* 2008;80:2539–2547.
34. Razumienko E, Ornatsky O, Kinach R, Milyavsky M, Lechman E, Baranov V, Winnik MA, Tanner SD. Element-tagged immunoassay with ICP-MS detection: evaluation and comparison to conventional immunoassays. *J. Immunol. Methods* 2008;336:56–63.
35. Kotecha N, Krutzik PO, Irish JM. Web-based analysis and publication of flow cytometry experiments. *Curr. Protoc. Cytom.* 2010;Chapter 10:Unit 10.17.
36. Vancaeyzeele C, Ornatsky O, Baranov V, Shen L, Abdelrahman A, Winnik MA. Lanthanide-containing polymer nanoparticles for biological tagging applications: nonspecific endocytosis and cell adhesion. *J. Am. Chem. Soc.* 2007;129:13653–13660.

INDEX

Flow Cytometry in Drug Discovery and Development, Edited by Virginia M. Litwin and Philip Marder